Rigid Flexibility

APPLIED LOGIC SERIES

VOLUME 34

SCOPE OF THE SERIES

Logic is applied in an increasingly wide variety of disciplines, from the traditional subjects of philosophy and mathematics to the more recent disciplines of cognitive science, computer science, artificial intelligence, and linguistics, leading to new vigor in this ancient subject. Kluwer, through its Applied Logic Series, seeks to provide a home for outstanding books and research monographs in applied logic, and in doing so demonstrates the underlying unity and applicability of logic.

The titles published in this series are listed at the end of this volume.

Rigid Flexibility

The Logic of Intelligence

by

Pei Wang
Temple University, Philadelphia, USA

Springer

A C.I.P. Catalogue record for this book is available from the Library of Congress.

ISBN-10 1-4020-5044-5 (HB)
ISBN-13 978-1-4020-5044-2 (HB)
ISBN-10 1-4020-5045-3 (e-book)
ISBN-13 978-1-4020-5045-3 (e-book)

Published by Springer,
P.O. Box 17, 3300 AA Dordrecht, The Netherlands.

www.springer.com

Printed on acid-free paper

Printed in the Netherlands

Contents

List of Tables

Preface

This book presents a research project aimed at the building of a "thinking machine," that is, a general-purpose artificial intelligence.

Artificial intelligence has a scientific and an engineering aspect. The former focuses on the explanation of "intelligence" displayed by the human mind, while the latter focuses on building a computer system that has such a nature. It is the most recently developed branch of a profound intellectual tradition, and is related to many problems studied in cognitive sciences, including philosophy, psychology, logic, linguistics, mathematics, neuroscience, and related disciplines. Ultimately, the goal is to understand notions like "intelligence," "cognition," "mind," and "thinking" well enough to reproduce them in computer systems.

Though the research field of artificial intelligence has existed for about half a century, we are still far from the goal of building a thinking machine. To a large degree, this is due to the complexity of the problem — since the mind is perhaps the most complicated phenomenon in the universe — as well as limitations of existing computer technology. However, there is also a possibility that the mainstream research in the field has been heading in the wrong direction.

The research reported in this book proposes a change in direction of the research of artificial intelligence, a "paradigm shift" *per se*. Instead of duplicating human behaviors or solving practical problems, this book proposes that the right thing to do in the research is to build systems that follow the same underlying *principle* as the human mind, that is, *to adapt to the environment and to work with insufficient knowledge and resources.*

In light of this opinion, the limitations of traditional theories are analyzed. Such theories include first-order predicate logic, model-theoretic semantics, probability theory, computability theory, and computational complexity theory. Each of these theories makes some explicit or implicit assumptions on the sufficiency of knowledge and/or resources, and will not work when these assumptions are not satisfied. The new theory introduced in this book is designed for a situation where no traditional theory can be applied, with the belief that this is what "intelligence" is about.

This research shows that it is possible to build a formal model, which is then be implemented in a computer, by standing firm on the assumption of insufficient knowledge and resources. Furthermore, the model uses a relatively simple mechanism to uniformly reproduce many cognitive faculties, including reasoning, learning, perceiving, remembering, categorizing, planning, predicting, problem solving, and decision making.

This book consists of four parts:

Part I introduces the philosophical and methodological foundation of the *Non-Axiomatic Reasoning System*, NARS for short, project. The major schools of thought in the field are analyzed, and a new working definition of "intelligence" is proposed, according to which it is the capability of a system to adapt to its environment and to work with insufficient knowledge and resources. The choice of the reasoning system framework for this project is justified. Finally, the major components of NARS are briefly and informally introduced.

Part II describes a formal model based on the above theory. This part contains the most technical results in NARS. First, a logic system, NAL, is defined in several phases. The logic uses a categorical language, an experience-grounded semantics, and extended syllogistic inference rules. Then, the memory structure and control mechanism of the system are introduced, which let the system operate adequately while the computational resources, time and space, are in short supply.

Part III compares the above model with related approaches on several topics, and discusses the corresponding properties of NARS.

The topics include knowledge representation, semantics and interpretation, various types of inference, learning and categorization, and the control of inference processes. It is shown that an outstanding feature of NARS is that it provides a unified solution to many problems in artificial intelligence and cognitive sciences.

Part IV summarizes the conclusions reached in this research so far, and outlines the plan for the next stage of the project.

This organization is not perfect. Usually, a new technical idea should be introduced together with the problem it aims, as well as with comparisons to other related ideas. In this book, however, each idea introduced in Part II is not fully discussed until Part III, after the whole system is described. This is because that each idea in NARS is typically motivated by more than one consideration, and the solution of a problem is typically provided by the cooperation of multiple components of the system. Consequently, to fully understand part of the system without mentioning the other parts is very hard, if not impossible. Given this nature, one possible reading strategy is to read Part II quickly in the first pass, just to get a big picture of the system, then to come back to the technical details during or after reading the remaining part of the book.

This research is highly interdisciplinary. Its theoretical foundation is rooted in philosophical and psychological studies; the formal model is mainly about logic and artificial intelligence; the implementation is carried out with the tools provided by computer science.

This book is aimed at general readers interested in mind, thinking, and the computer. The readers are expected to be moderately familiar with artificial intelligence, formal logic, computer science, and cognitive sciences.

NARS is an on-going project. For up-to-date information about its progress, please visit the project website,[1] which contains on-line demonstrations, working examples, related publications, as well as additional materials and links.

---

[1]Currently at `http://www.cogsci.indiana.edu/farg/peiwang/NARS/`.

Acknowledgment

I became interested in artificial intelligence (AI) in the early 1980s, when I was an undergraduate student in Peking University. At that time AI was a new topic in China, so few faculty was working on it, and no course was offered on the subject. As a result, I simply read what I could find on AI and related topics. This fact partially explains why I did not follow any established approach, but tried to put together a theory by myself from the beginning of my research. The atmosphere in Peking University in those years was extraordinary — just out of the "cultural revolution", the young generation was enthusiastically, seriously, and bravely challenging traditional theories and values in all domains, and attempting to build their own. The influence of the so-called "Spirit of Peking University" was decisive to me; without it, I would not have had the confidence to work on my ideas for more than twenty years without major signs of acceptance from the AI community. Therefore, to a large extent, this research is a product of the atmosphere of "Peking University in the 1980s" — I don't think I would pursue the same path at a different place or at a different time.

As I had very limited access to established researchers and reference materials, many of my ideas were inspired by discussions with friends. Among them I want to thank Chen Gang, Cai Shan, Sun Yongping, and in particular, Bai Shuo.[2]

After developing some preliminary ideas in my thesis for my bachelor's degree, I decided to pursue them further when I became a graduate student in the same university. Professor Xu Zhuoqun (also known as Hsu Cho-Chun) found my ideas promising, and agreed to be my adviser.

[2] All Chinese names in this acknowledgment are written in the traditional Chinese order, with the family name followed by the given name.

Without his support, it would have been difficult for me to finish my earliest research as a Master Thesis. It was also from his Operating System course that I got my idea of resource management. Another member of my advisory committee, Professor Sun Huaimin of Beihang University (BUAA), was the one who triggered my interest in symbolic logic. I also obtained help from the other committee members and faculty of several universities, including Li Wei, Lin Jianxiang, Ma Xiwen, Shi Chunyi, Shi Zhongzhi, and Wang Yutian.

After I got my master's degree in July 1986, I was offered the position of Lecturer by the department. I accepted it, because Peking University was so much like a home for me. During the summer vacation that year, a fellow student, Yan Yong, persuaded me to join the translation project of Douglas Hofstadter's *Gödel, Escher, Bach: an Eternal Golden Braid.* The translation process turned out to be very exciting, and I found many of my own ideas were presented in the book in a beautiful way. As a consequence, I wrote to Professor Hofstadter and then we exchanged some research papers, at which point both of us noticed a surprising degree of overlap in our research projects. Consequently, he arranged for me to join his research group at Indiana University, and I was also accepted by the Ph.D program in computer science and cognitive science at IU.

I left Peking University (after about twelve years), and arrived in the USA in May 1991. This time I was lucky again to have an open-minded adviser. Though Professor Hofstadter did not share all my opinions about intelligence, he still provided me the support I needed to continue my research. His opinions have always been stimulating to my thought. Thanks to his help, the four and half years I spent on the peaceful Bloomington campus were very productive and pleasant. The other members of the Center for Research on Concepts and Cognition were very helpful. They include David Moser, Robert French, Terry Jones, David Chalmers, Gary McGraw, Jim Marshall, and John Rehling. Helga Keller gave me enormous help in administrative affairs. Jeff Logan played a major role in the improvement of my English. Carol Hofstadter's kindness to my family is unforgettable for us. I also want to thank the other members of my advisory committee — David Leake, Gregory Rawlins, and James Townsend — for their valuable comments and suggestions.

After getting my Ph.D. at the end of 1995, I could not get a research position anywhere, so I went into industry. From early 1996 to Spring 1998, I worked in three companies, doing applied AI work, mostly in expert systems, and continued the work in NARS in my own time. In April 1998 I was attracted to a start-up company, Intelligenesis (renamed to Webmind later), by a recruiting advertisement that looked for people with "a passion for making computers think." After talking with the co-founder (also the Chairman and CTO), Ben Goertzel, we found enough overlap in our ideas for me to join the company. One major task I finished for the company was to design a customized version of NARS as the inference engine of a general-purpose AI system, Webmind, which integrated several techniques. Though I and Ben had been arguing many issues all the time, the cooperation was stimulating and pleasant overall. I was also benefited from discussions with Jeff Pressing, Karin Verspoor, and other colleagues in the company.

In April 2001, the company finally ran out of money in a tough economical season. In Summer 2001, with the help of Fan Wenfei, I got a teaching position at the Computer and Information Sciences Department of Temple University, where I have stayed until now. At Temple, I have been getting help from Robert Aiken, Frank Friedman, John Nosek, and other faculty members.

In the recent years, my research has benefited from the discussions with the following friends: Deng Lang, Hu Xiangen, Lin Fangzhen, Lin Yunqing, Lin Zuoquan, Ju Shier, Wang Hongbin, Yang Yingrui, Zhou Beihai, as well as with many people in various mailing lists and news groups.

I knew I needed to write such a book many years ago, simply because NARS addresses too many issues in such an interwoven manner, that it is very difficult to clearly explain any of them without touching the others. My Master Thesis [Wang, 1986] and Ph.D. Dissertation [Wang, 1995a] can be seen as immature versions of this book. The current manuscript also includes many materials in my other writings (journal articles, conference papers, book chapters, and technical reports), which are included in the Bibliography of this book. Thanks to the following publishers for their kind permission for me to use my previous publications in this book: Elsevier [Wang, 1994b, Wang, 1996a, Wang, 2004a, Wang, 2005], IEEE [Wang, 1996b], Shaker Publishing

[Wang, 2001b], Springer Science and Business Media [Wang, 1993a, Wang, 1994a, Wang, 2000c, Wang, 2004c, Wang, 2006b], and World Scientific Publishing [Wang, 1995b, Wang, 2004b, Wang, 2004d].

The first version of this book was finished in the Summer of 2003. Since then, it had been rejected by several publishers until it finally got into the hands of Dov Gabbay of the Applied Logic Series, and got favorable evaluations from two anonymous reviewers. Without them, I do not know how much longer it would take for this manuscript to be published. During this process, Jane Spurr of the Applied Logic Series and Lucy Fleet of Springer provide abundant help.

Thanks to Kevin Copple, Paul Fidika, Jordan Fultz, Edward Heflin, J. W. Johnston, Klaus Witzel, and Brad Wyble for making many comments and English corrections on earlier versions of this manuscript.

Finally, I would like to thank the help from my family members, especially, from my dear wife, Sun Hongyuan, with her understanding, support, and love.

Part I

Theoretical Foundation

Chapter 1

The Goal of Artificial Intelligence

Generally speaking, Artificial Intelligence (AI) is the creation of intelligence, as displayed by the human mind, in an artificial entity, especially, a computer system.

This chapter surveys the current state of the field of AI, albeit through my personal perspective.

1.1 To define intelligence

1.1.1 The field of AI

A key characteristic that distinguishes the human being from other currently known entities (animals, machines, and so on) is "intelligence" (similar terms include "mind," "cognition," and "thinking"). Whether this capability can be understood and reproduced in machines is a question that has been considered for a long time by philosophers, mathematicians, scientists, engineers, as well as by writers and movie makers. However, it is the modern digital computer that makes it possible to seriously test various answers to this question.

The electronic computer first appeared in the 1940s. Though initially the computer was used for numerical calculations, a mental activity which previously could only be accomplished by a human mind, soon people realized that they could carry out many other mental activities by manipulating various types of symbols or codes. Naturally, people began to wonder whether all mental activities could be carried out by computers, and if not, where does the border lie?

Roughly speaking, all attempts to answer the above questions belong to the study of "Artificial Intelligence" (AI), that is, to the attempts to produce "intelligence" in artifacts, especially, computer systems.

Toward this general goal, two motivations of AI research and development coexist:

- As a science, AI attempts to establish a theory of intelligence to explain human intellectual activities and abilities.

- As a technology, AI attempts to implement a theory of intelligence in computer systems to reproduce these activities and abilities and use them to solve practical problems.

In AI, the science aspect ("What is intelligence?") and the technology aspect ("How to reproduce intelligence?") are closely related to each other. Although different researchers may focus on different aspects of the research, a complete AI project typically consists of works on the following three levels of description:[1]

1. a *theory* of intelligence, as writings in natural languages such as English or Chinese,

2. a formal *model* of intelligence based on the above theory, as formulas and expressions in formal languages like the ones used in logic or mathematics,

3. a computer *system* implementing the above model, as programs in programming languages such as Lisp or Java. Optionally, some AI projects include works on computer hardware and special devices.

[1]Similar level distinctions are made by other authors [Marr, 1982, Newell, 1990], and a summary can be found in [Anderson, 1990, page 4]. The above level distinction differs from the others in that here it is mostly determined by the *language* in which the research results are presented, and is, therefore, mostly independent of the content of the AI approach under discussion.

Roughly speaking, the mapping between descriptions of a higher level and those of a lower level is one-to-many, in the sense that one theory may be represented in more than one model (though each model only represents one theory), and that one model may be implemented in more than one way (though each implementation only realizes one model).

Because of the nature of the field, AI is closely related to other disciplines. At the top level, AI borrows concepts and theories from the disciplines that study the various aspects of the human brain and mind, including neuroscience, psychology, linguistics, and philosophy. At the middle level, AI uses tools and models developed in mathematics, logic, and computer science. At the bottom level, AI depends on components and systems provided by computer technology, like programming language, software, and hardware.

1.1.2 The need for definition

Though the previous subsection provided a brief description of the field of AI, it does not answer a key question: What is the definition of artificial intelligence?

It is generally acknowledged that the forming of AI as a research field was signified by the Dartmouth Meeting in 1956. After half a century, there is a substantial AI community with thousands of researchers all over the world, producing many books, journals, conferences, and organizations. However, the current state of AI research activities are not bounded together by a common theoretical foundation or by a set of methods, but by a group of loosely related problems.

In the acronym "AI," the "A" part is relatively easier to define — by "artificial," we mean "artifacts," especially electric computing machinery. However, the "I" part is much harder, because the debate on the essence of intelligence has been going on since the existence of the related fields like psychology and philosophy, etc, not to mention AI, and there is still no sign of consensus.

Consider what the "founding fathers" of AI had in mind about the field:

> "AI is concerned with methods of achieving goals in situations in which the information available has a certain

> complex character. The methods that have to be used are related to the problem presented by the situation and are similar whether the problem solver is human, a Martian, or a computer program." [McCarthy, 1988]
>
> Intelligence usually means "the ability to solve hard problems". [Minsky, 1985]
>
> "By 'general intelligent action' we wish to indicate the same scope of intelligence as we see in human action: that in any real situation behavior appropriate to the ends of the system and adaptive to the demands of the environment can occur, within some limits of speed and complexity." [Newell and Simon, 1976]

The above statements clearly have something in common, but there are still differences among them. The same is also true for the definitions of intelligence in AI books and articles. In fact, almost everyone in the field has a personal opinion about how the word "intelligence" should be used. These opinions in turn influence the choice of research goals and methods, as well as serve as standards for judging other researchers' results.

Maybe it is too early to define intelligence. It is obvious that, after decades of study, we still do not know very much about it. There are more questions than answers. Any definition based on the current knowledge is doomed to be revised by future works. We all know that a well-founded definition is usually the *result*, rather than the *starting point*, of scientific research.

However, there are still reasons for us to be concerned about the definition of intelligence at the current time.

Inside the AI research community, the lack of a common definition of the key concept of the field is the root of many controversies and misunderstandings. Many debates can be reduced to the fact that different sides use the term "intelligence" to mean very different things, and therefore they propose very different conclusions for questions like "What is the best way to achieve AI," "How to judge whether a system is intelligent," and so on.

Outside the AI community, AI researchers need to justify their field as a scientific discipline. Without a relatively clear definition of intelligence, it is hard to say why AI is different from, for instance, computer science or psychology. Is there really something novel and special, or just a fancy label on old stuff?

More importantly, each researcher in the field needs to justify his/her research approach in accordance with such a definition. For a concept as complex as "intelligence," no direct study is possible, especially when an accurate and rigid tool, namely the computer, is used as the research medium. We have to specify the problem clearly and only then be in a position to try to solve it. In this sense, anyone who wants to work on AI is facing a two-phase problem: firstly, choosing a working definition of intelligence, and then, producing it on a computer.

A *working definition* is a definition that is concrete enough to allow a researcher to directly work with it. By accepting a working definition of intelligence, a researcher does not necessarily believe that it fully captures the concept "intelligence," but the researcher takes it as a goal to be sought after for the current research effort. Such a definition is not for an AI journal editor who needs a definition to decide what papers are within the field or a speaker of the AI community who needs a definition to explain to the public what is going on within the field — in those cases, what is needed is a "descriptive definition" obtained by summarizing the individual working definitions.

Therefore, the lack of a consensus on what intelligence is does not prevent each researcher from picking up (consciously or not) a working definition of intelligence. Actually, unless a researcher keeps a working definition, he/she cannot claim to be working on AI. It is a researcher's working definition of intelligence that relates the current research, no matter how domain-specific, to the larger AI enterprise.

By accepting a working definition of intelligence, a researcher makes important commitments on the acceptable assumptions and desired results, which bind all the concrete work that follows. Limitations in the definition can hardly be compensated by the research, and improper definitions will make the research more difficult than necessary, or lead the study away from the original goal.

To better understand the relationship between a working definition of intelligence and AI research, consider an analogy. Imagine a group of people that want to climb a mountain. They do not have a map, and the peak is often covered by clouds. At the foot of the mountain, there are several paths leading in different directions. When you join the group, some of the paths have been explored for a while, but no one has reached the top.

If you want to get to the peak as soon as possible, what should you do? It is a bad idea to sit at the foot of the mountain until you are absolutely sure which path is the shortest, because you know it only after all paths have been thoroughly explored. You have to try some path by yourself. On the other hand, taking an arbitrary path is also a bad idea. Although it is possible that you make the right choice from the beginning, it is more advisable to use your knowledge about mountains and to study other people's reports about their explorations, so as to avoid a bad choice in advance.

There are three kinds of "wrong paths": (1) those which lead nowhere, (2) those which lead to interesting places (even to unexpected treasures) but not to the peak, and (3) those which eventually lead to the peak but are much longer than some other paths. If the only goal is to reach the peak as soon as possible, a climber should use all available knowledge to choose the most promising path to explore. Although switching to another path is always possible, it is time consuming.

AI researchers face a similar situation in choosing a working definition for intelligence. There are already many such definitions, which are different but related to each other (so hopefully we are climbing the same mountain). As a scientific community, it is important that competing approaches are developed at the same time, but it does not mean that all of them are equally justified, or will be equally fruitful.

1.1.3 Criteria of a good definition

Before studying concrete working definitions of intelligence, we need to establish the general criteria for what makes one definition better than another.

The same problem of general criteria is encountered in other areas. For example Carnap tried to clarify the concept of "probability." The task "consists in transforming a given more or less inexact concept into an exact one or, rather, in replacing the first by the second," where the first may belong to everyday language or to a previous stage in the scientific language, and the second must be given with explicit rules for its use [Carnap, 1950].

According to Carnap, the second concept, or the *working definition* as it is called here, must fulfill the following requirements [Carnap, 1950]:

1. It is *similar* to the concept to be defined, as the latter's vagueness permits.

2. It is defined in an *exact* form.

3. It is *fruitful* in the study.

4. It is *simple*, as the other requirements permit.

Since these criteria seem suitable for our purpose, let us see what they mean concretely to the working definition of intelligence (here I change the names and order of the first two requirements):

Sharpness. The definition should draw a relatively sharp line between the systems with intelligence and the ones without it. Given the working definition, whether or how much a system is intelligent should be clearly decidable. For this reason, intelligence cannot be defined in terms of other ill-defined concepts, such as *mind, thinking, cognition, intentionality, rationality, wisdom, consciousness*, etc., though these concepts do have close relationships with intelligence. As well, the definition needs to answer the complement question: "What is not intelligent?" — The reason is simply if everything is intelligent, then the concept becomes empty.[2]

Faithfulness. The line drawn by the definition should not be an arbitrary one. Though "intelligence" has no precise meaning in everyday language, it does have some common usage with which the working definition should agree. For instance, normal human beings are intelligent, but most animals and machines (including ordinary computer systems) are either not intelligent at all or much less intelligent than human beings. For this reason, AI

[2]For this reason, to define intelligence using the recently fashionable term "agent" is also not a good idea, because the term is too vague and too outstretched.

should not be defined to have the same meaning as "computer science."

Fruitfulness. The line should not only be descriptive, but also be constructive. Given the nature of AI as both a science and a technology, the "what is it?" and the "how to do it?" parts are closely related. The working definition should provide concrete guidelines for the research based on it. For instance, what assumptions can be accepted, what phenomena can be ignored, what properties are desired, and so on. Most importantly, the working definition of intelligence should contribute to solving fundamental problems in AI. For this reason, we want to avoid various "sterile" definitions, which sound correct, but tell us little about how to build an intelligent system.

Simplicity. Although intelligence is surely a complex mechanism, the working definition should be as simple as possible. From a theoretical point of view, a simple definition can be explored in detail; from a practical point of view, a simple definition is easy to use.

For our current purpose, there is no "right" or "wrong" working definition for intelligence, but there are "better" and "not-so-good" ones, judged according to the above criteria. Though there is no evidence showing that in general the requirements cannot be satisfied at the same time, the four requirements may conflict with each other when comparing proposed definitions. For example, one definition is more fruitful, while another is simpler. In such a situation, some weighing and trade-off become necessary.

Especially, the requirement of "faithfulness" should not be understood as to mean that the working definition of intelligence should be determined according to an authoritative dictionary, or a poll among all the people. A working definition might even be counter-intuitive, if there is evidence showing that such a definition is faithful to the "deep meaning" of a concept. This is why we cannot argue that Einstein's concepts of "time" and "space" should be renamed because they conflict with our everyday usage of these terms. As Feyerabend said, "without a constant misuse of language there cannot be any discovery, any progress." [Feyerabend, 1993].

1.2 Various schools in AI research

With the above criteria in mind, we can evaluate the current AI approaches by analyzing their working definitions of intelligence. Since it is impractical to study each of the existing working definitions of intelligence one by one (there are simply too many of them), I will group them into several schools of thought and consider each school in turn. As usual, a concrete definition may belong to more than one school.

Stated previously, AI is the attempt of building computer systems that are "similar to the human mind." But in what sense are they "similar"? To different schools, the desired similarity may involve *structure, behavior, capability, function,* or *principle* of the systems. In the following, I discuss typical opinions in each of the five schools, to see where such a working definition of intelligence will lead research to.

1.2.1 To simulate the human brain

In the middle of all puzzles and problems about intelligence, there is one obvious and undoubtable fact, that is, the most typical example of intelligence we know today is produced by the human brain. Therefore, it is very natural to attempt to achieve AI by building a computer system that is as similar to a human brain as possible.

There are many researchers working on various kinds of "brain models" and "neurocomputational systems," though not all of them associate themselves with AI. However, there are people who believe that the best way to achieve AI is by looking into the brain, and some of them even argue that "the ultimate goals of AI and neuroscience are quite similar" [Reeke and Edelman, 1988]. Recent attempts in this direction include [Hawkins and Blakeslee, 2004, Hecht-Nielsen, 2005].

Though there is motive to identify AI with a *brain model*, few AI researchers take such an approach in a very strict sense. Even the "neural network" movement is "not focused on *neural modeling* (i.e., the modeling of neurons), but rather ... focused on *neurally inspired* modeling of cognitive processes" [Rumelhart and McClelland, 1986].

Why? One obvious reason is the daunting *complexity* of this approach. Current technology is still not powerful enough to simulate a huge neural network, not to mention the fact that there are still many mysteries about the brain.

Moreover, even if we were able to build a brain model at the neuron level to any desired accuracy, it could not be called a success for AI, though it would be a success for neuroscience. From the very beginning, and for a good reason, AI has been more closely related to the notion of a "model of mind", that is, a *high-level* description of brain activity in which biological concepts do not appear [Searle, 1980].

A high-level description is preferred, not because a low-level description is impossible, but because it is usually simpler and more general. When building a model, it is not always a good idea to copy the object or process to be modeled as accurately as possible, because a major purpose of modeling is often to identify the "essence" of the object or process, and to filter out unnecessary details. By ignoring irrelevant aspects, we gain insights that are hard to discern in the object or process itself. For this reason, an accurate duplication is not a model, and a model including unnecessary details is not a good model.

Intelligence (and the related notions like "thinking" and "cognition") is a complicated phenomena mainly observed only in the human brain at the current time. However, the very idea of "*artificial* intelligence" assumes that the same phenomena can be reproduced in something that is different from the human brain. This attempt to "get a mind without a brain", i.e., to describe mind in a medium-independent way, is what makes AI important and attractive. Even if we finally build an "artificial brain" which is like the human brain in all details, it still does not tell us much about intelligence and thinking in general. If one day we can build a system which is very different from the human brain in details, but we nevertheless recognize it as intelligent, then it will tell us much more about intelligence than a duplicated brain does.

If we agree that "brain" and "mind" are different notions, then a good model of the brain is not a good model of the mind, though the former is useful for its own sake, and may be helpful for the building of the latter.

1.2.2 To duplicate human behavior

For the people who believe that intelligence can be defined independently of the structure of the human brain, a natural alternative is to define it in terms of human intellectual behavior. After all, if a system behaves like a human mind, it should deserve the title of "intelligence" for both theoretical and practical reasons. From this standpoint, whether the system's internal structure is similar to the human brain is mostly irrelevant.

This view is perhaps best captured by Turing in his famous "Imitation Game," later known as the "Turing Test" [Turing, 1950]. According to this idea, if a computer is indistinguishable from a human in a conversation (where the physical properties of the system are not directly observable), the system has intelligence.

After half a century, "passing the Turing Test" is still regarded by many people as the ultimate goal of AI [Saygin et al., 2000]. There are some research projects targeting it, sometimes under the name of "cognitive modeling." In recent years, there are also many "chatbots" developed to simulate human behavior in conversation.

On the other hand, this approach to AI has been criticized from various directions:

Is it sufficient? Searle argues that even if a computer system can pass the Turing Test, it still cannot *think*, because it lacks the *causal capacity* of the brain to produce *intentionality*, which is a biological phenomenon [Searle, 1980]. However, he does not demonstrate convincingly why thinking, intentionality, and intelligence cannot have high-level (higher than the biological level) descriptions.

Is it possible? Due to the nature of the Turing Test and the resource limitations of present computer systems, it is unlikely for the system to have stored in its memory all possible questions and proper answers in advance, and then give a convincing imitation of a human being by searching its memory upon demand. The only realistic way to imitate human behavior in a conversation is to produce the answers in real time. To do this, it needs not only cognitive faculties, but also much prior "human experience" [French, 1990]. It must, therefore, have a "body" that feels human, and all human motivations, including biological ones. Simply put, it must be an "artificial person," rather than a computer system with artificial intelligence. Furthermore, to build such a system is not merely a technical problem, since acquiring human experience means that humans treat and interact with it as a human being.

Is it necessary? By using behavior as evidence, the Turing Test is a criterion solely for *human* intelligence, not for intelligence in general [French, 1990]. As a working definition of intelligence, such an approach can lead to good psychological models, which are valuable for many reasons, but it suffers from "human chauvinism" [Hofstadter, 1979]. We would have to say, according to this definition, that "extraterrestrial intelligence" cannot exist, simply because that human experience can only be obtained on the Earth. This strikes me as a very unnatural and unfruitful way to use concepts. Actually, Turing did not claim that passing the imitation test is a necessary condition for being intelligent. He just thought that if a machine could pass the test satisfactorily, we would not be troubled by the question [Turing, 1950]. However, this part of his idea is often ignored, and consequently his test is taken by many people as a sufficient and necessary condition of intelligence.

In summary, though "reproducing human (verbal) behavior" may still be a sufficient condition for being intelligent (as suggested by Turing), such a goal is difficult, if not impossible, to achieve presently. More importantly, it is not a necessary condition for "intelligence", if we want it to be a more general notion than "human intelligence."

1.2.3 To solve hard problems

For people whose main interest in AI is its practical application, whether a system is structured like a brain or behaves like a human does not matter at all, but what counts is what practical problems it can solve — after all, that is how the intelligence of a human being is measured. Therefore, according to this opinion, intelligence means the capability of solving hard problems.

This intuitive idea explains why early AI projects concentrated on typical and challenging intellectual activities, such as theorem proving and game playing, and why achievements on these problems are still seen as milestones of AI progress. For example, many people, both within the AI community and among the general public, regard the victory of IBM's supercomputer Deep Blue over the World Chess Champion Kasparov as a triumph of AI.

For similar reasons, many AI researchers devote their effort to building "expert systems" in various domains, and view this as the way to general AI. The relation between these systems and the notion of intelligence seems to be obvious — experts are more intelligent in their domains than the average person. If computer systems can solve the same problems, they should deserve the title of intelligence, and whether the solutions are produced in a "human manner" has little importance. The way Deep Blue plays chess is very different from the way a human player plays chess. But as far as it wins the game, why should we care? Similarly, the intelligence of an expert system is displayed by its capability to solve problems for which it was designed.

Compared to the previously discussed goals, e.g., to model the human brain or to pass the Turing Test, this kind of goals is much easier to achieve, though still far from trivial. As today, we already have some success stories in game playing, theorem proving, and expert systems in various domains.

Though this approach toward AI sounds natural and practical, it has its own trouble.

If intelligence is defined as "the capability to solve hard problems," then the next obvious question is "Hard for whom?" If we say "hard for human beings," then most existing computer systems are already intelligent — no human manages a database as well as a database management system, or substitutes a word in a file as fast as an editing program. If we say "hard for computers," then AI becomes "whatever hasn't been done yet," which has been dubbed "Tesler's Theorem" [Hofstadter, 1979] and the "gee whiz view" [Schank, 1991].

The view that AI is a "perpetually expanding frontier" makes it attractive and exciting, which it deserves, but tells us little about how it differs from other research areas in computer science — is it fair to say that the problems there are easy? If AI researchers cannot identify other commonalities of the problems they attack besides mere hardness, they will not be likely to make any progress in understanding and replicating intelligence.

This application-oriented movement has drawn in many researchers, produced many practically useful systems, attracted significant funding, and thus made important contributions to the development of the AI enterprise. However, though often profitable, these systems do not provide much insight into how the mind works. No wonder people ask, after learning how such a system works, "Where's the AI?" [Schank, 1991] — these systems look just like ordinary computer applications. Actually, many "AI systems" are indeed developed in the same way as ordinary software.

Nowadays AI researchers often complain that the field does not get the credit it deserves, since many AI research results have been used by other fields without the AI label. This seems to confirm that many people in AI are actually doing ordinary computer science and application, and therefore the results are just like the results obtained outside AI, so they do not need a fancy label. If someone insists that these works should be called AI simply because they solve problems that were previously solvable only by the human mind, then by the same token numerical calculating programs should be called AI, as well.

Beside the issues of label and credit, the real problem of this approach is that it fails to explain why ordinary computer systems are not intelligent. Many people enter AI to look for a fundamentally different way to build computer systems. To them, traditional computer systems are stupid, not because they cannot do anything (in fact, they can do many amazing things), but because they solve problems in a rigid manner. Therefore, whether a system is intelligent not only depends upon what it can do, but also upon how it does it. If an expert system is as brittle as a conventional computing system [Holland, 1986], it hardly deserves to be called "intelligent."

An interesting example is the victory of IBM's supercomputer Deep Blue alluded to earlier. While many people applaud this as a great achievement for AI [Newborn, 2002], the research team that developed the system never made such a claim [Campbell et al., 2002]. Instead, they made it clear that "although Deep Blue's speed and search capabilities enable it to play grandmaster-level chess, it is still lacking in general intelligence." [Campbell, 1997].

Human beings usually use their intelligence to play games, but it does not mean that a computer system must do the same. In theory, it is possible to find (or invent) a game that is simple enough for a supercomputer to perform an exhaustive search to find the best move, but still too complicated for a human mind to play in like manner. Such a game can still be seen as a testing of human intelligence, because intelligent players will play better after a while (given that there are recognizable patterns in the game). However, a simple brute-force search algorithm, e.g., minimax, will be the world champion, simply because it will always find the optimum solution. In this case, should the algorithm fit the criteria of "intelligence"?

1.2.4 To carry out cognitive functions

As an attempt to generalize the various concrete behavior and capability into domain-independent form, the current AI field is often seen as studying a set of cognitive functions, including searching, recognizing, categorizing, reasoning, planning, decision making, problem solving, learning, and so on. Furthermore, for the interaction between the system and its environment (including other systems), sensorimotor and natural language processing can also be seen as cognitive functions.

Each cognitive function is typically specified as a *computation* process that starts with given input data, and after some processing generates the desired output data (plus certain side-effects inside and/or outside the system). The goal of the research is to find the most efficient *algorithm* to carry out a given function. Finally, the algorithm is implemented into a computer system, which can then handle the problem for us [Marr, 1982].

This approach has produced, and will continue to produce, information-processing tools in the form of software packages and even specialized hardware, each of which can carry out a function that is similar to certain mental skills of human beings, and therefore can be used in various domains for practical purposes. However, this kind of success does not justify the claim that it is the right way to study AI. To define intelligence as a "toolbox" of cognitive functions has the following weaknesses:

- When specified in isolation, a formalized function is often quite different from its "natural form" in the human mind. For example, to study analogy without perception leads to distorted cognitive models [Chalmers et al., 1992].

- Having any particular cognitive function is not enough to make a system intelligent. For example, problem-solving by exhaustive search is usually not considered intelligence, and many unintelligent animals have excellent perceptual capability.
- Even if we can produce all the desired functions, it does not mean that we can easily integrate them into one system, because different functions may be developed under different assumptions, which prevent the tools from being integrated. According to the past experience in building integrated systems, "Component development is crucial; connecting the components is more crucial" [Roland and Shiman, 2002].

The basic problem with the "toolbox" approach is: without a "big picture" in mind, the study of a cognitive function in an isolated, abstracted, and often distorted form simply does not contribute much to our understanding of intelligence.

A common counterargument runs something like this: "Intelligence is very complex, so we have to start from a single function to make the study tractable." For many systems with weak internal connections, this is often a good choice, but for a system like the mind the situation may be just the opposite. When the so-called "functions" are actually phenomena produced by a complex-but-unified mechanism, reproducing all of them together (by duplicating the mechanism) is simpler than reproducing only one of them. For example, we can grow a tree, but we cannot generate a leaf *alone*, although a leaf is much simpler than a tree. Intelligence may be such a phenomenon.

As Piaget said: "Intelligence in action is, in effect, irreducible to everything that is not itself and, moreover, it appears as a total system of which one cannot conceive one part without bringing in all of it."[Piaget, 1963] This opinion does not deny that intelligence includes many distinguishable functions carried out by distinct mechanisms, but it stresses the close relations among the functions and processes, which produce intelligence as a whole. If intelligence is a toolbox, where is the hand that use the tools?

1.2.5 To develop new principles

In the cognitive sciences, especially, AI, psychology, and philosophy, there are some researchers who believe that intelligence (or cognition) are governed by a small set of general and simple principles. According to this opinion, all behaviors, capabilities, and functions of the human mind can be explained as produced by the application of these principles in concrete situations [Chater and Oaksford, 1999].

Typically, these principles are represented as some kind of "rationality," formed by the evolution process as the best adaptation strategy in a certain sense. Here are some examples:

> *Bounded rationality* [Simon, 1983]: "Within the behavioral model of bounded rationality, one doesn't have to make choices that are infinitely deep in time, that encompass the whole range of human values, and in which each problem is interconnected with all the other problems in the world."
>
> *Type II rationality* [Good, 1983]: "Type II rationality is defined as the recommendation to maximize expected utility allowing for the cost of theorizing. It involves the recognition that judgments can be revised, leading at best to consistency of *mature* judgments."
>
> *Minimal rationality* [Cherniak, 1986]: "We are in the finitary predicament of having fixed limits on our cognitive resources, in particular, on memory capacity and computing time."
>
> *General principle of rationality* [Anderson, 1990]: "The cognitive system operates at all times to optimize the adaptation of the behavior of the organization."
>
> *Limited rationality* [Russell and Wefald, 1991a]: "Intelligence was intimately linked to the ability to succeed as far as possible given one's limited computational and informational resources."

According to these ideas, an AI theory should establish a few principles to derive all the functions and behaviors, then a formal model of intelligence should be formulated according to such a normative theory (such as logic, probability theory, and so on), which always "does the right thing," according to the underlying principles.

If such an approach works, we will eventually have a well-justified theory of AI, in which all functionalities are based on a consistent foundation.

Like the previous approaches, this approach has its problems. Though it is not a new idea that the human mind, like many other objects of scientific research, can eventually be explained by a small set of principles, none of the "principles" proposed so far has successfully achieved this goal yet. Although "haven't found such principles" does not prove "no such principles can exist," it often leaves people feeling that way. In AI, to date, there have been too many promises of success only to be followed by failure for people to believe in any new "silver bullet" that can solve the AI problem with one shot. Consequently, they would rather believe that given the complexity of the domain, AI must be treated as a collection of concrete problems that have to be handled one by one.

A much more serious challenge to this kind of approach is the existence of many well-documented psychological phenomena, showing that people often violate the proposed normative theories, such as predicate logic [Wason and Johnson-Laird, 1972] and probability theory [Tversky and Kahneman, 1974]. Any rationality-base theory must try to explain these phenomena [Anderson, 1990].

1.3 AI as a whole

1.3.1 Relations among the goals

From the previous discussion, we can see that instead of currently having one common research goal, the field of AI has a set of different, but related, goals pursued through different schools of thought. As Nilsson said: "AI shows all the signs of being in what the late Thomas Kuhn called a pre-paradigmatic, pre-normal-science stage." [Hearst and Hirsh, 2000].

Each of these goals reflects a particular aspect of our current usage of the word "intelligence," defines the term in a relatively sharp and simple way, and has been producing interesting results. In this sense, all of them guide legitimate scientific research, and contribute to our understanding of intelligence, as well as to the progress of each other. Since they have different goals, they can and should co-exist for a long time.

However, at a more general level, these schools do compete — as the best way to build a "thinking machine." There is no contradiction here. When these schools are evaluated as research goals in their own right, each of them is valuable in a particular way. But if they are evaluated as paths to the common goal of AI, as introduced at the beginning of the chapter, they are not equally good. Of course, which one is better is a controversial issue.

From the above presentation, one point I want to make is that there is no "natural" or "self-evident" definition of intelligence, and nobody in the field can escape from the responsibility of choosing a school to work within. Many people claim that they are not interested in philosophical debates, and they simply choose the natural or obvious problem to work on, but in reality they have made the choice unconsciously, guided by their intuition or non-academic factors, such as personal background, adviser expertise, practical need, grant source, publication possibility, current fashion, and so on. After the initial choices (which are typically made early in their career), they gradually get accustomed to them, and spend most of their time in solving the problems specified by the school, without considering more fundamental questions like whether they are the right problems to work on.

Some people may think that the different schools are aimed at different "parts" of intelligence, like in the parable of "The Blind Men and the Elephant," and the best way is to "integrate" them. However, here the situation is different. These schools are generally *incompatible* (though they have small overlaps here or there), and therefore cannot be fully integrated into a consistent theory on intelligence, or be satisfied together by a computer system.

There are many systems that use techniques developed in different schools of thought, but these "integrated" or "hybrid" systems are often justified by what they can do, rather than by a consistent theory on intelligence. This is the case because different schools usually make different design decisions. For example, the most efficient way to solve a problem in a computer is often very different from the way used in the human mind. Which one is more "intelligent"? Well, it depends on that you mean by "intelligence."

It is fine to set up a "major" goal, and, at the same time, to achieve other "minor" goals as much as possible. Even in this case, school conflicts exist, necessitating compromise and trade-offs to make progress.

The multi-school nature of the current AI field causes much confusion, because people often use the requirement of one school to judge the results produced by another school, which usually does not provide a fair conclusion. Also, the answers to many general questions on AI depend on which school is referred to. Examples of these questions include "When can we get an intelligent system?" "Can an AI be more intelligent than a human being?" "Can an intelligent system be creative or original or conscious?" "Can an intelligent system run out of human control?" and so on — their answers are different in different schools.

1.3.2 Different opinions on unified AI

Should AI be addressed as one problem, or a collection of loosely related problems that can be handled one by one separately? Again, the answer to this question depends on the interpretation of the concept of "intelligence," and there are very different opinions.

The majority of the current AI community believes in a "divide-and-conquer" approach toward AI. Many researchers claim that their research will contribute to the whole AI enterprise by focusing on a particular aspect of intelligence. Usually, there is an implicit assumption under this kind of claims, that is, when all these particular solutions are finally put together, we will have an AI, a "thinking machine."

However, as mentioned previously, such an assumption is hard to justify. It may be true that a complicated problem should be cut into pieces of smaller problems and solved one by one, but if everyone cuts the problem in his/her own way, and only works on a small piece of the problem obtained in this manner, we have no reason to believe that the solutions can later be put together to form a solution to the original problem [Brooks, 1991].

This is well illustrated by considering the present state of research in the field of AI. Browsing a journal with AI in its title or attending a conference with AI in its name, it is all too common to find articles or presentations that have very little to do with each other. In fact, nowadays few people even mention the relation between their current research and the big picture of AI.

Many people seem comfortable with this situation. They think that the idea of a "thinking machine" or something like that belongs to science fiction only, and that few people are pursuing the goal only means that the field has become mature. They do not care whether their systems are really "intelligent," which is just a label that can be attached or removed according to context for convenience purposes.

Of course, there are still attempts to unify the AI field. Newell is one of the few people who actually tried to build a unified AI theory. In [Newell, 1990], he argued for the need of unified theories, and discussed what such a theory should include. Though his theory is well known, and his project, Soar, is still alive, this kind of work does not attract many followers these days.

For the people who feel uncomfortable about the fragmented status of the field, one response is to find a unified way to *describe* the problems and solutions. The most recent attempt in this category is to uniformly describe the field within the framework of *intelligent agent* [Nilsson, 1998, Russell and Norvig, 2002]. Though this effort improves the coherence of AI textbooks, it is far from enough in unifying the techniques covered under the umbrella of "agent."

People who still associate themselves to the original AI goal find the current situation disappointing. As Minsky said [Stork, 1997b]:

> The bottom line is that we really haven't progressed too far toward a truly intelligent machine. We have collections of dumb specialists in small domains; the true majesty of general intelligence still awaits our attack.
>
> We have got to get back to the deepest questions of AI and general intelligence and quit wasting time on little projects that don't contribute to the main goal.

Wolfram made a similar comment [Stork, 1997a]:

> Nobody's trying more fundamental stuff. Everyone assumes it's just too difficult. Well, I don't think there's really any evidence of that. It's just that nobody has tried to do it. And it would be considered much too looney to get funded or anything like that.

Though there is no evidence showing the impossibility of unified AI, the past experience does make the AI community turn away from such a goal. In this atmosphere, only two types of people continue to pursue the "thinking machine" dream: the well-established researchers, and the people at the margin or even the outside of the AI community. Though these two groups of people have opposite status in many attributes, they have one thing in common, that is, they don't care too much about what the others say, and they can keep their research going even if the majority of the AI community dislikes it.

1.3.3 AGI projects

The "Thinking Machine Dream" mentioned above goes with many names, such as "Unified AI," "Strong AI," "Real AI," "Hard AI," "AGI (Artificial General Intelligence)," "Human-Level Intelligence," and so on. I'll use the term AGI in this book, though this choice does not really make any difference, since few of these terms are accurately defined. The common thesis behind these terms is the belief that intelligence is a unified mechanism that should be described and developed as a whole, independent of any application domain. Even if the development must be carried out step by step, an overall plan should be drawn first to guide the process.

For such a project, one crucial issue is to have a theoretical foundation with sufficient width to support all kinds of functions and capabilities. Though there has been a very small number of people doing this kind of research, they still belong to different schools of thought, as described previously.

In the following I will discuss some representative AGI projects, though it is not an attempt of reviewing all such projects exhaustively.

To many people, the capability to solve various types of problems is at the core of intelligence. The first attempt to build a general-purpose system for this task is the General Problem Solver (GPS) [Newell and Simon, 1976]. By analyzing "protocols" observed in the problem-solving processes of human beings, Newell and Simon represented them as state-space search, and used "means-ends analysis" to lead the search process. According to this approach, problem solving is treated as finding a sequence of actions that transforms the initial state into a final state, step by step. When there are multiple alternatives at a state, the difference between the next state and a final state is used as a *heuristic* to estimate the distance from the former to the latter. Although few people still believe that all problem solving can and should be handled in this way, the search is still referred by some as "the most fundamental method of all" [Newell, 1990].

Another ambitious attempt to make a breakthrough in AI is the Fifth Generation Computer Systems (FGCS) project of Japan, initiated in the early 1980's. The belief behind the project, roughly speaking, was that the bottleneck of AI was in the von Neumann computer architecture, which was initially designed for *sequential calculation*. On the contrary, the key in AI should be *parallel inference*. To build a machine that is "as smart as a person," we should turn from "sequential calculation on data" to "parallel inference on knowledge," and to build a proper computer system to support the latter is the key of AI [Feigenbaum and McCorduck, 1983]. This approach caused quite a splash in the AI community and consequently received a lot of attention from governments and companies all around the world. Today, however, there haven't been any remarkable advances as a result of FGCS, not to mention any breakthrough in AI research, although parallel inference engines have been built as scheduled. On the contrary, now this project is rarely mentioned, as if it never existed.

No matter what the reasons are, none of the historical AGI projects has delivered the results initially promised. This partly explains why there are few such projects being actively worked on.

For these on-going projects, in the following I directly cite their project websites, with a brief description.

In the mainstream AI, there are three well-known projects that can be categorized as AGI.

Cog (http://www.ai.mit.edu/projects/humanoid-robotics-group/cog/)
This project is based on the belief that intelligence should come out of a robot that directly interacts with the physical world.

CYC (http://www.cyc.com)
This project was initially the American response to FGCS. Instead of focusing on the inference engine, this project puts most of its efforts in the building of a huge knowledge base that holds "common sense."

Soar (http://sitemaker.umich.edu/soar)
This system can be seen as a follow up of GPS. It attempts to provide a unified model of cognition in the framework of state-space search, which is implemented as a production system.

The next group of projects have smaller scopes in their goals, though they also are in the direction of AGI, to various degrees.

ACT-R (http://act-r.psy.cmu.edu/)
This is not an AI project, but a psychological model of human cognition. Nevertheless, it is still closely related to AGI. The basic architecture is also a production system (like Soar).

OSCAR (http://www.u.arizona.edu/~pollock/)
This is an architecture for rational agents based upon a philosophical theory of rational cognition. The core technique is defeasible inference.

SNePS (http://www.cse.buffalo.edu/sneps/)
This is an attempt to unify knowledge representation, reasoning, and natural-language processing, using a semantic network. The research has begun to integrate sensorimotor capability into the system.

Finally, there is a group of projects that are ambitious and take more radical paths, though they have not got much attention from the mainstream AI community.

AIXI (http://www.idsia.ch/~marcus/ai/index.htm)
A mathematical theory of universal induction, based on probability theory and computation theory.

a2i2 (`http://adaptiveai.com/project/index.htm`)
A connectionist AGI system, which is embodied, and learns from its interaction with the environment.

CAM-Brain (`http://www.cs.usu.edu/\%7Edegaris/cam/index.html`)
An artificial brain consisting of roughly a million modules of cellular automata based neural circuits, which grow and evolve.

Novamente (`http://www.agiri.org/engine.htm`)
An integrated AGI system with multiple techniques, include probabilistic reasoning, genetic programming, and so on.

These projects, as well as some other AGI approaches, are described in [Goertzel and Pennachin, 2006].

In summary, past research on unified artificial intelligence has not produced encouraging results; currently there is only a small number of people involved in AGI work; wherein the on-going AGI projects are based on very different opinions.

Chapter 2

A New Approach Toward AI

In this book, a new approach toward AI will be presented. This chapter first informally introduces the basic ideas.

2.1 To define AI

As described in the previous chapter, different schools of AI research are based on different working definitions of "intelligence." Therefore, I start by clarifying my definition, which encapsulates the goal of the research.

2.1.1 Information System

The working definition of intelligence, no matter what it is, should distinguish one type of system from another type of system. More concretely, here I want to distinguish one type of *information system* from another type.

The concept "information," like "intelligence," is also a concept used differently by different people. In this book, this concept is used to set the "platform" or "background" for the discussion on intelligence, so I only state my working definition for it, that is, what I mean by it, without a detailed discussion about why it is better than alternative definitions.

An information system, or information-processing system, is a system whose internal activities and interactions with its environment can be described *abstractly* — that is, without specifying the concrete entity and process (hardware) that carries out the activities and interactions.

Usually, such a system has certain *tasks* (also called *goals*) to carry out, given by the environment or generated by the system itself. To do this, the system takes various *actions* (also called *operations*), guided by its *knowledge* (also called *beliefs*) about how the actions and the tasks are related. Any internal activity costs the system some *resources*, especially, processing time and memory space.

The *environment* of such a system may be the physical world (if the system has sensorimotor capability), or other information-processing systems (human or computer). In either case, the interactions are specified by the *experiences* (or *stimuli*) and the *behaviors* (or *responses*) of the system, which can be described as streams of input and output information, respectively. For the system, recognizable patterns of input and producible patterns of output constitute its *interface language*.

According to this definition, all human beings and computer systems, as well as many animals and automatic control systems, can be described as information systems.

To call a system an "information system" means to describe the system at an abstract level, and many low-level details will be omitted from the description. A computer server is an information system to a remote user, who does not care about its size, color, and weight. However, to a worker who is moving the server from one room to another, it is no longer suitable to treat it as an information system. If you throw a ball to a friend as a prearranged signal for something, your action is information transformation, and where the ball goes does not matter. However, it would be silly to call the ball-throwing "information transformation" in a baseball game — it is not wrong, but contributes little to our understanding of the game.

Even if a entity or process cannot be treated as an information system, it often can be "modeled" or "simulated" in an information system. It means that the system can be described at an abstract level, and another system can be built that has the same high-level description, which contains essential features of the system to be modeled, though these two systems are completely different at a lower level. For example, a hurricane can be modeled in a computer, so that its movement can be predicted. However, "being a kind of movement of air" is a defining property of hurricane, which is not in the computer simulation. In this sense, a model is not the entity or process to be modeled, and they are only similar at a high level of description.

However, if the entity or process being modeled is such a system that all of its major properties are shown at the "information-processing" level, then we no longer call the above procedure "modeling" or "simulating," but call it "reproducing," "replicating," or "implementing." For example, arithmetic calculation are manipulations of symbolic entities and relations. Whether it is done by stones, abacus, or pen and paper usually has little importance to the result. When such a process is carried out by a computer, we do not say that the arithmetic calculation is "simulated" in the computer — unlike a hurricane, the calculation in a computer is genuine.

Sensitive readers would have realized why I start the discussion about AI with information system. Actually, to ask "To what extent can intelligence be produced by a computer?" is the same as to ask "To what extent can intelligence be described as information processing?" and the latter question will be answered by this book, using the information-processing terminology introduced above.

2.1.2 A working definition of intelligence

Following the preparation of the previous subsection, I propose here a working definition of intelligence:

> *Intelligence is the capability of an information system to adapt to its environment while operating with insufficient knowledge and resources.*

Here the meanings of "information system," "environment," "knowledge," and "resources" have been clarified previously. However, the two major components of the definition: "adaptation" and "insufficient knowledge and resources" remain to be explained.

In terms of behavior change, we can distinguish three types of systems:

Instinctive system: The behaviors of the system remains the same, and do not change over time.

Erratic system: The behaviors of the system change, but not as a function of its experience.

Adaptive system: The behaviors of the system change according to its experience. It attempts to improve its performance in carrying out the tasks, under the assumption that its future experience will be similar to its past experience.

On the other hand, we can roughly distinguish three kinds of environment, in terms of its interaction with a system in it:

Constant environment: The environment is deterministic, and the results of system behaviors never change. In this kind of environment, the best way to carry out a task is by an instinctive system specially built for the task. An adaptive system may be able to learn the behavior, but it is less efficient.

Random environment: The environment is completely unpredictable. In this situation, all systems are equally bad in the long run.

Stable environment: The environment may change, though not randomly, and the results of system behaviors are usually predictable, but with exceptions from time to time. Adaptive systems work better in this kind of environment, so long as its adaptation-rate is not too slow compared to the speed of change in the environment.

In this sense, an adaptive system is not necessarily better than a non-adaptive system. Actually, if a problem can be handled without adaptation (i.e., can be solved "mechanically"), it is better to do it that way. Adaptation is needed only when there is no predetermined solution available.

Just being "adaptive" is not enough for being "intelligent." A complex system can be called "adaptive" only because a few parameters in it can be tuned by itself according to its experience. Intelligence requires more than that, and this is why I have another component in the working definition of intelligence.

Insufficient knowledge and resources means that the system works under the following restrictions:

Finite: The information-processing capability of the system's hardware is fixed.

Real-time: All tasks have time constraints attached to them.

Open: No constraint is put on the content of the experience that the system may have, as long as they are representable in the interface language.

Not all information-processing systems take the insufficiency of knowledge and resources into full consideration. Non-adaptive systems, for instance, simply ignore new knowledge in their interactions with their environment. As for artificial adaptive systems, most of them are not finite, real-time, and open, in the following senses:

1. Though all concrete systems are finite, many theoretical models (for example, Turing Machine) neglect the fact that the requirements for processor time and/or memory space may go beyond the supply capability of the system [Hopcroft and Ullman, 1979].

2. Most current AI systems do not consider time constraint at run time. Most real-time systems can handle time constraint only if they are essentially deadlines [Strosnider and Paul, 1994].

3. In most systems, various explicit or implicit constraints are imposed on what a system can experience. For example, only questions that can be answered by retrieval and deduction from current knowledge are acceptable, new knowledge cannot conflict with previous knowledge, and so on.

Many computer systems are designed under the assumption that their knowledge and resources, though *limited* or *bounded*, are nevertheless *sufficient* to fulfill the tasks that they will be called upon to handle. When facing a situation where this assumption fails, such a system simply panics, and asks for external intervention, usually from a human manager of the system.

For a system to work under the *Assumption of Insufficient Knowledge and Resources*, hereforth known as *AIKR*, it should have mechanisms to handle the following situations:

> A processor is required when all processors are occupied;
>
> A piece of memory is required when all memory is already full;
>
> A task comes up when the system is busy with something else;
>
> A task comes up with a time constraint, so exhaustive processing is not affordable;
>
> New knowledge conflicts with previous knowledge;
>
> A question is presented for which no sure answer can be deduced from available knowledge;
>
> etc.

For traditional computing systems, these situations usually either require human intervention, or simply cause the system to reject the task or knowledge involved. However, for a system designed under AIKR, these are *normal situations*, and should be managed smoothly by the system itself.

The two main components in the working definition, *adaptation* and *insufficient knowledge and resources*, are related to each other. An adaptive system must have some insufficiency in its knowledge and resources, for otherwise it would never need to change at all. On the other hand, without adaptation, a system may have insufficient knowledge and resources, but make no attempt to improve its capability. Such a system acts, for all intents and purposes, as if its knowledge and resources were sufficient.

According to the above definition, intelligence is indeed a "highly developed form of mental adaptation" [Piaget, 1960]. This assertion is consistent with the usages of the two words in natural language: we are willing to call many animals, computer systems, and automatic control systems "adaptive," but not "intelligent," because the latter has a higher standard than the former.

When defining intelligence, many authors ignore the complementary question: what is unintelligent? If everything is intelligent, then this concept is empty. If every computer system is intelligent, it is better to stay within the theory of computation. Even if we agree that intelligence, like almost all properties, is a matter of degree, we still need criteria to indicate what makes a system more intelligent than another. An unintelligent system is not necessarily incapable or gives only wrong results. Actually, most ordinary computer systems and many animals can do something that human beings cannot. However, these abilities do not earn the title "intelligent" for them. What is missing in these capable-but-unintelligent systems?

According to the working definition of intelligence introduced previously, an *unintelligent* system is one that does not adapt to its environment. Especially, in artificial systems, an unintelligent system is one that is designed under the assumption that it only works on problems for which the system has sufficient knowledge and resources.

An intelligent system is not always "better" than an unintelligent system for practical purposes. Actually, it is the contrary: when a problem can be solved by both of them, the unintelligent system is usually better, because it guarantees a correct solution. As Hofstadter said, for tasks like adding two numbers, a "reliable but mindless" system is better than an "intelligent but fallible" system [Hofstadter, 1979].

2.1.3 Comparison with other definitions

According to the classification of AI schools in the previous chapter, the above working definition of intelligence belongs to the attempts that treat intelligence as being defined by some underlying *principles*.

Furthermore, it is an attempt to attack the whole AI problem, rather than part of it. Therefore we can also classify it as an "AGI," using the term introduced in the previous chapter. Personally, however, I would rather stay with the term "AI," because to me, "intelligence" is a domain-independent capability, so it is redundant to say "general intelligence." However, since many other people do not use the term in this way, I do not mind to use AGI or similar terms to stress the "general purpose" nature of the related systems.

How is my working definition of intelligence different from the others discussed in the previous chapter?

- In the following chapters, I will show that a system developed on such a foundation has many cognitive *functions*, but they are better thought of as emergent phenomena than as well-defined tools used by the system.

- By learning from its experience, the system potentially can acquire the *capability* to solve hard problems — actually, "hard" problems are exactly those for which the system has insufficient knowledge and resources. However, this capability is not built into the system, and thus, without proper training, no capability is guaranteed, and acquired capability can even be lost.

- Because the human mind also follows the above principles, such a system can be expected to behave similarly to human beings, but the similarity would exist at a more abstract level than that of concrete *behavior*. Due to the fundamental difference between human experience and the experience of an AI system, the system will not accurately reproduce masses of psychological data or guarantee to pass a Turing Test.

- Although the internal *structure* of the system has some properties in common with a description of the human mind at a certain level, it is not an attempt to simulate a biological neural network or the brain as a whole.

To be sure, what has been proposed in my definition is not entirely new to the AI community. Few would deny that adaptation, or learning, is important for intelligence (though many people are still working on "AI" projects that have no learning capability — to them, learning is something that can be added into the picture at a later time). Moreover, "insufficient knowledge and resources" is the focus of many subfields of AI, such as heuristic search, reasoning under uncertainty, real-time planning, and machine learning. Besides the various types of "rationality" listed in the previous chapter, similar attitudes toward intelligence can be found in the following quotations:

> Medin and Ross: Much of intelligent behavior can be understood in terms of strategies for coping with too little information and too many possibilities. [Medin and Ross, 1992]
>
> Michalski: By "intelligence," we mean a set of capabilities that let a system with limited resources (energy, time, and memory) operate under limited input information (incomplete, uncertain, inconsistent, or incorrect). [Hearst and Hirsh, 2000]

Given all that has already been said, what is *new* in my approach? As far as the working definition is concerned, this approach is new in the following aspects:

1. An explicit and unambiguous definition of intelligence as "adaptation under insufficient knowledge and resources."
2. A further clarification of the phrase "with insufficient knowledge and resources" as meaning *finite*, *real-time*, and *open*.
3. The design of all formal and computational aspects of an AGI project keeping the two previous definitions foremost in mind.

Though many AI systems are designed according to some kind of "bounded rationality", and assume some restrictions in knowledge and resources, few of them can be said to be based on AIKR. We will see detailed analysis on this topic in Part III, where my approach is compared with other approaches.

2.2 Intelligent reasoning systems

As stated in the previous chapter, a typical AI project consists of an informal theory, a formal model, and a computer implementation. The previous working definition belongs to the theory level. Now let us see how to use it to choose a proper formal language to build a model.

2.2.1 Different traditions in formalization

Formalization means, in the current context, the process of describing the status and activities of a system in a formal (artificial, symbolized) language.

In the current AI research, there are different traditions of formalization. The major ones are the following:

Dynamic system: In this tradition, the state of a system at a given time is specified by the values of a fixed set of attributes. Intuitively, it corresponds to a point in a multi-dimensional space, where each dimension corresponds to an attribute, and the coordinate of the point on that dimension corresponds to the system's value on the attribute. In this representation, the change of state (caused either by the system itself or by an outside factor) is a trajectory line in the space, indicating how one state follows another. The regularity of the system is represented by equations that describe the possible trajectory lines. AI inherits this type of formalization from dynamics, system theory, and cybernetics, and uses it in pattern recognition systems, connectionist models, and so on.

Reasoning system: In this tradition, the state of a system at a given time is specified by a set of sentences in a formal language. Each sentence represents a belief of the system, or a piece of knowledge about the environment. The change of state can either be caused by the system's inference activity (i.e., using a fixed set of rules to derive new sentences from existing ones), or by its communication activity (i.e., the input and output of the sentences). The regularity of the system is represented by the set of inference rules. AI inherits this type of formalization from mathematical logic, and uses it in various types of "knowledge-based systems."

Computing system: In this tradition, the state of a system at a given time is specified by a data structure, in which individual data items are organized together. The change of state is caused by the execution of a program, which modifies the data structure, and produces certain side effects. The regularity of the system is specified by algorithms, which are abstract representation of the programs. AI inherits this type of formalization from computer science, and uses it in searching, learning, and many other techniques.

All the three traditions are very powerful, and in principle they can emulate one another, in the sense that a virtual machine specified in one tradition can be implemented by another virtual machine specified in a different tradition. In this sense, they have equivalent expressive and processing capacity. It is also possible to build a hybrid system, in which multiple formalization traditions are integrated.

However, for a given problem, one tradition may be more natural, convenient, and efficient than the others. Therefore they are not always equivalent under practical considerations for a given problem.

I choose to formalize my working definition of intelligence in the framework of a reasoning system, mainly based on the following considerations:

- It is a general-purpose system. Working in such a framework keeps us from being bothered by domain-specific properties, and also prevents us from cheating by using domain-specific tricks.

- It uses a rich formal language, especially compared to the "language" used in multi-dimensional space, where a huge number of dimensions are needed to represent a moderately complicated situation.

- Since the activities of a reasoning system consists of inference steps, it allows a natural combination of the *rigidness* (i.e., justifiability) of each individual step and the *flexibility* (i.e., context-dependency) of the inference processes, especially compared to the algorithm-governed processes, where the linkage from one step to the next is fixed, and the process usually cannot stop in the middle.

- Compared with cognitive activities like low-level perception and motor control, reasoning is at a more abstract level, and is one of the cognitive skills that collectively make human beings so qualitatively different from other animals.

- As will be displayed by this book, the notion of "reasoning" can be extended to cover many cognitive functions, including learning, searching, categorizing, planning, decision making, and so on.

- Most research on reasoning systems is carried out within a school based on assumptions directly opposed to AIKR. By "fighting in the backyard of the rival," we can see more clearly what kinds of effects the new ideas have.

In summary, I believe that an *intelligent reasoning system* provides a suitable framework for the study of intelligence, though being a reasoning system is neither necessary nor sufficient for being intelligent. Here I will not justify the above claims, but leave that task to the end of the book, after the whole formal model is described and discussed.

2.2.2 Reasoning systems and logics

An automatic (computerized) reasoning system is an information-processing system that consists of the following major (conceptual) components:

1. *a language*, defined by a formal grammar, for the (external) communication between the system and its environment, and for the (internal) knowledge representation within the system;[1]

2. *a semantics* of the language that provides the principles to determine the meanings of the words and the truth values of the sentences in the language;

3. *a set of inference rules* that is defined formally, and can be used to match questions with knowledge, to infer conclusions from premises, to derive sub-tasks from tasks, and so on;

4. *a memory* that systematically stores tasks, beliefs, and so on, as well as provides a work place for inferences;

5. *a control mechanism* that is responsible for resource management, including to choose premises and inference rules in each inference step, and to allocate memory space.

[1] It is possible to have the two functions be accomplished by two different languages, with a translation mechanism in between.

The first three components are usually referred to as a *logic*, or the *logical part* of the reasoning system, and the last two as the *control part* of the system.

Before showing how an intelligent reasoning system is designed, let us first see its opposite — that is, a reasoning system designed under the assumption that its knowledge and resources are *sufficient* to answer the questions asked by its environment (so no adaptation is needed). By definition, such a system has the following properties:

1. No new knowledge is necessary. All the system needs to know to answer the questions is already there at the very beginning, expressed by a set of *axioms*.

2. The axioms are *true*, and will remain true, in the sense that they correspond to the actual situation of the environment.

3. The system answers questions by applying a set of formal rules to the axioms. The rules are sound and complete (with respect to the valid questions), therefore they guarantee correct answers for all questions.

4. The memory of the system is so big that all axioms and intermediate results can always be stored within it, and so effective that any content can be retrieved faithfully whenever needed.

5. There is an algorithm that can carry out any required inference in finite time, and it runs so fast that it can satisfy all time requirements attached to the questions.

This is the type of system dreamed of by Leibniz, Boole, Hilbert, and many others. It is usually referred to as a "decidable axiomatic system." The attempt to build such systems has dominated the study of logic for a century, and has strongly influenced the research of AI. Many researchers believe that such a system can serve as a model of human thinking.

However, if intelligence is defined as "to adapt under insufficient knowledge and resources," what we want is the *contrary*, in some sense, to an axiomatic system, though it is still *formalized* or *symbolized* in a technical sense. That is why *Non-Axiomatic Reasoning System*, NARS for short, is chosen as the name for the intelligent reasoning system to be introduced in this book.

2.3 Major design issues of NARS

Before going into detailed formal descriptions of NARS in the chapters of Part II, this section provides an informal introduction to the major design issues of the system. For each topic, the problem is summarized first, then the solution provided in NARS is briefly described. Full discussions and comparisons with other approaches will be given in Part III of the book, because the NARS solutions to many problems are provided by multiple components of the system.

2.3.1 Validity and rationality

A central issue of NARS is: when a system has to work with insufficient knowledge and resources, what is the criteria of *validity* or *rationality*? This problem is obviously related to Hume's problem of induction [Hume, 1748] — if the future is different from the past, how can we predict the former by the latter?

This issue needs to be addressed, because the aim of NARS is to provide a normative model for intelligence in general, not a descriptive model of human intelligence. It means that what the system does should be "the right thing to do," that is, can be *justified* against certain simple and intuitively attractive principles of validity or rationality.

In traditional logic, a "valid" or "sound" inference rule is one that never derives a *false* conclusion (that is, it will be contradicted by the future experience of the system) from *true* premises [Copi, 1982]. However, such a standard cannot be applied to a system that has to work under AIKR, since by definition, such a system has no way to guarantee the infallibility of its conclusions. On the other hand, it does not mean that every conclusion is equally valid.

As discussed previously, since intelligence is a special kind of adaptation, and in an adaptive system the behavior is determined by the assumption that the future experience will be similar to the past experience, in NARS a "valid conclusion" is one that is most consistent with the evidence in the past experience, and a "valid inference rule" is one whose conclusions are supported by the evidence provided by the premises used to derive them.

Furthermore, restricted by insufficient resources, NARS cannot exhaustively check every possible conclusion to find the best conclusion for every given task. Instead, it has to settle down with the best it can find with available resources.

In this sense, NARS can also be called an "adaptive reasoning system," whose central principle of rationality is "to predict the (unknown) future according to the (experienced) past, and to satisfy the (potentially infinite) resource request with the (actually finite) resources supply."

The following components are designed according to this principle.

2.3.2 Semantics

As was stated earlier, semantics studies how the items in a language are related to the environment in which the language is used.

Model-theoretic semantics is the dominant theory in the semantics of formal languages. For a language **L**, a model **M** consists of the relevant part of some domain described in another language **ML**, and an interpretation **I** that maps the items in **L** onto the objects in the domain (labeled by words in **ML**). **ML** is referred to as a "meta-language," which can be either a natural language, like English, or another formal language.

Given the above components, the *meaning* of a term in **L** is defined as its image in **M** under **I**, and whether a sentence in **L** is *true* is determined by whether it is mapped by **I** onto a "state of affairs" that holds in **M**. For a reasoning system, valid inference rules are those that always derive true conclusions from true premises.

With insufficient knowledge and resources, what relates the language **L**, used by a system **R**, to the environment is not a *model*, but the system's *experience*. For a reasoning system like NARS, the experience of the system is a stream of sentences in **L**, provided by a human user or another computer.

In such a situation, the basic semantic notions of "meaning" and "truth" still make sense. The system may treat terms and sentences in **L**, not solely according to their syntax (shape), but in addition taking into account their relations to the environment, according to the system's experience. Therefore, What we need is an *experience-grounded semantics*.

Under AIKR, NARS should not (and cannot) use "true" and "false" as the only truth values of sentences. To handle conflicts in experience properly, we need to determine what counts as positive evidence in support of a sentence, and what counts as negative evidence against it, and in addition we need some way to measure the *amount* of evidence in terms of some fixed unit. In this way, a truth value will simply be a numerical summary of available evidence.

Similarly, the meaning of a term (or word) is defined by the role it plays in the experience of the system, that is, by its relations with other terms, according to the experience of the system.

As was mentioned above, "experience" in NARS is represented in **L**, too. Therefore, in **L** the truth value of a sentence, or the meaning of a word, is defined by a set of sentences, also in **L**, with their own truth values and meanings — which seems to have led us into a circular definition or an infinite regress.

The way out of this seeming circularity in NARS is "bootstrapping." In the following, I will first define a very simple subset of the language, with its semantics. Then, I will use it to define the semantics of the whole language.

As a result, the truth value of statements in NARS uniformly represents several types of uncertainty, such as randomness, fuzziness, and ignorance. The semantics specifies how to understand sentences in the language, and provides justifications for the inference rules.

2.3.3 Grammar and inference rules

When presenting NARS, I take a path that is opposite to the usually accepted one. Instead of first defining a language formally, then attaching a semantics to it, I analyze the desired semantics first (guided by my working definition of intelligence), then analyze the language that can support such a semantics. The advantage of such an approach is argued in [Ellis, 1993].

From the previous discussion, we can see that what NARS needs is a language in which the meaning of a term is represented by its relation with other terms, and the truth value of a sentence is determined by available evidence. For these purposes, the concept of (positive or negative) evidence should be naturally introduced into the language. Unfortunately, the most popular formal language used in First-Order Predicate Logic does not satisfy the requirement, as revealed by the "Confirmation Paradox" [Hempel, 1943].[2] A traditional rival to predicate logic is known as *term logic*. Such logics, exemplified by Aristotle's Syllogistic, have the following features: [Bocheński, 1970, Englebretsen, 1981]

1. A typical sentence is *categorical*, which consists of a *subject term* and a *predicate term*, related by a *copula* intuitively interpreted as "to be."

2. A typical inference rule is *syllogistic*, which takes two sentences that share a common term as premises, and from them derives a conclusion formed by the other two terms.

Traditional term logic has been criticized for its poor expressive power. In NARS, this problem is solved by introducing various types of *compound terms* into the language, to represent *sets*, *intersections*, *differences*, *products*, *images*, *statements*, and so on.

The inference rules in this extended term logic carry out inheritance-based inference. Basically, each of them indicates how to use one item as another one, according to the experience of the system. Different rules correspond to different combinations of premises, and use different truth-value functions to calculate the truth values of the conclusions from those of the premises, justified according to the semantics of the system.

The inference rules in NARS uniformly carry out *choice, revision, deduction, abduction, induction, exemplification, comparison, analogy, compound term composition and decomposition*, and so on.

2.3.4 Inference control

Under AIKR, NARS cannot guarantee to process every task perfectly — with insufficient knowledge, the best way to carry out a task is

[2]This issue will be discussed in detail in Section 9.2.2.

unknown; with insufficient resources, the system cannot exhaustively try all possibilities.

Since NARS still needs to do its best in this situation, its solution is to let the items and activities in the system compete for the limited resources. Again, the validity of the resource allocation policy is justified according to the past experience of the system (rather than its future experience), and the aim is to satisfy the goals of the system as much as possible.

In the system, different data items (task, belief, or concept) have different *priority values* attached, according to which the system's resources are distributed. These values are determined according to the past experience of the system, and are adjusted according to the change of situation.

A special data structure is developed to implement a probabilistic priority queue with a limited storage. Using it, each access to an item takes roughly a constant time, and the accessibility of an item depends on its priority value. When no space is left, items with low priority will be removed.

The memory of the system contains a collection of concepts, each of which is identified by a term. Within the concept, all the tasks and beliefs (i.e., pieces of knowledge) that have the term as subject or predicate are collected together.

The running of NARS consists of individual inference steps. In each step, a concept is selected probabilistically (according to its priority), then within the concept a task and a belief are selected (also probabilistically), and the applicable inference rules take the task and the belief as premises to derive new tasks and beliefs, which are added into the memory.

The system runs continuously, and interacts with its environment all the time, without stopping at the beginning and ending of each task. The processing of a task is interwoven with the processing of other existing tasks, so as to give the system a dynamic and context-sensitive character.

Part II

Non-Axiomatic Reasoning System

Chapter 3

The Core Logic

The logic implemented in NARS is called *Non-Axiomatic Logic*, hereforth *NAL*. The formal language used in NARS is called *Narsese*.

In the following, both Narsese and NAL will be defined layer by layer, each of which adds new grammar rules and inference rules into the system. Overall, there will be eight layers, and the language used in NAL-n is Narsese-n (n = 1, ..., 8).

In this chapter, the simplest non-axiomatic logic, NAL-1, is defined. Narsese-1 is a simple language, in which each statement consists of two atomic terms, linked by an inheritance relation. NAL-1 has inference rules defined on this language.

To provide a proper semantics for this logic, an idealized version of the logic, NAL-0, is introduced first. This logic is not non-axiomatic, but will be used as part of the meta-language of NAL. Set theory and first-order predicate logic are also used as parts of the meta-language of NAL.

3.1 NAL-0: binary inheritance

NAL-0 is a simple binary deductive logic[1]. It is not actually "non-axiomatic," but we need it to define the semantics of NAL.

[1] It was formerly called Inheritance Logic in [Wang, 1994b, Wang, 1995a].

3.1.1 Language: term and inheritance

NAL-0, like all members of the NAL family, is a "term logic." This type of logic is different from predicate logics, because of its usage of *categorical* sentences and *syllogistic* inference rules [Bocheński, 1970, Copi, 1982, Englebretsen, 1996]. Therefore, it is also called "categorical logic" or "syllogistic logic."[2]

First, let us define the smallest unit of Narsese, "term."

Definition 1 *A* term, *in the simplest form, is a string of letters in an alphabet. In such a form, it is also called an* atomic term.

The default alphabet in this book is the alphabet of English plus digits 0 to 9, and most terms we use as examples are common English words, such as "bird", "animal", and "water" (we also allow hyphenated terms). It is easy for NAL to use words in another natural language (such as Chinese) as terms, because the following design does not depend on the choice of the alphabet or characters.

Later in this book, we will see that a term is the name of a concept in NARS. The above definition only gives the simplest form of terms, and more complicated forms will be introduced later.

Definition 2 *The* inheritance relation, *"$\rightarrow$", is a relation from one term to another term, and defined by* being reflexive and transitive. *An* inheritance statement *consists of two terms related by the inheritance relation. In the inheritance statement "$S \rightarrow P$", S is the* subject term *and P is the* predicate term.

In a statement, the two terms can be the same.

According to this definition, for any term X, "$X \rightarrow X$" is always true (reflexivity). Also, if "$X \rightarrow Y$" and "$Y \rightarrow Z$" are true, so is "$X \rightarrow Z$" (transitivity). On the other hand, if there is a relation defined among terms, which is both reflexive and transitive, and has no additional property, then it is the same as the inheritance relation defined in NAL.

[2] A detailed comparison between predicate logics and term logics will be given in Chapter 10.

The inheritance relation is neither symmetric nor anti-symmetric. That is, for different X and Y, given "$X \rightarrow Y$", whether "$Y \rightarrow X$" is also true cannot be determined.

NAL is a "term logic," partially because its sentences are "categorical," with a "subject-copula-predicate" format. The inheritance relation is a kind of "copula," and intuitively, it indicates that S is a *specialization* of P, and P is a *generalization* of S. It roughly corresponds to "S is a kind of P" in English. For example, "$apple \rightarrow fruit$" says that "Apple is a kind of fruit."

The inheritance relation defined above is closely related to many well-known relations, such as "belongs to" (in Aristotle's syllogisms), "subset" (in set theory), "IS-A" (in semantic networks) [Brachman, 1983], "inheritance assertion" (in inheritance systems [Touretzky, 1986]), and "inheritance" in object-oriented programming (such as "extends" in Java).

What makes the inheritance relation in NAL different from the other relations are:

1. It is a relation between two *terms* (not between sets, classes, or concepts).

2. The relation is defined completely (no more, no less) by the two properties, *reflexivity* and *transitivity*.

Now we use the above notions to define the corresponding version of Narsese.

Definition 3 Narsese-0 *is a formal language whose sentences are inheritance statements.*

Therefore, the grammar of Narsese-0 is given in Table 3.1.

3.1.2 Semantics: truth and meaning

Now let us establish a semantics for Narsese-0, by defining the notions of "truth" and "meaning" in the language.

For a reasoning system implementing the logic NAL-0, the language Narsese-0 is used for both internal knowledge representation and external communication. The initial knowledge of the system, obtained from the environment, is defined as its "experience."

<*sentence*>	::=	<*statement*>
<*statement*>	::=	<*term*><*copula*><*term*>
<*copula*>	::=	$\rightarrow$
<*term*>	::=	<*word*>
<*word*>	:	a string in a given alphabet

Table 3.1: The Grammar of Narsese-0

Definition 4 *The system's* experience, K, *is a non-empty and finite set of sentences in Narsese-0. In each statement in* K, *the subject term and the predicate term are different.*

For example, we can have $K = \{apple \rightarrow fruit,\ fruit \rightarrow plant\}$. As a set, K has no duplicated elements, and there is no order among its elements.

Definition 5 *Given experience* K, *the system's* beliefs, K^*, *is the transitive closure of* K, *excluding statements whose subject and predicate are the same term.*

Therefore, K^* is also a non-empty and finite set of sentences in *Narsese-0*, which includes K, as well as the sentences derived from K according to the transitivity of the inheritance relation. For the above K, $K^* = \{apple \rightarrow fruit,\ fruit \rightarrow plant,\ apple \rightarrow plant\}$. K^* can be generated from K in finite steps using an ordinary closure-generating algorithm.

In NARS, the words "belief" and "knowledge" are usually treated as exchangeable with each other.[3] Therefore, K^* can also be called the *knowledge base* of the system (as in some previous publications on NARS).

Now we can define the "truth value" of a statement and the "meaning" of a term, with respect to a given K.

Definition 6 *The* truth value *of a statement in NAL-0 is either* true *or* false. *Given experience* K, *the truth value of a statement is true if it is in* K^*, *or has the form of* $T \rightarrow T$, *otherwise it is false.*

[3]A justification of this decision will be given in Section 7.4.1.

Therefore we have two types of truth in NAL-0: *empirical* and *literal* (or call them *synthetic* and *analytic*, respectively). The former is "true according to experience," and the latter is "true by definition." Given the above definitions, truth in these two categories have no overlap.

In the following, we will call true statements "positive knowledge," and false statements "negative knowledge." All analytic truths are positive knowledge (and all of them fall into the pattern "$T \rightarrow T$"). Synthetic knowledge may be positive or negative. In NAL-0, negative knowledge are implicitly represented: they are not sentences in Narsese-0, but propositions in its meta-language. The amount of positive knowledge, (i.e., number of beliefs in K^*) increases monotonically with the increase of the experience K, but that is not the case for negative knowledge, which is implicitly defined by the former as "statements that not known to be true."

Therefore, here "true" is "derivable from experience," or "have the relation"; while "false" is "haven't found the relation," but not "have an anti-relation." NAL-0 accepts the "closed world" assumption, where "lack known relation" is treated as "no relation." It is necessary here, because if truth value is a summary of experience, then "having an unknown truth value" makes no sense.

For a term T that does not appear in K, all statements having T in them are false, except "$T \rightarrow T$." For example, given the above experience K, "$orange \rightarrow fruit$" is false. Though "$\neg(orange \rightarrow fruit)$" is not a valid sentence of Narsese-0, it is a valid proposition in the meta-language of NAL-0 (by taking statements of NAL-0 as propositions in propositional logic).

To clarify how a particular term T is related to other terms according to the experience of the system, the *extension* and *intension* of T are defined as the sets of its known specializations and generalizations, respectively:

Definition 7 *Given experience K, let the set of all terms appearing in K to be the* vocabulary *of the system, V_K. Then, the* extension *of a term T is the set of terms $T^E = \{x \mid x \in V_K \land x \rightarrow T\}$. The* intension *of T is the set of terms $T^I = \{x \mid x \in V_K \land T \rightarrow x\}$.*

Obviously, both T^E and T^I are determined with respect to K, so they can also be written as T^E_K and T^I_K. In the following, the simpler notions are used, with the experience K implicitly assumed.

Since "extension" and "intension" are defined in a symmetric way in NAL, for any result about one of them, there is a dual result about the other. Each belief of the system reveals part of the intension for the subject term and part of the extension for the predicate term. For example, "*apple* $\rightarrow$ *fruit*" indicates that "*apple*" is in the extension of "*fruit*," and "*fruit*" is in the intension of "*apple*."

Theorem 1 *For any term* $T \in V_K$, $T \in (T^E \cap T^I)$. *If* T *is not in* V_K, $T^E = T^I = \{\}$, *though "*$T \rightarrow T$*" is still true.*

Since all the theorems in this book are easy to prove, I will leave the detailed proofs to the interested reader, and only explain their implications.

The above theorem states that any given term in V_K has a non-empty extension and a non-empty intension — both of them contain at least the term itself.

Definition 8 *Given experience* K, *the* meaning *of a term* T *consists of its extension and intension.*

Therefore, the meaning of a term is its relation with other terms, according to the experience of the system. A term T is "meaningless" to the system, if $T^E = T^I = \{\}$ (that is, it has never got into the experience of the system), otherwise it is "meaningful." The larger the extension and intension of a term are, the "richer" its meaning is.

Theorem 2 *If both* S *and* P *are in* V_K, *then* $(S \rightarrow P) \equiv (S^E \subseteq P^E) \equiv (P^I \subseteq S^I)$.

Here "$\equiv$" is the "if and only if" in propositional logic. This theorem says that "$S \rightarrow P$" is true if and only if the extension of S is fully contained in the extension of P, and also if and only if the intension of P is fully contained in the intension of S. In other words, the statement "There is an inheritance relation from S to P" is equivalent to both "P inherits the extension of S" and "S inherits the intension of P." This is the reason that "$\rightarrow$" is called an "inheritance" relation. Here I change the intuitive meaning of the word "inheritance" to indicate the situation where the two terms get different things from each other.

If "$S \rightarrow P$" is false, it means that the inheritance is incomplete — either $(S^E - P^E)$ or $(P^I - S^I)$ is not empty. However, it does not mean that S and P share no extension or intension.

Theorem 3 $(S^E = P^E) \equiv (S^I = P^I)$.

That is, the *extensions* of S and P precisely coincide if and only if their *intensions* precisely coincide. This means that in NAL-0 the extension and intension of a term are mutually determined. Consequently, one of the two uniquely determines the meaning of a term.

NAL-0 is a logic that from given experience determines the truth values of statements and meaning of terms, and this is what I call "experience-grounded semantics." I will come back to this topic again and again in this book.

3.1.3 Inference: rules and properties

As we can see from the above definitions, NAL-0 only has one inference rule, justified by the transitivity of the inheritance relation. As given in Table 3.2, this rule takes two statements as premises, and derive one statement as conclusion. This rule can be used to exhaust all beliefs according to a given experience.

$J_2 \setminus J_1$	$M \rightarrow P$	$P \rightarrow M$
$S \rightarrow M$	$S \rightarrow P$	
$M \rightarrow S$		$P \rightarrow S$

Table 3.2: The Inference Rule of NAL-0

There is also a "matching rule" in the system, which derives no new belief, but matches questions to answers.

Definition 9 *For different terms S and P, a* question *that can be answered with NAL-0 has one of the following three forms: (1) $S \rightarrow P?$,*

(2) $S \rightarrow ?$, and (3) $? \rightarrow P$. An empirical truth $S \rightarrow P$ is an answer to any of the three. If no such an answer can be found in K^*, "NO" is answered.

The first form asks for an *evaluation* of a given statement, while the other two ask for a *selection* of a term with a given relation with another term.

If there are more than one answers to (2) and (3), any of them is equally good. Literal truth "$T \rightarrow T$" is a trivial answer to such a question, so it is not allowed.

The matching rule is shown in Table 3.3.

$J \backslash Q$	$S \rightarrow P?$	$S \rightarrow ?$	$? \rightarrow P$
$S \rightarrow P$	$S \rightarrow P$	$S \rightarrow P$	$S \rightarrow P$

Table 3.3: The Matching Rule of NAL-0

In NAL-0 the user cannot ask the system "What is not T?," because any term not appearing in K may become a (trivial) answer for this question.[4]

NAL-0 is consistent (since negative knowledge is implicitly represented), sound (since all derived statements are true), complete (since all truths are either literal, or in K^*), and decidable (since K^* can be generated in finite steps from a given K, and it can also be searched in finite time). Because in NAL-0 the time-space cost of inference is ignored, we do not need to worry about how the sentences are stored, and how the premises are chosen in each inference step.

NAL-0 is a term logic, with categorical statements, experience-grounded semantics, and syllogistic inference rules. However, it is not really a "non-axiomatic" logic, because it ignores the assumption of insufficient knowledge and resources. Though NAL-0 looks simple (and even trivial) by itself, its importance in defining the NAL family will be shown by the following sections.

[4]In NAL, there is a way to ask "What instances of S are not instances of P?," and it will be introduced in the next chapter.

3.2 The language of NAL-1

As mentioned in the previous chapter, a central issue in NARS is to treat "truth" as a matter of degree. For that purpose, we first define the concept of "evidence," then define "truth value" as a function of available evidence.

What is defined in Narsese-0 can be called "complete inheritance" (of extension/intension), and it can be naturally extended to the situation of "incomplete inheritance," and the concept of evidence will be introduced in the process.

3.2.1 Evidence and its measurement

As shown by a previous theorem, an inheritance statement is equivalent to a statement about the inclusion of extension (or intension) between two terms. Furthermore, such an inclusion can be seen as a summary of a set of inheritance statements. Based on this observation, "evidence" of an inheritance statement is defined as the following.

Definition 10 *For an inheritance statement "$S \rightarrow P$," its* evidence *are terms in S^E and P^I. Among them, terms in $(S^E \cap P^E)$ and $(P^I \cap S^I)$ are* positive evidence, *and terms in $(S^E - P^E)$ and $(P^I - S^I)$ are* negative evidence.

Here the related extensions and intensions are sets of terms, and "$\cap$" and "$-$" are the "intersection" and "difference" of sets, respectively, as defined in set theory.

Concretely, for a statement "$S \rightarrow P$" and a term M, if both "$M \rightarrow S$" and "$M \rightarrow P$" are true, it is positive evidence for the statement; if "$M \rightarrow S$" is true but "$M \rightarrow P$" is false, it is negative evidence. Symmetrically, if both "$P \rightarrow M$" and "$S \rightarrow M$" are true, it is positive evidence for the statement; if "$P \rightarrow M$" is true but "$S \rightarrow M$" is false, it is negative evidence.

Evidence is defined in this way, because as far as a term in positive evidence is concerned, the inheritance statement is correct; as far as a term in negative evidence is concerned, the inheritance statement is incorrect.

According to this definition, what counts as a piece of evidence is a *term*, not a *statement*. However, whether a given term M is positive or negative evidence for the statement "$S \rightarrow P$" is determined by two statements, one between M and S, and another between M and P.

Now we can rephrase the definition of truth value in NAL-0 in terms of "evidence": "$S \rightarrow P$" is true in NAL-0 if and only if according to the experience of the system, there is no negative evidence for the statement (that is, all available evidence is positive).

When a system has to answer questions with insufficient knowledge and resources, to only indicate whether there is (positive or negative) evidence is usually too rough as a summary of experience. When a statement has both positive and negative evidence, the system often needs to balance them, and takes into account the influence of future evidence. To do this, it is not enough to qualitatively indicate the existence of a certain type of evidence — we need to quantitatively *measure* evidence. For an adaptive system, though past experience is never sufficient to accurately predict future situations, the amount of evidence does matter for the system's decision, and the beliefs based on more evidence should be preferred.

Since according to the previous definition, terms in the extension or intension of a given term are equally weighted, the amount of evidence can be simply measured by the size of the corresponding set.

Definition 11 *For "$S \rightarrow P$," the amount of positive, negative, and total evidence is, respectively,*

$$\begin{aligned} w^+ &= |S^E \cap P^E| + |P^I \cap S^I| \\ w^- &= |S^E - P^E| + |P^I - S^I| \\ w &= w^+ + w^- \\ &= |S^E| + |P^I| \end{aligned}$$

For example, an observed black raven is a piece of positive evidence for "Raven is a kind of black-thing" ($w = w^+ = 1$), and an observed non-black raven is a piece of negative evidence for it ($w = w^- = 1$). Here we assume the observations have no uncertainty.

Amount of evidence captures the idea that an inheritance statement can be seen as a summary of some other inheritance statements.

An important feature of the above definition of evidence is that the "extensional factor" and the "intensional factor" are merged. From the amounts of evidence of a statement alone, there is no way to tell how much of it comes from extensional comparison or intensional comparison of the two terms. I will explain why this is desired later.

3.2.2 Truth value: frequency and confidence

Because all the operations in the system are based on available evidence, w^+ and w^- contain all the information about the uncertainty of the statement, as far as the current discussion is concerned. However, when represented in this way, the information is inconvenient for certain purposes, especially when we talk about beliefs where uncertainty is not obtained by directly counting evidence.

When comparing competing beliefs and deriving new conclusions, we usually prefer *relative measurements* to *absolute measurements*, because the evidence of a premise usually cannot be directly used as evidence for the conclusion. Also, it is often more convenient for the measurements to take values from a finite interval, while the amount of evidence has no upper bound. This point will become more clear later.

In principle, all intervals of real number can be mapped into the interval [0, 1], and this interval corresponds to notions like "ratio," "proportion," or "percentage," which are naturally used to represent "approximation" and "discount" in our daily life. Also, [0, 1] is a natural extension of the binary truth values, traditionally represented as {0, 1}. For these reasons, I use it for the uncertainty measurements in NARS.

A natural relative measurement for uncertainty is the *frequency*, or *proportion*, of positive evidence among all available evidence. In NAL, the "frequency" of a statement is defined as

$$f = w^+/w$$

If the system has observed 100 ravens, and 90 of them are black, but the other 10 are not, the system sets $f = 0.9$ for "Raven is a kind of black thing." When $w = 0$ (and therefore $w^+ = 0$), f is defined to be 0.5.

Although f is a natural and useful measurement, it is not enough for our current purpose. Intuitively, we have the feeling that the uncertainty evaluation $f = 0.9$ is uncertain itself. For a simple example, let us consider the following two situations: (1) the system only knows 10 ravens, and 9 of them are black, and (2) the system knows 10000 ravens, and 9000 of them are black. Though in both situations we have $f = 0.9$, the first case is obviously "more uncertain" than the second. Because here the uncertainty is about the statement "The frequency for ravens to be black is 0.9," we are facing a *higher-order* uncertainty, which is the uncertainty of an evaluation about uncertainty.

As mentioned previously, in NARS the uncertainty in a statement appears as the result of insufficient knowledge. Specially, the first-order uncertainty, measured by frequency, is caused by *known* negative evidence, and the higher-order uncertainty is caused by *potential* negative evidence. For the second measurement, we are looking for a function of w, call it c for *confidence*, that satisfies the following conditions:

1. Confidence c is a continuous and monotonically increasing function of w. (More evidence, higher confidence.)

2. When $w = 0$, $c = 0$. (Without any evidence, confidence is minimum.)

3. When w goes to infinity, c converges to 1. (With infinite evidence, confidence is maximum.)

There are infinite functions satisfying the above requirements, therefore we need more intuition to pick up a specific one.

Many functions with value range [0, 1] can be naturally interpreted as a *proportion* of a certain amount in a total amount. Following this path, when comparing available evidence to potential evidence, we might want to define c as the ratio of "the amount of evidence the system has obtained" to "the amount of evidence the system will obtain." Obviously, the first item is w, but for a system that is always open to new evidence, the second item is infinity, therefore the ratio is always 0. When compared with an infinite "future," the difference among the various finite "past" cannot be perceived. Therefore, it makes little sense to talk about an infinite future.

However, it makes perfect sense to talk about the *near* future. What the system needs to know, from the value of w, is how *sensitive* a frequency will be to new evidence; then the system can use this information to make a choice among competing beliefs. If we limit our attention to a future of fixed horizon, we can represent the information in w in a *ratio* form.

Let us introduce a positive number k, whose value can be metaphorically thought of as the distance to the (temporal) horizon, in the sense that k is the number of times we will still test the given inheritance statement. With this new notion of "horizon," measured by k, we can define a new measurement — confidence, in terms of the amount of total evidence w.

Now we get the relation between c and w in NAL:

$$c = w/(w + k)$$

where k is a positive parameter indicating the evidence to be collected in the "near future." Obviously, this function satisfies the three requirements listed previously.

In this way, the frequency and confidence of a statement are independent of each other, in the sense that, from the value of one, the other's value cannot be determined, or even estimated or bounded (except the trivial case where $c = 0$ implies $f = 0.5$).

For a specific system, k should remain fixed to make the system's behaviors consistent, but different systems can have different values for k. In this book, the default value of k is 1 (and we will discuss the choice of k later). Under such a definition, confidence indicates the ratio of the current amount of evidence to the amount of evidence the system will have after it gets new evidence with a unit amount. The more the system already knows about a statement, the less the new evidence will contribute (relatively), therefore the more confident, or the less ignorant, the system is, on the given statement. When $w = 1$, $c = 0.5$, and the new evidence will double the amount of available evidence; When $w = 999$, $c = 0.999$, and the new evidence will have little effect on the system's belief.

Together, f and c form the truth value of a statement in NAL, and they are defined by the amount of evidence.

Definition 12 *The truth value of a statement consists of a pair of real numbers in [0, 1]. One of the two is called* frequency*, computed as* $f = w^+/w$ *(or 0.5 if* $w = 0$*); the other is called* confidence*, computed as* $c = w/(w + k)$*, where* k *is a positive number.*

From a given truth value, the amount of positive, negative, and total evidence can be uniquely determined. Therefore, the "truth value" representation of uncertainty is functionally equivalent to the "amount of evidence" representation.[5]

3.2.3 Frequency interval

Interestingly, there is a third way to represent the uncertainty of a statement in NAL: as an interval of the frequency of success.

Given the above definition of frequency, after the coming of evidence of the amount k, the new f value will be in the interval

$$[w^+/(w + k),\ (w^+ + k)/(w + k)]$$

This is because the current frequency is w^+/w, so in the "best" case, when all evidence in the near future is positive, the new frequency will be $(w^+ + k)/(w + k)$; in the "worst" case, when all evidence in the near future is negative, the new frequency will be $w^+/(w + k)$.

Let us define this interval formally.

Definition 13 *The* lower frequency *of a statement,* l*, is* $w^+/(w + k)$*; the* upper frequency *of a statement,* u*, is* $(w^+ + k)/(w + k)$*. The* frequency interval *of the statement is* $[l, u]$.

This measurement has certain intuitive aspects in common with other interval-based approaches [Bonissone, 1987, Kyburg, 1988]. For example, the *ignorance* about where the frequency will be (in the near future) can be represented by the *width* of the interval, $i = u - l$. In NAL, i happens to be $1 - c$, so *ignorance* and *confidence* are complementary to each other.

It is important to remember that in NAL the interval $[l, u]$ indicates the range in which the frequency will lie in the *near* future, rather than

[5] The relations between this representation and other representations of uncertainty, such as probability and fuzziness, will be discussed in detail in Chapter 8.

in the *remote* future beyond that. According to the definition of truth value, with the coming of new evidence for a given statement, its confidence value monotonically increases, and eventually converges to 1, but its frequency may increase or decrease, and does not necessarily converge at all. For this reason, the frequency interval cannot be interpreted as indicating where the frequency will eventually be.

The interval representation of uncertainty provides a mapping between the "accurate representation" and the "inaccurate representation" of uncertainty, because "inaccuracy" corresponds to willingness to change a value within a certain range.

Within the system, it is necessary to keep an accurate representation of the uncertainty for statements, but it is often unnecessary for communication purposes. To simplify communication, uncertainty is often represented by a verbal label. In this situation, the truth value corresponds to the relative ranking of the label in the label set.

If in a language there are only N words that can be used to specify the uncertainty of a statement, and all numerical values are equally possible, the most informative way to communicate is to evenly divide the [0, 1] interval into N section: [0, 1/N], [1/N, 2/N], ..., [(N-1)/N, 1], and use a label for each section.

For example, if the system has to use a language where "false," "ambivalent" and "true" are the only valid words to specify truth value, and it is allowed to say "I don't know," then the most reasonable approach for input is to map the three words into [0, 1/3], [1/3, 2/3], and [2/3, 1], respectively, and ignore all "I don't know." For output, all conclusions whose confidence is lower than 1/3 become "I don't know," and for the others, one of the three words is used, according to the section in which the frequency of the conclusion falls.

A special situation of this is to use a single number, with its accuracy, to carry out both frequency and confidence information. In such a situation, "The frequency of statement S is 0.9" is different from "The frequency of statement S is 0.900" — though both give the same frequency, they give different confidence value. In the former case, the interval is [0.85, 0.95], so the confidence is $1 - (0.95 - 0.85) = 0.9$. In the latter case, the interval is [0.8995, 0.9005], so the confidence is $1 - (0.9005 - 0.8995) = 0.999$.

With the interval representation of uncertainty, NARS gains some flexibility in its communication. Though within the system, every belief is attached with numerical uncertainty measurement, in communications it is not necessary when accuracy is not required.

3.2.4 Relations among representations of uncertainty

Now we have three functionally equivalent ways to represent the uncertainty of a statement:

1. as a pair of *amounts of evidence* $\{w^+, w\}$, where $0 \leq w^+ \leq w$ (they do not have to be integers);

2. as a truth-value $<f, c>$, where both f and c are real numbers in $[0, 1]$, independent of each other;

3. as a *frequency interval* $[l, u]$, where $0 \leq l \leq u \leq 1$.

To avoid confusion, three types of brackets ("{}," "<>," and "[ ]") are used in this book for the three forms of uncertainty, respectively. Again, each of them is calculated with respect to certain part of the system's experience, which is implicitly assumed.

Formulas for inter-conversion among the three truth-value forms are displayed in Table 3.4.

This table can be easily extended to include w^- (the amount of negative evidence) and i (degree of ignorance). In fact, any valid (not inconsistent or redundant) assignments to any two of the eight measurements (for example, setting $w^+ = 3.5$ and $i = 0.1$, or setting $f = 0.4$ and $l = 0.3$) will uniquely determine the values of all the others. Therefore, the three forms of uncertainty measurement can even be used in a mixed manner.

Having several closely related forms and interpretations for uncertainty has the following advantages:

1. It gives us a better understanding of what uncertainty of statement really means in NARS, since we can explain it in different ways. The mappings also give us interesting relations among the various uncertainty measurements.

to \ from	$\{w^+, w\}$	$<f, c>$	$[l, u]$
$\{w^+, w\}$		$w^+ = k\frac{fc}{1-c}$ $w = k\frac{c}{1-c}$	$w^+ = k\frac{l}{u-l}$ $w = k\frac{1-(u-l)}{u-l}$
$< f, c >$	$f = \frac{w^+}{w}$ $c = \frac{w}{w+k}$		$f = \frac{l}{1-(u-l)}$ $c = 1 - (u - l)$
$[l, u]$	$l = \frac{w^+}{w+k}$ $u = \frac{w^++k}{w+k}$	$l = fc$ $u = 1 - c(1 - f)$	

Table 3.4: The Relations Among Forms of Truth-Value

2. It provides a user-friendly interface. If the environment of the system consists of human users, the uncertainty of a statement can be expressed in different ways, such as, "I've tested it w times, and in w^+ of them it was true," or "Its past success frequency was f, and the confidence was c," or "I'm sure that its success frequency will remain in the interval $[l, u]$ in the near future." We can maintain a single form as the internal representation (in the current implementation, it is the truth-value form), and, using the mappings in the above table, translate it into/from the others in the interface of the system when necessary.

3. It makes the designing of inference rules easier. For each rule, there should be a function that calculates the truth value of the conclusion from the truth values of the premises, with different rules of course equipped with different functions. As we will see in the following, for some rules it is easier to choose a function if we directly deal with truth values, while for other rules we may prefer to convert truth values into amounts of evidence, or frequency intervals.

4. It facilitates the comparison between measurements in NARS and the uncertainty measurements of various other approaches, because different forms capture different intuitions about uncertainty.[6]

[6]These comparisons will be left to Chapter 8.

Given experience K (as a finite set of binary inheritance statements), for an inheritance relation "$S \rightarrow P$" derived from it, w is always finite. Also, since the system has no need to keep statements for which there is no evidence, w should be larger than 0. For uncertainty represented in the other two forms, these translate into $0 < c < 1$ and $l < u$, $u - l < 1$, respectively.

Beyond the above normal values of uncertainty, there are two limit cases useful for the interpretation of uncertainty and the design of inference rules:

Null evidence: This is represented by $w = 0$, or $c = 0$, or $u - l = 1$, and of course means that the system knows nothing at all about the statement.

Full evidence: This is represented by $w = \infty$, or $c = 1$, or $l = u$. It means that the system already knows everything about the statement — no future modification of the uncertainty value is possible.

Though the above values never appear in actual beliefs of the system, they play important role in system design.

3.2.5 Narsese-1 and experience

Now let me summarize the grammar of Narsese-1 in Table 3.5. We can see that it is similar to that of Narsese-0, except that a binary "statement" plus its truth value becomes a multi-valued "judgment."

In the interface of the system, the other two types of uncertainty representation can also be used in place of the truth value of a judgment, though within the system they will be translated to (from) truth value. Also, truth values corresponding to "null evidence" and "full evidence" are not allowed to appear in the interface (or within the system), though they are used in the meta-language, as limit points, when the inference rules are determined.

Now we can treat Narsese-0 (defined in Table 3.1) as a subset of Narsese-1. In Narsese-1, "$S \rightarrow P < 1, 1 >$" indicates that the inheritance is complete (and negative evidence can be practically ignored), so it is identical to "$S \rightarrow P$" in Narsese-0.

<table>
<tr><td><<i>sentence</i>></td><td>::=</td><td><<i>judgment</i>> | <<i>question</i>></td></tr>
<tr><td><<i>judgment</i>></td><td>::=</td><td><<i>statement</i>><<i>truth-value</i>></td></tr>
<tr><td><<i>question</i>></td><td>::=</td><td><<i>statement</i>>?</td></tr>
<tr><td></td><td></td><td>| ? <<i>copula</i>><<i>term</i>> | <<i>term</i>><<i>copula</i>>?</td></tr>
<tr><td><<i>statement</i>></td><td>::=</td><td><<i>term</i>><<i>copula</i>><<i>term</i>></td></tr>
<tr><td><<i>copula</i>></td><td>::=</td><td>$\rightarrow$</td></tr>
<tr><td><<i>term</i>></td><td>::=</td><td><<i>word</i>></td></tr>
<tr><td><<i>truth-value</i>></td><td>:</td><td>a pair of real number in$[0, 1] \times [0, 1]$</td></tr>
<tr><td><<i>word</i>></td><td>:</td><td>a string in a given alphabet</td></tr>
</table>

Table 3.5: The Grammar of Narsese-1

In this way, the semantics of Narsese-1 is defined by a subset of the language, Narsese-0. Given the experience of the system K in Narsese-0, the binary inheritance language, the truth value of a judgment in Narsese-1, with subject and predicate in V_K, can be determined by comparing the meaning of the two terms. All these judgments form the beliefs of the system, K^*.

Similarly, we extend the concept of "meaning." For a system whose beliefs are represented in Narsese-1, the meaning of a term still consists of the term's extensional and intensional relations with other terms, as in NAL-0. The only difference is that the definition of extension and intension is modified as follows:

Definition 14 *A judgment "$S \rightarrow P < f, c >$" states that S is in the extension of P and that P is in the intension of S, with the truth value of the judgment specifying their degrees of membership.*

Consequently, extensions and intensions in NAL-1 are no longer ordinary sets with well-defined boundaries (as in NAL-0). They are similar to fuzzy sets [Zadeh, 1965], because terms belong to them to different degrees. What makes them different from fuzzy sets is how the "membership" is measured (in NAL, two numbers are used) and interpreted (in NAL, it is experience-grounded).[7]

[7]This issue will be discussed in detail in Section 8.2.

Given any set of statements of Narsese-0 as the experience of a NAL-1 system, the truth values of judgments and the meanings of terms can be determined. In this way, Narsese-0 is used as a meta-language of Narsese-1. At the same time, the former is a subset of the latter. Therefore, the experience-grounded semantics for Narsese is established in a "bootstrapping" manner.

However, since NAL-1 is used with insufficient knowledge and resources, its *actual* experience is a stream of sentences in Narsese-1, not a set of statements in Narsese-0 (as the idealized experience used in the above semantics).

There are several important differences between the "idealized experience" and the "actual experience" of the system.

- A judgment in the idealized experience has truth value $<1, 1>$, while a judgment in the actual experience has truth value $<f, c>$, where c is less than 1.

- The idealized experience is a set of sentences, which is available altogether to the system at the beginning, while the actual experience is a stream of sentences, coming to the system one at a time.

- For a given statement, in the idealized experience each piece of evidence is equally weighted, while in the actual experience, pieces of evidence make different contributions to its truth value, depending on several factors (to be described later).

- When *defining* the truth value of a judgment, the whole idealized experience is considered, while when *calculating* the truth value of a judgment, an inference rule only takes part of the actual experience into account.

- The idealized experience is used at design time to define truth value (of statements) and meaning (of terms), as well as to justify the inference rules, while the actual experience is used at run time by the inference rules to derive new statements (or terms) with their truth value (or meaning), or to modify the existing ones.

For example, if the system has a belief "$S \rightarrow P$ $<0.75, 0.80>$," then from the relationship between the truth value and the amount of evidence (and assuming $k = 1$), we get $w = 4$, $w^+ = 3$. Therefore, the system believes the statement "$S \rightarrow P$" to such an extent, *as if* it had tested the statement 4 times in idealized situations (by checking common elements of the extensions or the intensions of the two terms), in which the relation had been confirmed 3 times, and disproved 1 time. This does not imply, of course, that the system actually got the truth value by carrying out such tests — such absolute certainty can never be obtained in real life. Indeed, the system may have checked the relation more than four times in less-than-ideal situations (i.e., with results represented by judgments whose confidence values are less than 1), or the conclusion may have been derived from other beliefs, or even directly provided by the environment. But no matter how the truth value $< 0.75, 0.80 >$ is generated in practice (there are infinitely many ways it could arise), it can always be *understood* in a unique way, as stated above.

For any approach to extend a binary logic to a multi-valued logic, there is always the question for the meaning of the numerical truth value that need to be answered to make everything else meaningful, while "numerical statements are meaningful insofar as they can be translated, using the mapping conventions, into statements about the original qualitative structure" [Krantz, 1991]. In other words, "ideal experience" is being used in NAL as an "ideal meter-stick" to measure degrees of certainty. Like all measurements, though its unit is *defined* in an idealized situation, it is not *used* only in idealized situations — when we say that a cord is "3 meters long," we do not mean that we have compared it with three end-to-end meter-sticks.

Clearly, the actual experience of NARS is much more complex than the ideal experience as defined in the semantics, but it does not prevent us from saying that the truth value of a judgment summarizes its evidential support, and that the meaning of a term is derived from its experienced relations with other terms.

3.3 The inference rules of NAL-1

Now we can define inference rules for NAL-1, whose premises and conclusions are judgments of Narsese-1, and whose validity is justified according to the experience-grounded semantics.

3.3.1 Revision rule

In NAL, *revision* indicates the inference step in which evidence from different sources is combined. For example, assuming the system's previous uncertainty for "Ravens are black" is $<9/10, 10/11>$ (we know that it corresponds to "10 ravens observed, and 9 of them are black" when $k = 1$), now a new judgment comes, which is "Ravens are black $< 3/4, 4/5 >$" (so it corresponds to "4 ravens are observed, and 3 of them are black"). If the system can determine that no evidence is repeatedly counted in the two sources, then the uncertainty of the revised judgment should be $< 6/7, 14/15 >$ (corresponding to "14 ravens observed, and 12 of them are black").

Formally, the revision rule is defined in Table 3.6, where S can be any statement. The two premises may be conflicting to each other (when the two frequency values are very different), though this is not necessarily the case. Conflicting or not, the information in the two should be summarized into the conclusion.

$J_2 \setminus J_1$	$S < f_1, c_1 >$
$S < f_2, c_2 >$	$S < F_{rev} >$

Table 3.6: The Revision Rule

Since in this case the evidence of either premise is also evidence for the conclusion, and there is no overlapping evidence between the two premises, we have

$$w^+ = w_1^+ + w_2^+, \ w = w_1 + w_2$$

Then, according to the relationship between truth value and amount of evidence, we get the truth-value function for the revision rule:

$$F_{rev} : \ f = \frac{f_1 c_1(1-c_2)+f_2 c_2(1-c_1)}{c_1(1-c_2)+c_2(1-c_1)}, \quad c = \frac{c_1(1-c_2)+c_2(1-c_1)}{c_1(1-c_2)+c_2(1-c_1)+(1-c_1)(1-c_2)}$$

This function has the following properties:

- The order of the premises does not matter.

- As a weighted average of f_1 and f_2, f is usually a "compromise" of them, and is closer to the one that is supported by more evidence.

- The value of c is never smaller than either c_1 or c_2, that is, the conclusion is supported by no less evidence than either premise.

- If $c_1 = 0$ and $c_2 > 0$, then $f = f_2$ and $c = c_2$, that is, a judgment supported by null evidence cannot revise another judgment.

- If $c_1 = 1$ and $c_2 < 1$, then $f = f_1$ and $c = c_1$, that is, a judgment supported by full evidence cannot be modified by empirical evidence.

Because actual confidence values are always in $(0, 1)$, the last two cases do not actually appear at run time, but serve as limit situations. Also because of this reason, it does not matter for the above function has undefined value when $c_1 = c_2 = 0$ and $c_1 = c_2 = 1$.

This definition is compatible with our intuition about evidence and revision — revision is nothing but to reevaluate the uncertainty of a statement by taking new evidence into account. Revision is not updating, where old evidence is thrown away.[8] A high w means that the system already has much evidence for the statement, therefore its confidence is high and its ignorance is low, and consequently the judgment is relatively insensitive to new evidence. All these properties are independent to the decisions on how w is divided into w^+ and w^-, as well as to how they are actually measured (so these decisions may change from situation to situation without invalidating the revision rule).

What happens in revision is similar to what Keynes said: "As the relevant evidence at our disposal increases, the magnitude of the probability of the argument may either decrease or increase, according as the new knowledge strengthens the unfavorable or the favorable evidence; but *something* seems to have increased in either case — we have a more substantial basis upon which to rest our conclusion." [Keynes, 1921].

It needs to be clarified that here "revision" refers to the operation by which the system summarize two (maybe conflicting) beliefs.

[8]This issue will be discussed with more details in Section 8.3.1.

In this operation the conclusion always has a higher confidence. However, generally speaking, in NARS it is possible for the system to lose its confidence in a belief. This can be caused by the "forgetting" or "explaining away" of previously available evidence. This issue will be discussed later, after other relevant components of NARS are introduced.

Now the remaining issue in revision is how to recognize and handle the "overlapping evidence" situation. For that, we need to record, for each judgment, the fragments of experience its truth value is based on.

Definition 15 *If J is an input judgment that appears in the system's experience, with a unique serial number N, it is based on the fragment of experience $\{N\}$. If J is derived from premises $J_1, \cdots, J_n$, which are based on the fragments of experience $K_1, \cdots, K_n$, respectively, then J is based on fragment $K_1 \cup \cdots \cup K_n$.*

The serial numbers will be generated by the program that implements NAL. If the same judgment appears twice in the system's experience, each occurrence will have its own serial number, and the two occurrences will later be treated as different pieces of evidence. On the contrary, if one occurrence produces multiple copies in the system, they will all have the same serial number, and be treated as the same piece of evidence.

The system is designed in this way, because for an adaptive system, what really matters is to predict whether a given statement will be true next time. For someone who lives on a small island with a black swan, "Swan is black" should have a higher frequency than "Swan is white" — though the person has the knowledge that most swans in the world are white, "black swan" appears more often in his/her personal experience. Of course, we do not want to count the same observation more than once, but different observations of the same swan should be treated as multiple pieces of evidence. Therefore, accurately speaking, in NARS the truth value attached to "Swan is black" is not about *how many swans (in the world) are black*, but about *how often a black swan (in the system's experience) is encountered.*

If judgments J_1 and J_2 are based on fragments of experience K_1 and K_2, respectively, and K_1 and K_2, as sets of serial numbers, have no common elements, then the evidence supporting the two judgments do not overlap with each other (that is, no piece of evidence is used to calculate the truth values of both premises). If the two judgments are about the same statement, then they can be used by the revision rule as premises to derive a (summarized) conclusion.

With insufficient resources, NARS cannot maintain a complete record of the supporting experience for each judgment, because it may ask for time and space that the system cannot afford. Therefore the "overlapping-evidence recognition problem" cannot be completely solved by a system with insufficient resources.

Obviously, this limitation holds also for human beings: we could not possibly remember all evidence that supports each judgment we make. Nevertheless, NARS needs to be able to handle this problem somehow, which is not limited to revision only; otherwise, as Pearl points out, "a cycle would be created where any slight evidence in favor of A would be amplified via B and fed back to A, quickly turning into a stronger confirmation (of A and B), with no apparent factual justification." [Pearl, 1988].

The NARS strategy for dealing with this problem is to record only a *constant-sized fragment* of the experience supporting each judgment, and to use such fragments to determine approximately whether two judgments are based on overlapping evidence. As mentioned above, each input judgment is automatically assigned a unique serial number when accepted by the system. In each inference step, the conclusion is assigned a list of serial numbers constructed by interleaving its parents' (the premises') serial-number lists, and then truncating that list at a certain length.

For example, suppose the maximum length for serial-number lists is 4. In this case, if two judgments have a parent or grandparent judgment in common, their serial-number lists will overlap. Now the revision rule is applied only if the two premises' serial-number lists have no common elements, meaning that they are related, if at all, more than two "generations" ago. This mechanism is only an approximation to the perfect solution to the problem, of course.

Though not perfect, it is a reasonable solution when resources are insufficient, and "reasonable solutions" are exactly what we expect from a non-axiomatic system. It is also similar to the strategy of the human mind, since we usually have impressions about where our judgments come from, but such impressions are far from complete and accurate. Also, there is no guarantee that we never repeatedly using the same evidence to adjust our degree of belief.

3.3.2 Choice rule

What should NARS do when two conflicting judgments $S < f_1, c_1 >$ and $S < f_2, c_2 >$ are based on overlapping evidence?

Ideally, we would like to record the precise contribution of each input judgment, and then to subtract the amount of the overlapping evidence from the truth value of the conclusion, so that nothing is counted more than once. Unfortunately, this is impossible, because the experience recorded for each judgment is incomplete, as has just been explained. Furthermore, to find out the contribution of a given input judgment to the overall conclusion is very difficult, and simply impossible given incomplete records.

Nevertheless, NARS needs to be able to handle this situation. For example, the two conflicting judgments may be candidate answers to an evaluative question. If it is impossible to combine them, then NARS needs to make a choice between the two. In the current situation, the *choice* rule is very simple: the judgment having a *higher confidence* (no matter what its *frequency* is) is taken as the better answer, the idea being that if an *adaptive* system must make a choice between conflicting judgments, the one based on more experience has higher priority.

To make a choice between two competing answers for a selective question is more complicated. Let us say that the system is asked the selective question "$S \rightarrow ?$," meaning that it should come up with a term T that is a "typical element" in the intension of S (not S itself, of course). Ideally, the best answer would be provided by a judgment "$S \rightarrow T < 1, 1 >$." But of course this is impossible, because confidence can never reach 1 in NARS. Therefore, we have to settle for the best answer the system can find under the constraints of available knowledge and resources.

Suppose the competing answers are "$S \rightarrow T_1 < f_1, c_1 >$" and "$S \rightarrow T_2 < f_2, c_2 >$." Which one would be better? Let us consider some special cases first:

1. When $c_1 = c_2$, the two answers are supported by the same amount of evidence. For example, both come from statistical data of 100 samples. Obviously, the answer with the *higher frequency* is preferred, since that statement has more positive evidence than the other.

2. When $f_1 = f_2 = 1$, all available evidence is positive. Now the answer with the *higher confidence* is preferred, since it is more strongly confirmed by experience.

3. When $f_1 = f_2 = 0$, all available evidence is negative. Now the answer with the *lower confidence* is preferred, since it is less strongly refuted by the experience. Of course such an answer is still a bad one because of its negative nature, but it may be the best (the least negative) answer the system can find for the question.

From these special cases, we can see that to set up a *general rule* to make a choice among competing judgments, we need somehow to combine the two numbers in a truth value into a single measurement. The current situation is different from the previous one. "$S \rightarrow T_1 < f_1, c_1 >$" and "$S \rightarrow T_2 < f_2, c_2 >$" do not conflict with each other — they have different contents — but they compete for being the "best supported intensional relation of S."

In NARS, an *expectation* measurement, e, is defined on every judgment for this purpose. Different from truth value (which is used to record past experience), expectation is used to predict future experience. "$e = 1$" means that the system is absolutely sure that the statement will always be confirmed by future experience; "$e = 0$" means it will always be refuted; and "$e = 0.5$" means the system considers it equally likely to encounter a piece of positive or a negative evidence.

To calculate e from $< f, c >$, we can see that under the assumption that the system makes extrapolations from its (past) experience, it would be natural to use f as e's "first-order approximation." However, such a *maximum-likelihood estimate* is not good enough when c is small [Good, 1965]. For example, if a hypothesis has been tested only once, it would not make sense to set one's expectation to 1 (if the test was a success) or to 0 (if the test was a failure).

Intuitively, e should be more "conservative" (i.e., closer to 0.5, the "no-preference point") than f, to reflect the fact that the future may be different from the past. Here is where the confidence c affects e — the more evidence the system has accumulated, the more confident the system is (indicated by a larger c) that its predicted frequency e should be close to its experienced frequency f. Therefore, it is natural to define

$$F_{exp}: \; e = c(f - 0.5) + 0.5.$$

In particular, when $c = 1$ (full evidence), $e = f$; when $c = 0$ (null evidence), $e = 0.5$. Alternatively, this equation can be rewritten as $c = (e - 0.5)/(f - 0.5)$ (when $f \neq 0.5$), showing that c indicates the ratio of e's and f's distances to 0.5.

To express the definition of e in the other two forms of uncertainty leads to interesting results.

When the uncertainty is represented as a frequency interval, from the inter-conversion formulas in Table 3.4, we get

$$e = (l + u)/2$$

Thus e is precisely the *expectation of the future frequency* — that is, the midpoint of the interval in which the frequency will lie, in the near future.

When the uncertainty is represented as amounts of evidence, from the mappings in Table 3.4 we get

$$e = (w^+ + k/2)/(w + k)$$

which is a continuum (i.e., a family) of functions with k as a parameter. This formula turns out to be closely related to what has been called the "beta-form based continuum" (with positive and negative evidence weighted equally) [Good, 1965], and the "λ-continuum" (with the "logical factor," or prior probability, being 1/2) [Carnap, 1952]. Though interpreted differently, the three continuum share the same formula and make identical predictions. All three continua have *Laplace's law of succession* as a special case (when $k = 2$), where the probability of success on the next trial is estimated by the formula $(w^+ + 1)/(w + 2)$.

Now we can see how the choice of the parameter k can influence the behavior of a system. Let us compare a system A_1 with $k = 1$ and a system A_2 with $k = 10$. The problem is to make a choice between two competing answers "$S \rightarrow P_1 \; \{w_1^+, w_1\}$" and "$S \rightarrow P_2 \; \{w_2^+, w_2\}$" (where the truth values are represented as weights of evidence). It is easy to see that when $w_1 = w_2$ or $w_1^+/w_1 = w_2^+/w_2$, the two systems make the same choice. It is only when a system needs to make a choice between a higher f and a higher c that the value of k will matter. For example, let us suppose that $w_1^+ = w_1 = 2, w_2^+ = 5$, and $w_2 = 6$. In this situation, in A_1, $e_1 = (2+0.5)/(2+1) \approx 0.83$, $e_2 = (5+0.5)/(6+1) \approx 0.79$, and thus A_1 will choose the first answer (since all of its evidence is positive); in A_2, $e_1 = (2+5)/(2+10) \approx 0.58$, $e_2 = (5+5)/(6+10) \approx 0.63$, and thus A_2 will choose the second answer (since it is more fully tested, and its frequency is not much lower than that of the other alternative).

Therefore, k is one of the "personality parameters" of the system, in the sense that it indicates a certain systematic preference or bias, for which there is no "optimal value" in general. The larger k is, the more "conservative" the system is, in the sense that the system always makes smaller adjustments when e is reevaluated according to new evidence, than a system having a smaller value of k. This parameter was called the "flattening constant" by Good ([Good, 1965], where he also tried to estimate its value according to certain factors that are beyond our current consideration), and was interpreted by him as a way to choose a prior probability distribution. The same parameter was interpreted by Carnap as the "relative weight" of the "logical factor" [Carnap, 1952].

In summary, the choice rule is formally defined in Table 3.7, where $S_1 < f_1, c_1 >$ and $S_2 < f_2, c_2 >$ are two competing answers to a question, and $S < F_{cho} >$ is the chosen one. When S_1 and S_2 are the same statement, the one with a higher *confidence* value is chosen, otherwise the one with a higher *expectation* value is chosen.

$J_2 \setminus J_1$	$S_1 < f_1, c_1 >$
$S_2 < f_2, c_2 >$	$S < F_{cho} >$

Table 3.7: The Choice Rule

3.3.3 Truth-value functions in general

A typical inference rule in NAL has the following format:

$$\{premise_1 <f_1, c_1>, premise_2 <f_2, c_2>\} \vdash conclusion <f, c>$$

and a truth-value function calculates $< f, c >$ from $< f_1, c_1 >$ and $< f_2, c_2 >$. Alternatively, it can be put into a table (as we have seen previous) where each row and column corresponds to a premise.

The previously defined revision rule is the only inference rule in NAL whose premises and conclusion contain the same statement. Consequently, the evidence of the premises can be directly treated as evidence of the conclusion, and the conclusion, based on accumulated evidence, has a higher confidence value than the premises.

In the other inference rules, the premises and the conclusion are judgments about different statements, so each of them has its own evidence space, and the evidence of a premise cannot be directly used as evidence of the conclusion. Even if a premise and a conclusion have overlapping evidence spaces, evidence in the premise will be counted less in the conclusion. This is the case because according to the semantics of NAL, "amount of evidence" actually measures evidence that *directly* supports the statement. When indirect evidence is recognized, it is turned into direct evidence with a reduced amount (and the detail differs from rule to rule). Consequently, in every inference rules in NAL, except revision, the conclusion always has a lower confidence value than the premises.

I have introduced several uncertainty measurements, and most of them take values from the [0, 1] interval. Even the amount of evidence, which is not defined with this range in general, corresponds to this interval when it is limited to a piece of evidence within a unit amount. Since they cannot be easily interpreted as "probability" as defined in probability theory and statistics, we cannot directly apply an existing theory to guide their calculation in the truth-value functions attached to various inference rules.[9]

The approach used in NARS is to see the values in [0, 1] as extended Boolean values, 0 and 1, and to handle their calculation by extending the Boolean operators, namely "*not*," "*and*," and "*or*."

[9]Arguments for this conclusion will be provided in Chapter 8.

The extended "*and*" and "*or*" are often called *Triangular norm* (T-norm) and *Triangular conorm* (T-conorm), respectively. They are functions defined on real numbers in [0, 1], being commutative and associative, and monotonic in each variable. T-norm has boundary conditions satisfying the truth tables of the Boolean operator "*and*," and T-conorm those of "*or*." [Bonissone and Decker, 1986, Dubois and Prade, 1982, Schweizer and Sklar, 1983].

In this book, these two functions are directly written as $and(x_1, x_2)$ and $or(x_1, x_2)$. Because each is commutative and associative, they can be extended to take an arbitrary number of arguments:

$$and(x_1, \ldots, x_n) = and(and(x_1, \ldots, x_{n-1}), x_n),$$

$$or(x_1, \ldots, x_n) = or(or(x_1, \ldots, x_{n-1}), x_n).$$

The usage of T-norm and T-conorm in NARS is different from that in other approaches [Bonissone and Decker, 1986, Dubois and Prade, 1982], where they are only used to determine the degree of certainty of the *conjunction* and *disjunction* of two propositions, respectively. In NARS, the T-norm function $y = and(x_1, \ldots, x_n)$ is used when a quantity y is *conjunctively determined* by two or more other quantities $x_1, \ldots, x_n$ — that is, $y = 1$ if and only if $x_1 = \cdots = x_n = 1$, and $y = 0$ if and only if $x_1 = 0$ or ... or $x_n = 0$; similarly, the T-conorm function $y = or(x_1, \ldots, x_n)$ is used when a quantity y is *disjunctively determined* by two or more other quantities $x_1, \ldots, x_n$ — that is, $y = 1$ if and only if $x_1 = 1$ or ... or $x_n = 1$, and $y = 0$ if and only if $x_0 = \cdots = x_n = 0$. These functions are not directly about the *conjunction* or *disjunction* of Narsese statements.[10]

Intuitively, a variable y is conjunctively determined by variables x_1, ..., x_n when all the x's are its *necessary factors*, or numerically, if y is never bigger than any of them. Similarly, y is disjunctively determined by $x_1, \ldots, x_n$ when all the x's are its *sufficient factors*, or numerically, it is never smaller than any of them. In this way, T-norm and T-conorm are applied in situations where a quantity is determined by several factors, where we wish the boundary condition to be satisfied, and where no one factor is more important than any of the others.

---

[10]In Chapter 5, we will see that they are still used for the conjunctions and disjunctions of statements. They are just not *merely* used in that situation.

There are an infinite number of ways of numerically satisfying the prescribed conditions on T-norm and T-conorm. For our purpose, it is desired for them to be continuous and strictly increasing, so that any upward (downward) change in any argument will cause an upward (downward) change in the function value. In [Schweizer and Sklar, 1983] it is proved that all functions satisfying the above conditions are isomorphic to (i.e., can be represented as a monotonic transform of) the "probabilistic" operators:

$$and(x, y) = xy; \quad or(x, y) = x + y - xy.$$

It is also shown in [Bonissone and Decker, 1986] that only a small finite subset of the infinite set of possible T-norms and T-conorms will produce significantly different results, if we limit our concern to the "finest level of distinction among different quantifications of uncertainty." Among those representative operators in the small subset, the above pair is the only continuous and strict T-norm and T-conorm. These results show that the above T-norm and T-conorm have not been chosen arbitrarily for NARS; although in principle there are other pairs satisfying our requirements, they are usually more complex, and are not significantly different from the above pair.

The above choice is also justifiable in another way. We call quantities *mutually independent* of each other, when given the values of any of them, the remaining ones cannot be determined, or even bounded approximately. This type of mutual independence among arguments is assumed by the probabilistic operators, but not by other representative operators, such as the "min/max" pair used in fuzzy logic [Bonissone and Decker, 1986].

Obviously, to use the probabilistic operators when the mutual independence does not hold (e.g., $x = y$ or $x = not(y)$) leads to counterintuitive results. In the following, the T-norm and T-conorm are only used when the "mutual independence" condition is satisfied. As far as the two premises are not based on overlapping evidence, f_1, c_1, f_2, and c_2 satisfy this requirement, because given the values of any three of them, the value of the last one cannot be determined, or even bounded.

It should be mentioned that though the T-norm and T-conorm used in NARS share intuition and mathematical forms with probabilistic formula, they should not been understood as $and(x, y) = P(x \text{ and } y)$ and

$or(x, y) = P(x \text{ or } y)$, simply because x and y are usually not random variables with probability distribution function P.

As usual, the "*not*" operator on the extended Boolean variable is defined as

$$not(x) = 1 - x$$

In NAL, the truth-value function for most of the inference rules (with the previously defined revision and choice as exceptions) are built by the following steps:

1. To treat all the uncertainty values involved as Boolean variables whose value are either 0 or 1. According to the definition of these uncertainty measurements and the semantics of Narsese, the uncertainty values of the conclusion is determined for each combination of those of the premises.

2. To represent the uncertainty values of the conclusion as Boolean expressions of the the uncertainty values of the premises that satisfy the above boundary conditions. Usually there are infinitely many functions that satisfy the restriction, and the ones accepted are those that are simple and have natural interpretations.

3. To replace the *and*, *or*, and *not* operator in the Boolean function by the T-norm ($and(x, y) = x * y$), T-conorm ($or(x, y) = 1 - (1 - x)(1 - y)$), and Negation ($not(x) = 1 - x$) functions, respectively, so as to get a general function on [0, 1].

4. To rewrite the uncertainty functions as truth-value functions (if they are not already in that form), according to the relationship between truth value and the other uncertainty measurements.

These are the conceptual steps of the design procedure. When truth-value functions are introduced in the following descriptions, the first two steps are often merged, and the last step is often taken implicitly. Since the result of the above Step 2 may be not unique, the above approach of building truth-value functions is not a mathematical proof of the function obtained, and with the progress of the research, the functions have been modified in different versions of NARS in the past, and it may still happen in the future, if negative evidence for the design of these functions is found. As everything else in this theory, these functions are just "the best we can get according to available evidence."

3.3.4 Syllogistic rules

In term logics, when two judgments share exactly one common term, they can be used as premises in an inference rule that derives an inheritance relation between the other two (unshared) terms. Altogether, there are four possible combinations of premises and conclusions, corresponding to the four figures of Aristotle's Syllogisms [Aristotle, 1989], three of which are also discussed by Peirce [Peirce, 1931]. They are listed in Table 3.8.

$J_2 \setminus J_1$	$M \rightarrow P\ <f_1, c_1>$	$P \rightarrow M\ <f_1, c_1>$
$S \rightarrow M\ <f_2, c_2>$	$S \rightarrow P\ <F_{ded}>$	$S \rightarrow P\ <F_{abd}>$
$M \rightarrow S\ <f_2, c_2>$	$S \rightarrow P\ <F_{ind}>$	$S \rightarrow P\ <F_{exe}>$

Table 3.8: The Syllogistic Rules of NAL-1

The four rules in the table are explained in the following:

1. $\{M \rightarrow P <f_1, c_1>,\ S \rightarrow M <f_2, c_2>\} \vdash S \rightarrow P <f, c>$
   This rule corresponds to Aristotle's *first figure* and Peirce's *deduction.*

2. $\{P \rightarrow M <f_1, c_1>,\ S \rightarrow M <f_2, c_2>\} \vdash S \rightarrow P <f, c>$
   This rule corresponds to Aristotle's *second figure* and Peirce's *abduction* (and he also called it *hypothesis*).

3. $\{M \rightarrow P <f_1, c_1>,\ M \rightarrow S <f_2, c_2>\} \vdash S \rightarrow P <f, c>$
   This rule corresponds to Aristotle's *third figure* and Peirce's *induction.*

4. $\{P \rightarrow M <f_1, c_1>,\ M \rightarrow S <f_2, c_2>\} \vdash S \rightarrow P <f, c>$
   This rule corresponds to the *fourth figure* of Aristotle's Syllogistic [Bocheński, 1970].

In NAL, the first three rules are named using Peirce's words. The fourth rule is called *exemplification*. The truth-value functions in the table are named by three letters after the corresponding rule. They are built according to the general procedure introduced previously.

Obviously, each pair of premises also derives a judgment "$P \rightarrow S$," whose truth value can be determined by one of the four functions.

The *deduction* rule in NAL-1 extends the "rule of transitivity" in NAL-0. For the frequency of the conclusion, f, it is 1 if and only if both premises have frequency 1. As for the confidence of the conclusion, c, it reaches 1 only when both premises have truth values $< 1, 1 >$. Therefore, the Boolean function we get for deduction is

$$f = and(f_1, f_2), \quad c = and(f_1, c_1, f_2, c_2)$$

which leads to truth-value function

$$F_{ded}: \; f = f_1 f_2, \quad c = f_1 c_1 f_2 c_2$$

The deduction rule is symmetric to the premises, that is, their order does not matter.

In NARS, *abduction* is the inference that, from a shared element M of the *intensions* of S and P, determines the truth value of "$S \rightarrow P$," and *induction* is the inference that, from a shared element M of the *extensions* of S and P, determines the truth value of "$S \rightarrow P$." Therefore, derived from the duality of extension and intension, we have a duality of abduction and induction in NAL.

In both cases, the premises provide a piece of positive evidence with a unit amount if and only if both of them have truth-value $< 1, 1 >$, which can be represented as Boolean function

$$w^+ = and(f_1, c_1, f_2, c_2)$$

For the total amount of evidence, in abduction we get

$$w = and(f_1, c_1, c_2)$$

and in induction we get

$$w = and(c_1, f_2, c_2)$$

Please note that in the above representation we are mixing two forms of uncertainty measurement: in the premises, the truth values are used, while in the conclusion, the amounts of (positive/total) evidence are used. Also, in these two rules, the two premises play different roles, and their order matters.

After rewriting the result as truth-value functions, for abduction, it is

$$F_{abd}: \ f = f_2, \ c = f_1 c_1 c_2 / (f_1 c_1 c_2 + k)$$

and for induction, it is

$$F_{ind}: \ f = f_1, \ c = c_1 f_2 c_2 / (c_1 f_2 c_2 + k)$$

In the process of designing truth-value function for induction (and abduction), a crucial point is to see that when the premises are $\{M \rightarrow P, M \rightarrow S\}$, it is the term M (as a whole) that is taken as evidence, and the amount of evidence it can provide is less than 1. A mistake easy to make here is to think of M as a set of evidences for "$S \rightarrow P$," and think of the number of instances in M as the amount of evidence of the conclusion. That interpretation is inconsistent with the semantics of Narsese.

In term logics, "conversion" is an inference from a single premise to a conclusion by interchanging the subject and predicate terms of the premise [Bocheński, 1970]. Now we can see conversion, defined in Table 3.9, as a special case of abduction by taking "$P \rightarrow S < f_0, c_0 >$" and "$S \rightarrow S < 1, 1 >$" (a tautology) as premises, and "$S \rightarrow P < f, c >$" as conclusion.

Using F_{abd}, we obtain the truth-value functions for the conversion rule

$$F_{cnv}: \ f = 1, \ c = f_0 c_0 / (f_0 c_0 + k)$$

$\{S \rightarrow P \ < f_0, c_0 >\} \vdash P \rightarrow S \ < F_{cnv} >$

Table 3.9: The Conversion Rules of NAL-1

We could also derive this same result by seeing conversion as a special case of induction with "$P \rightarrow P < 1, 1 >$" and "$P \rightarrow S < f_0, c_0 >$" as premises.

Similarly we can get the truth-value functions for *exemplification*. This rule takes the same premises as the deduction rule, but in its conclusion the most general term in the premises, P, becomes the subject, while the most specific term in the premises, S, becomes the predicate. As in the case of the conversion rule, no negative evidence for the conclusion can be collected in this way, and

$$w = w^+ = and(f_1, c_1, f_2, c_2)$$

Therefore the truth-value function is

$$F_{exe}: \; f = 1, \; c = f_1 c_1 f_2 c_2 / (f_1 c_1 f_2 c_2 + k)$$

As mentioned above, from "$M \rightarrow P < f_1, c_1 >$" and "$M \rightarrow S < f_2, c_2 >$," NARS can directly get "$S \rightarrow P < f_1, \; c_1 f_2 c_2 / (c_1 f_2 c_2 + k) >$" by induction. Now there is also an indirect way to derive "$S \rightarrow P$" from the same premises: via conversion, the second premise yields "$S \rightarrow M < 1, c_2 f_2 / (c_2 f_2 + k) >$"; then, deductively combining this judgment with the first premise, NARS arrives at the conclusion "$S \rightarrow P < f_1, \; f_1 c_1 f_2 c_2 / (c_2 f_2 + k) >$." Compared with the direct result, this indirect conclusion has the same frequency value, but a lower confidence value. Similarly, abduction can be replaced by conversion-then-deduction, and exemplification by conversion-then-deduction or deduction-then-conversion, but all of them give lower confidence values, compared to the ones produced by the above rules.

These results show that each application of a syllogistic rule in NARS will cause some information loss (while preserving other information, of course), and therefore *direct* conclusions will always be more confident. On the other hand, the fact that exactly the same frequency value is arrived at by following different inference pathways shows that the truth-value functions defined above have not been coined individually in *ad hoc* ways, but are closely related to each other, since all of them are based on the same semantic interpretation of the truth value.

In general, from the same set of premises, different sequences of inference steps may assign different truth values to the same statement. According to the experience-grounded semantics, the truth value assigned to a statement reflects the evidence collected in a certain way, as specified by the rule used in this step of inference. Therefore, it is normal if different paths lead to different judgments. As far as each step is justified according to the semantics, all the judgments are valid, and the system usually just chooses the most confident one.

By comparing the inference rules of NAL-1, we can get the following conclusions:

- Both frequency and confidence contribute to inference, but in different ways.

- Revision is the only rule where the confidence of the conclusion may be higher than those of the premises.

- The confidence of a syllogistic conclusion is never higher than the confidence of either premise, that is, confidence "declines" in syllogistic inference.

- In general, confidence declines much slower in deduction than in induction and abduction.[11] In deduction, if both premises have a confidence value of 1, the conclusion may also have a confidence value of 1. In induction and abduction, however, the confidence of the conclusion has an upper bound $1/(1 + k)$, far less than 1. So, by saying that "Induction and abduction are more uncertain when compared with deduction," what is referred to is not the "first-order uncertainty," f (inductive and abductive conclusions can have a frequency of 1 when all available evidence is positive), but the "higher-order uncertainty," c.

Here we can see another function of the personality parameter k: to indicate the relative confidence of abductive/inductive conclusions.

[11] This conclusion does not apply to situations where the confidence values of the premises are all very low. However, the conclusions produced in those situation usually have little impact on the system.

Intuitively speaking, all intelligent systems (human and computer) need to maintain a balance between the strictness of deduction and the tentativeness of induction and abduction. Comparatively speaking, a system with a small k relies more on abduction and induction, while a system with a large k relies more on deduction. There is no single "optimal value" for such a parameter, at least for our current discussions.

3.3.5 Backward inference

The inference rules introduced before (except the choice rule) are for *forward inference*, since each of them takes a pair of judgments as premises (except the conversion rule, which takes a single premise), and derive a new judgment as conclusion. *Backward inference*, on the other hand, happens when a judgment and a question are taken as premises. We already discussed a special case of backward inference, that is, the choice rule. This rule is used to decide whether a question can be directly answered by a judgment, as well as to select an answer among candidates.

Formally, the backward inference rules for questions are determined by the following principle: A question Q and a judgment J will give rise to a new question Q' if and only if an answer for Q can be derived from an answer for Q' and J, by applying a forward inference rule.

For example, the system is asked to decide the truth value for "$goose \rightarrow swimmer$," that is, whether a goose swims. The system does not have a direct answer for it, but it has a belief "$goose \rightarrow bird < 1, 0.9 >$." From the question and the belief, a backward inference rule produces a derived question "$bird \rightarrow swimmer$," because from an answer to this question and the belief, the system can derive an answer to the original question by deduction.

Defined in this way, it is easy to get backward inference rules from forward inference rules. For example, for a given forward-inference rule table, first we take the conclusions in the table as questions (Q), one premise (J_2) as a judgment (J), and the other premise (J_1, without truth value) as the derived question. After renaming the terms and rearranging the order, we get a backward-inference rule table, in which some terms in the questions can be a "?," indicating a query for terms satisfying given condition.

$J \setminus Q$	$M \rightarrow P$	$P \rightarrow M$
$S \rightarrow M$	$S \rightarrow P$	$S \rightarrow P$
$M \rightarrow S$	$S \rightarrow P$	$S \rightarrow P$

Table 3.10: The Backward Syllogistic Rules of NAL-1

For the forward syllogistic rules in Table 3.8, the corresponding backward-inference rules are in Table 3.10.

This table turns out to be identical to Table 3.8, if the truth-value functions and the question/judgment difference are ignored. This elegant symmetry reveals an implicit property of the syllogistic rules of NARS — that is, for any three judgments J_1, J_2, and J_3, if J_3 can be derived from J_1 and J_2 by a syllogistic rule, then from J_3 and J_1 J_2' can be derived, which has the same statement as J_2 (their truth values may be different). Intuitively, the three inheritance relations constitute a triangle from any two sides of which the third side can be derived. Such a property does not give rise to infinite loops in the system, because if J_3 is really derived from J_1 and J_2, it must share serial numbers with each of the two, which prevents the system from taking J_3 and J_1 (or J_2) as premises in further inferences.

In NAL-1, if a question cannot be directly answered by the choice rule, backward inference is used to recursively "reduce" the question into derived questions, until all of them have direct answers. Then these answers, together with the judgments contributed in the previous backward inference, will derive an answer to the original question by forward inference.

3.3.6 NAL-1 summary

Now we have completed the description of a Non-Axiomatic Logic, NAL-1, with its formal language, semantics, and inference rules. Let me use an example to show what this logic can do.

To make the description simple, for the initial knowledge we give frequency value 1 to positive judgments, frequency value 0 to negative judgments, and confidence value 0.9 to every judgment. At the beginning, the following judgments are given to the system as experience:

$$
\begin{aligned}
&(1)\ swan \rightarrow bird < 1, 0.9 >\\
&(2)\ swan \rightarrow swimmer < 1, 0.9 >\\
&(3)\ seagull \rightarrow bird < 1, 0.9 >\\
&(4)\ seagull \rightarrow swimmer < 1, 0.9 >\\
&(5)\ robin \rightarrow bird < 1, 0.9 >\\
&(6)\ robin \rightarrow swimmer < 0, 0.9 >\\
&(7)\ goose \rightarrow bird < 1, 0.9 >\\
&(8)\ dolphin \rightarrow swimmer < 1, 0.9 >
\end{aligned}
$$

With (1) and (2) as premises, the induction rule derives

$$(9)\ bird \rightarrow swimmer < 1, 0.45 >$$

Similarly, from (3) and (4), by induction the system gets

$$(10)\ bird \rightarrow swimmer < 1, 0.45 >$$

Since (9) and (10) are derived from distinct bodies of evidence, they can be used as premises of the revision rule to get

$$(11)\ bird \rightarrow swimmer < 1, 0.62 >$$

Given the symmetry of the premises, following the same path the system can get another conclusion

$$(12)\ swimmer \rightarrow bird < 1, 0.62 >$$

However, this symmetry does not apply to negative judgments. From (5) and (6), by induction the system gets

$$(13)\ bird \rightarrow swimmer < 0, 0.45 >$$

but its symmetric conclusion gets a confidence value 0. Again, applying the revision rule to (11) and (13), the result is

$$(14)\ bird \rightarrow swimmer < 0.67, 0.71 >$$

Therefore, the inductive conclusions are just like statistical conclusions, except that they are revised incrementally, under the influence of their confidence values.

What makes NAL different from a purely statistical inference system is that in it different types of inference are unified, and therefore the conclusions are not statistical in the traditional sense anymore. From (7) and (14), by deduction the system get

$$(14)\ goose \rightarrow swimmer < 0.67, 0.43 >$$

but the conclusion does not mean that "Sixty-seven percent of geese can swim." Similarly, from (8) and (14), by abduction the system gets

$$(12)\ dolphin \rightarrow bird < 1, 0.3 >$$

but the conclusion is not based on any observed dolphin which is also a bird. Instead, the evidential support in the conclusion comes from the experienced common property of the two terms.

Chapter 4

First-Order Inference

In this chapter, three new layers of the logic, NAL-2, NAL-3, and NAL-4, will be defined, which introduce terms with internal structures and variants of the inheritance relation into NAL. Consequently, the expressive and inferential power of the logic will be increased, step by step. At the end of the chapter, we will get a complete First-Order Non-Axiomatic Logic.

4.1 Compound terms

In NAL-1, each term is "atomic," and named by a word, which is simply a unique identifier without internal structure. Obviously, Narsese-1 can only express simple statements.

To represent more complicated experience, "compound terms" are needed.

Definition 16 *A* compound term $(op\ c_1\ \cdots\ c_n)$ *is a term formed by one or more terms* $c_1, \cdots, c_n$*, called its* component(s)*, with a* term operator*,* op*. The order of the components usually matters.*

Sometimes we prefer the "infix" format of a compound term, that is, to write $(op\ c_1\ \cdots\ c_n)$ as $(c_1\ op\ \cdots\ op\ c_n)$. When introducing term operators with two or more components in the following, usually they are only defined with two components, and the general case (for both the above prefix representation and the infix representation) is translated into the two-component case by the following definition.

Definition 17 *If* $c_1 \cdots c_n$ $(n > 2)$ *are terms and* op *is a term operator defined as taking two arguments, both compound terms* $(op\ c_1 \cdots c_n)$ *and* $(c_1\ op \cdots op\ c_n)$ *are defined recursively as* $(op\ (op\ c_1 \cdots c_{n-1})\ c_n)$.

In NAL, the term operators are predefined as part of the grammar of Narsese, with determined (experience-independent) meaning. The meaning of a compound term has two parts, a *literal* part and an *empirical* part, where the former is determined by its definition and other literal truths about the term, while the latter comes from the system's experience when the compound term is used as a whole. In NAL, though empirical statements are all uncertain (i.e., with frequency in [0, 1] and confidence in (0, 1)), literal truths remain binary, so their truth values are omitted in the following description.

To indicate the syntactic complexity of a compound term, a notion of "level" is recursively defined as the following.

Definition 18 *Each term in NAL is on a certain* level *according to its syntactical complexity. If a term is atomic, then it is on level 1. If a term is a compound, then it is one level higher than the highest level of its components.*

Defined in this way, "level" is a syntactic concept, and it has nothing to do with semantics. Terms of different levels can have inheritance relations between each other.

All compound terms can be used by the inference rules defined in NAL-1. When doing so, their internal structures are ignored. In the following, three extensions of NAL-1 are defined, layer by layer, each of which processes some special types of compound term.

4.2 NAL-2: sets and variants of inheritance

NAL-1 is extended into NAL-2 by introducing new copulas to enrich the system's expressing capacity, so as to move Narsese closer to natural languages.

The copula in a term logic intuitively corresponds to the "to be" in English. However, even such a rough mapping cannot be simply established, because as a copula, "*to be*" has multiple usages, for example:

type: "Birds *are* animals."

element: "Tweety *is* a bird."

attribute: "Ravens *are* black."

identification: "The morning star *is* Venus."

The inheritance relation defined in NAL-1 can be used for the first case, but not for the others directly, though it is closely related to them. To introduce these new relations into NAL, the grammar, semantics, and inference rules all need to be extended.

4.2.1 Similarity

A symmetric inheritance relation *similarity* is written as "$\leftrightarrow$."

Definition 19 *The similarity statement "$S \leftrightarrow P$" is defined by two inheritance statements as* $S \leftrightarrow P \equiv (S \rightarrow P) \wedge (P \rightarrow S)$.

Therefore, the similarity relation is reflexive, symmetric, and transitive.

It follows that an inheritance relation between two terms is implied by a similarity relation between them.

Theorem 4 $(S \leftrightarrow P) \supset (S \rightarrow P)$

Here "$\supset$" is the *implication* operator defined in propositional logic. The expressions in the theorem are not statements in Narsese, but in its meta-language.

Two terms related by the similarity relation are in both the extension and the intension of each other. Here the extension and intension of a term are defined as before, that is, by the "$\rightarrow$" relation, not the new "$\leftrightarrow$" relation.

Theorem 5 $(S \leftrightarrow P) \equiv (S \in (P^E \cap P^I)) \equiv (P \in (S^E \cap S^I))$

Theorem 6 $(S \leftrightarrow P) \equiv (S^E = P^E) \equiv (S^I = P^I)$

That is, "$S \leftrightarrow P$" means "S and P have the same meaning." Or, we can say that the two terms are *identical.*

Two compounds terms are identical if they have the same term operator, and their corresponding components are identical pair by pair. Especially, if each of them has exactly one component, then the above "if" becomes "if and only if."

Definition 20 *The meaning of two compound terms are related in the following way:*

$$((c_1 \leftrightarrow d_1) \wedge \cdots \wedge (c_n \leftrightarrow d_n)) \supset ((op\ c_1\ \cdots\ c_n) \leftrightarrow (op\ d_1\ \cdots\ d_n))$$

$$(c \leftrightarrow d) \equiv ((op\ c) \leftrightarrow (op\ d))$$

To extend the relation to the situations of "incomplete similarity," the evidence of a similarity relation is defined like the evidence of an inheritance relation. For similarity statement "$S \leftrightarrow P$," its positive evidence is in $(S^E \cap P^E)$ and $(P^I \cap S^I)$, and its negative evidence is in $(S^E - P^E)$, $(P^E - S^E)$, $(P^I - S^I)$, and $(S^I - P^I)$. In this way, in general "similarity" is a matter of degree, measured by a truth value (defined in NAL-1, as a function of the amount of evidence). In the following, I reserve the word "identical" for the special situation where a similarity statement has truth value $<1, 1>$.

Corresponding to the syllogistic rules in NAL-1, in NAL-2 there are three combinations of inheritance and similarity, corresponding to *comparison*, *analogy*, and another form of *deduction*, respectively, as indicated by the names of truth-value functions in Table 4.1. To make the table (as well as the following inference tables) simpler, the truth values of the premises are omitted.

$J_2 \setminus J_1$	$M \rightarrow P$	$P \rightarrow M$	$M \leftrightarrow P$
$S \rightarrow M$		$S \leftrightarrow P < F_{com} >$	$S \rightarrow P < F'_{ana} >$
$M \rightarrow S$	$S \leftrightarrow P < F_{com} >$		$P \rightarrow S < F'_{ana} >$
$S \leftrightarrow M$	$S \rightarrow P < F_{ana} >$	$P \rightarrow S < F_{ana} >$	$S \leftrightarrow P < F_{dd2} >$

Table 4.1: The Syllogistic Rules of NAL-2

In the inference table, F'_x indicate the truth-value function obtained by exchanging $< f_1, c_1 >$ and $< f_2, c_2 >$ in function F_x, where x is the indicator of the inference rule (such as *ana* for analogy, *abd* for abduction, and *ind* for induction). Such a function is needed for every inference rule whose truth-value function is not symmetric with respect to the two premises.

In NARS, *comparison* refers to the inference rule by which a similarity judgment is obtained by comparing the inheritance relations of two terms to a third term. It is easy to see that the situation here is like the cases of abduction and induction, except that now the conclusion is symmetric. Using the same procedure as in NAL-1, we first build Boolean functions among the variables as the following:

$$w = and(or(f_1, f_2), c_1, c_2), \; w^+ = and(f_1, c_1, f_2, c_2)$$

which lead to the truth-value function

$$F_{com}: \; f = \frac{f_1 f_2}{f_1 + f_2 - f_1 f_2}, \; c = \frac{c_1 c_2 (f_1 + f_2 - f_1 f_2)}{c_1 c_2 (f_1 + f_2 - f_1 f_2) + k}$$

When $f_1 = f_2 = 0$, we define f to be 0 for the sake of continuity.

From an inheritance judgment J_1 and a similarity judgment J_2, NAL does a certain type of *analogy* by replacing a term in J_1 by a similar term provided by J_2. The situation here is quite similar to that of deduction as defined in NAL-1. The difference is, when the similarity judgment goes to extreme to become an identity judgment, the conclusion should have the truth-value of the inheritance judgment. Therefore, the confidence should depend more on f_2 and c_2, and not on f_1 anymore. Under this consideration NAL uses the following truth-value function:

$$F_{ana}: \; f = f_1 f_2, \; c = c_1 f_2^2 c_2^2$$

If the two premises are both similarity relations, the inference here is based on the transitivity of the similarity relation. It can be treated as deduction going in both directions. However, in this case, if one similarity judgment goes to extreme to become an identity judgment, the conclusion should have the truth-value of the other similarity judgment. Therefore, the following truth-value function is used:

$$F_{dd2}: \; f = f_1 f_2, \; c = c_1 c_2 (f_1 + f_2 - f_1 f_2)$$

The above two truth-value functions are introduced as variants of the deduction function, rather than obtained by directly analyzing the truth-value relationship between the premises and the conclusion.

4.2.2 Sets, instance and property

With the inheritance relation defined in NAL-0, terms can form an inheritance hierarchy, with respect to their level of generality/specificity.

- If "$T_1 \rightarrow T_2$" is true, and "$T_2 \rightarrow T_1$" is false, then T_1 is more specific than T_2, and T_2 is more general than T_1.
- If both "$T_1 \rightarrow T_2$" and "$T_2 \rightarrow T_1$" are true (that is, "$T_1 \leftrightarrow T_2$" is true), then T_1 and T_2 are on the same level of generality/specificity.
- If both "$T_1 \rightarrow T_2$" and "$T_2 \rightarrow T_1$" are false, then T_1 and T_2 cannot be compared with respect to generality/specificity.

For various purposes, we often need to define the boundary of such a hierarchy, by treating certain terms as at the most specific or the most general level. In NAL-2, two kinds of such compound terms, "extensional set" and "intensional set," are introduced.

Definition 21 *If T is a term, the* extensional set *with T as the only component, $\{T\}$, is also a term, and its meaning is defined by*

$$(\forall x)((x \rightarrow \{T\}) \equiv (x \leftrightarrow \{T\})).$$

That is, a compound term with such a form is like a set defined by a sole element. The compound therefore has a special property: all terms in the extension of $\{T\}$ must be identical to it, and no term can be more specific than it (though it is possible for some terms to be more specific than T).

Theorem 7 *For any term T, $\{T\}^E \subseteq \{T\}^I$.*

On the other hand, $\{T\}^I$ is not necessarily included in $\{T\}^E$.

To name a term like this means to treat its extension as including an *individual*. In a natural language, T often corresponds to a proper name, and $\{T\}$ corresponds to a category with a single instance indicated by that proper name. For example, "Tweety is a bird" can be represented as "$\{Tweety\} \rightarrow bird$" (but not "$Tweety \rightarrow bird$," which means "Tweety is a kind of bird").

An *instance relation*, "$\circ\!\!\rightarrow$," is another way to represent the same information.

Definition 22 *The instance statement "$S \circ\!\!\rightarrow P$" is defined by the inheritance statement "$\{S\} \rightarrow P$."*

So "Tweety is a bird" can also be represented as "$Tweety \circ\!\!\rightarrow bird$."

The intuitive meaning of "$\circ\!\!\rightarrow$" is similar to the membership relation ("$\in$") in set theory, but in NAL this relation is no longer primary or necessary (since it is defined by other notions).[1]

Theorem 8 $((S \circ\!\!\rightarrow M) \wedge (M \rightarrow P)) \supset (S \circ\!\!\rightarrow P)$.

However, "$S \rightarrow M$" and "$M \circ\!\!\rightarrow P$" does not imply "$S \circ\!\!\rightarrow P$."

Theorem 9 $(S \circ\!\!\rightarrow \{P\}) \equiv (S \leftrightarrow P)$.

"$T \circ\!\!\rightarrow \{T\}$" follows as a special case. On the other hand, the statement "$T \circ\!\!\rightarrow T$" is not a literal truth, though may be an empirical one.

According to the duality between extension and intension, we can define another special compound term and the corresponding copula.

Definition 23 *If T is a term, the* intensional set *with T as the only component, $[T]$, is also a term, and its meaning is defined by*

$$(\forall x)(([T] \rightarrow x) \equiv ([T] \leftrightarrow x)).$$

That is, a compound term with such a form is like a set defined by a sole attribute. The compound therefore has a special property: all terms in the intension of $[T]$ must be identical to it, and no term can be more general than it (though it is possible for some terms to be more general than T).

Theorem 10 *For any term T, $[T]^I \subseteq [T]^E$.*

On the other hand, $[T]^E$ is not necessarily included in $[T]^I$.

---

[1]This issue will be discussed in detail in subsection 10.1.3.

To name a term like this means to treat its intension as having an *attribute*. In a natural language, T often corresponds to an adjective, and $[T]$ corresponds to a category with that adjective as the defining property. For example, "Ravens are black" can be represented as "$raven \rightarrow [black]$" (but not "$raven \rightarrow black$").

A *property relation*, "$\multimap$," is another way to represent the same information.

Definition 24 *The property statement "$S \multimap P$" is defined by the inheritance statement "$S \rightarrow [P]$."*

So "Ravens are black" can also be represented as "$raven \multimap black$."

This relation can be used when we characterize terms by a set of primary properties. It can also be directly used in inference.

Theorem 11 $(S \rightarrow M) \wedge (M \multimap P) \supset (S \multimap P)$.

However, "$S \multimap M$" and "$M \rightarrow P$" does not imply "$S \multimap P$."

Theorem 12 $([S] \multimap P) \equiv (S \leftrightarrow P)$.

"$[T] \multimap T$" follows as a special case. On the other hand, the statement "$T \multimap T$" is not a literal truth, though may be an empirical one.

An *instance-property* relation, "$\circ\!\!\multimap$," is defined by combining "$\circ\!\!\rightarrow$" and "$\multimap$."

Definition 25 *The instance-property statement "$S \circ\!\!\multimap P$" is defined by the inheritance statement "$\{S\} \rightarrow [P]$."*

Intuitively, it states that an instance S has a property P. This relation is not really necessary, and it is just a way to simplify a statement.

So "Tweety is yellow" can also be represented as "$Tweety \circ\!\!\multimap yellow$" (but not "$Tweety \multimap yellow$," "$Tweety \circ\!\!\rightarrow yellow$," or "$Tweety \rightarrow yellow$").

Theorem 13 $(S \circ\!\!\multimap P) \equiv (\{S\} \multimap P) \equiv (S \circ\!\!\rightarrow [P])$

$$
\begin{array}{rcl}
<copula> & ::= & \leftrightarrow \mid \circ\!\!\rightarrow \mid \rightarrow\!\!\circ \mid \circ\!\!\rightarrow\!\!\circ \\
<term> & ::= & \{<term>\} \mid [<term>]
\end{array}
$$

Table 4.2: The New Grammar Rules of Narsese-2

$S \leftrightarrow P$	$\{S\} \leftrightarrow \{P\}$
$S \leftrightarrow P$	$[S] \leftrightarrow [P]$
$S \rightarrow \{P\}$	$S \leftrightarrow \{P\}$
$[S] \rightarrow P$	$[S] \leftrightarrow P$
$S \circ\!\!\rightarrow P$	$\{S\} \rightarrow P$
$S \rightarrow\!\!\circ P$	$S \rightarrow [P]$
$S \circ\!\!\rightarrow\!\!\circ P$	$\{S\} \rightarrow [P]$

Table 4.3: The Equivalence Rules of NAL-2

4.2.3 NAL-2 summary

In summary, while all the grammar rules of Narsese-1 are still valid in NAL-2, there are additional grammar rules of Narsese-2, as listed in Table 4.2.

Beside the syllogistic rules in Table 4.1, the previous definitions give the equivalence rules of NAL-2 in Table 4.3, where the two statements in the same row can replace each other. That is, a judgment with one statement can derive another judgment with the other statement, with the same truth value.

Since each new copula is defined in terms of the inheritance relation "$\rightarrow$," its semantics and relevant inference rules can be derived from those in NAL-1.

Please note that the extension and intension of a term are still defined by the inheritance relation, not by the new relations derived from it. Therefore,

- "$S \circ\!\!\rightarrow P$" says that the extension of P include $\{S\}$ (not S) as an element;

- "$S \rightarrow\!\!\circ P$" says that the intension of S include $[P]$ (not P) as an element.

To simplify the implementation of the system, relations "$\circ\!\!\rightarrow$," "$\rightarrow\!\!\circ$," and "$\circ\!\!\rightarrow\!\!\circ$" are only used in the input/output interface, and within the system they are translated into "$\rightarrow$." Therefore we do not really need to introduce inference rules for them. The same thing cannot be done to "$\leftrightarrow$." Though the "$\leftrightarrow$" relation is defined in terms of the "$\rightarrow$" relation, the system usually cannot translate a "$\leftrightarrow$" judgment into an equivalent "$\rightarrow$" judgment. Therefore, NAL-2 uses five copula in its interface language, but only keep two of them ("$\rightarrow$" and "$\leftrightarrow$") in its internal representation, without losing any expressive and inferential power.

As an example of inference in NAL-2, we start with the following judgments:

$$(1)\ Tweety \circ\!\!\rightarrow bird < 1, 0.9 >$$
$$(2)\ Tweety \circ\!\!\rightarrow\!\!\circ\ yellow < 1, 0.9 >$$
$$(3)\ Tweety \leftrightarrow Birdie < 1, 0.9 >$$

Using the equivalence rules, they derive the following judgments, respectively:

$$(4)\ \{Tweety\} \rightarrow bird < 1, 0.9 >$$
$$(5)\ \{Tweety\} \rightarrow [yellow] < 1, 0.9 >$$
$$(6)\ \{Tweety\} \leftrightarrow \{Birdie\} < 1, 0.9 >$$

From (4) and (5), by induction the system derives

$$(7)\ bird \rightarrow [yellow] < 1, 0.45 >$$

From (4) and (6), by analogy the system derives

$$(8)\ \{Birdie\} \rightarrow bird < 1, 0.73 >$$

which can be displayed as

$$(9)\ Birdie \circ\!\!\rightarrow bird < 1, 0.73 >$$

4.3 NAL-3: intersections and differences

In NAL-3, compound terms are composed by combining the extension or intension of existing terms in certain way.

4.3.1 Intersections

Definition 26 *Given terms* T_1 *and* T_2, *their* extensional intersection, $(T_1 \cap T_2)$, *is a compound term defined by*

$$(\forall x)((x \rightarrow (T_1 \cap T_2)) \equiv ((x \rightarrow T_1) \wedge (x \rightarrow T_2))).$$

From right to left, the equivalence expression defines the extension of the compound, i.e., "$(x \rightarrow T_1) \wedge (x \rightarrow T_2)$" implies "$x \rightarrow (T_1 \cap T_2)$"; from left to right, it defines the intension of the compound, i.e., "$(T_1 \cap T_2) \rightarrow (T_1 \cap T_2)$" implies "$(T_1 \cap T_2) \rightarrow T_1$" and "$(T_1 \cap T_2) \rightarrow T_2$."

As an example, "Ravens are black birds" can be represented as "*raven* $\rightarrow$ ([*black*] $\cap$ *bird*)," where the predicate term is an extensional intersection of the term [*black*] and the term *bird*.

Theorem 14

$$(T_1 \cap T_2)^E = T_1^E \cap T_2^E, \; (T_1 \cap T_2)^I = T_1^I \cup T_2^I$$

In the above expressions, the "$\cap$" sign is used in two different senses. On the right-side of the first expression, it indicates the ordinary intersection of sets, but on the left-side of the two expressions, it is the new intersection operator of terms. Though these two senses are intuitively similar, they are not the same, because the term operator is related to both the extension and the intension of a term.

As the common extension of the two terms, the compound term $(T_1 \cap T_2)$ inherits properties of both T_1 and T_2. That is why its intension is the union of the intensions of its two components.

The above definition and theorem specify the *literal* meaning of the compound terms, which is formed when a compound is built from its components. Later, the meaning of the compound may become more or less different, as the result of new experience and inference activity.[2]

The intensional intersection of terms is defined symmetrically.

Definition 27 *Given terms* T_1 *and* T_2, *their* intensional intersection, $(T_1 \cup T_2)$, *is a compound term defined by*

$$(\forall x)(((T_1 \cup T_2) \rightarrow x) \equiv ((T_1 \rightarrow x) \wedge (T_2 \rightarrow x))).$$

[2]This issue is discussed in detail in subsection 11.1.5.

From right to left, the equivalence expression defines the intension of the compound, i.e., "$(T_1 \rightarrow x) \wedge (T_2 \rightarrow x)$" implies "$(T_1 \cup T_2) \rightarrow x$"; from left to right, it defines the extension of the compound, i.e., "$(T_1 \cup T_2) \rightarrow (T_1 \cup T_2)$" implies "$T_1 \rightarrow (T_1 \cup T_2)$" and "$T_2 \rightarrow (T_1 \cup T_2)$."

Intuitively, the intensional intersection of two terms is defined by the common properties of the terms.

Theorem 15

$$(T_1 \cup T_2)^I = T_1^I \cap T_2^I, \ (T_1 \cup T_2)^E = T_1^E \cup T_2^E$$

Now we can see that the duality of *extension* and *intension* in NAL corresponds to the duality of *intersection* and *union* in set theory — *intensional intersection* corresponds to *extensional union*, and *extensional intersection* corresponds to *intensional union*.[3]

From the definitions, it is obvious that both intersections are symmetric to their components:

Theorem 16

$$\begin{array}{l}(T_1 \cap T_2) \leftrightarrow (T_2 \cap T_1)\\(T_1 \cup T_2) \leftrightarrow (T_2 \cup T_1)\end{array}$$

The relation between the compounds to their components is captured by the following theorem:

Theorem 17

$$\begin{array}{l}(T_1 \cap T_2) \rightarrow T_1\\T_1 \rightarrow (T_1 \cup T_2)\end{array}$$

Both operators can be extended to take more than two arguments. Since "$\cap$" and "$\cup$" are both associative and symmetric, the order of their components does not matter. A special situation for these two operators is when the components are the same.

---

[3]Some people may think that it is more natural to call $(T_1 \cup T_2)$ "extensional union" than "intensional intersection," because the symbol used for the term operator is the union operator in set theory. I do it the other way to stress its relation with the extensional intersection, as well as with the "difference" terms to be introduced in the following.

Theorem 18

$$(T \cup T) \leftrightarrow T$$
$$(T \cap T) \leftrightarrow T$$

The following implications are derived from the definition of the compound terms.

Theorem 19

$$\begin{array}{rcll} T_1 \rightarrow M & \wedge & \neg((T_1 \cup T_2) \rightarrow M) & \supset & \neg(T_2 \rightarrow M) \\ \neg(T_1 \rightarrow M) & \wedge & (T_1 \cap T_2) \rightarrow M & \supset & T_2 \rightarrow M \\ M \rightarrow T_1 & \wedge & \neg(M \rightarrow (T_1 \cap T_2)) & \supset & \neg(M \rightarrow T_2) \\ \neg(M \rightarrow T_1) & \wedge & M \rightarrow (T_1 \cup T_2) & \supset & M \rightarrow T_2 \end{array}$$

These implications will be turned into inference rules in the next chapter, which can be used to "decompose" a compound term to get conclusions about its components.

The two intersection operators keep inheritance and similarity relations between terms:

Theorem 20

$$\begin{array}{lcl} S \rightarrow P & \supset & (S \cup M) \rightarrow (P \cup M) \\ S \rightarrow P & \supset & (S \cap M) \rightarrow (P \cap M) \\ S \leftrightarrow P & \supset & (S \cup M) \leftrightarrow (P \cup M) \\ S \leftrightarrow P & \supset & (S \cap M) \leftrightarrow (P \cap M) \end{array}$$

In these propositions, M can be any term in V_K. The same is assumed for the implications introduced later.

If a term is taken as a set, then all the above theorems can be proved in set theory. However, in NAL even if sets can be defined (as in NAL-2), not all terms are sets, and the above theorems should not be understood as equivalent to their counterparts in set theory, but as (partially) isomorphic to them.

4.3.2 Differences

The above two compound terms are defined by restricting the extension or intension of a term with an additional *positive* statement. In the following we do the same thing, but with a *negative* statement.

Definition 28 *If T_1 and T_2 are different terms, their* extensional difference, *$(T_1 - T_2)$, is a compound term defined by*

$$(\forall x)((x \rightarrow (T_1 - T_2)) \equiv ((x \rightarrow T_1) \wedge \neg(x \rightarrow T_2))).$$

From right to left, the equivalence expression defines the extension of the compound, i.e., "$(x \rightarrow T_1) \wedge \neg(x \rightarrow T_2)$" implies "$x \rightarrow (T_1 - T_2)$"; from left to right, it defines the intension of the compound, i.e., "$(T_1 - T_2) \rightarrow (T_1 - T_2)$" implies "$(T_1 - T_2) \rightarrow T_1$" and "$\neg((T_1 \cap T_2) \rightarrow T_2)$."

Given this definition, "Penguins are birds that cannot fly" can be represented as "$penguin \rightarrow (bird - [flying])$," where the predicate term is a extensional difference of the term *bird* and the term $[flying]$.

Obviously, $(T_2 - T_1)$ can also be defined, but it will be different from $(T_1 - T_2)$.

Theorem 21

$$(T_1 - T_2)^E = T_1^E - T_2^E, \ (T_1 - T_2)^I = T_1^I$$

Intuitively, $(T_1 - T_2)$ is $(T_1 \cap (non\text{-}T_2))$, where $(non\text{-}T_2)$ is a term whose extension includes all terms in V_K that are not in the extension of T_2. However, the intension of this term is empty, because such a definition specifies no common (affirmatively specified) property for the terms in its extension. Since no additional property can be assigned to the compound, the intension of $(T_1 - T_2)$ is the same as that of T_1. For the same reason, in Narsese there is no term defined as the negation of another term, though we can talk about $(non\text{-}T)$ in the meta-language.[4]

Symmetrically, intensional differences can be defined.

Definition 29 *If T_1 and T_2 are different terms, their* intensional difference, *$(T_1 \ominus T_2)$, is a compound term defined by*

$$(\forall x)(((T_1 \ominus T_2) \rightarrow x) \equiv ((T_1 \rightarrow x) \wedge \neg(T_2 \rightarrow x))).$$

From right to left, the equivalence expression defines the intension of the compound, i.e., "$(T_1 \rightarrow x) \wedge \neg(T_2 \rightarrow x)$" implies "$(T_1 \ominus T_2) \rightarrow x$"; from left to right, it defines the extension of the compound, i.e., "$(T_1 \ominus T_2) \rightarrow (T_1 \ominus T_2)$" implies "$T_1 \rightarrow (T_1 \ominus T_2)$" and "$\neg(T_2 \rightarrow (T_1 \ominus T_2))$."

[4]Though in Narsese there is no negated term, there are negated statements, which will be defined in the next chapter.

Theorem 22

$$(T_1 \ominus T_2)^I = T_1^I - T_2^I, \ (T_1 \ominus T_2)^E = T_1^E$$

Intuitively, intensional difference is used to relax the requirement of a category by removing some positive properties. As a result, the original instances remains.

The relation between the difference compounds to their first component is captured by the following theorem:

Theorem 23

$$\begin{array}{l} (T_1 - T_2) \rightarrow T_1 \\ T_1 \rightarrow (T_1 \ominus T_2) \end{array}$$

The relation between the difference compounds to their second component is captured by the following theorem:

Theorem 24

$$\begin{array}{l} M \rightarrow (T_1 - T_2) \supset \neg(M \rightarrow T_2) \\ (T_1 \ominus T_2) \rightarrow M \supset \neg(T_2 \rightarrow M) \end{array}$$

Please notice the difference between the above two theorems: while in the former the result can be used to address both the extension and the intension of the compound, in the latter it is only about one of the two aspects.

Unlike the intersection operators, the difference operators cannot take more than two arguments, though we can still use both prefix and infix formats to represent them, so as to be consistent with other compound terms. Also, neither $(T - T)$ nor $(T \ominus T)$ is a valid term.

According to the literal meaning of the compound terms, there are the following implications involving extensional/intensional differences:

Theorem 25

$$\begin{array}{rcccr} T_1 \rightarrow M & \wedge & \neg((T_1 \ominus T_2) \rightarrow M) & \supset & T_2 \rightarrow M \\ \neg(T_1 \rightarrow M) & \wedge & \neg((T_2 \ominus T_1) \rightarrow M) & \supset & \neg(T_2 \rightarrow M) \\ M \rightarrow T_1 & \wedge & \neg(M \rightarrow (T_1 - T_2)) & \supset & M \rightarrow T_2 \\ \neg(M \rightarrow T_1) & \wedge & \neg(M \rightarrow (T_2 - T_1)) & \supset & \neg(M \rightarrow T_2) \end{array}$$

The two difference operators keep inheritance and similarity relations between terms, though may reverse the direction of the relation:

Theorem 26

$$
\begin{array}{lcl}
S \rightarrow P & \supset & (S - M) \rightarrow (P - M) \\
S \rightarrow P & \supset & (M - P) \rightarrow (M - S) \\
S \rightarrow P & \supset & (S \ominus M) \rightarrow (P \ominus M) \\
S \rightarrow P & \supset & (M \ominus P) \rightarrow (M \ominus S) \\
S \leftrightarrow P & \supset & (S - M) \leftrightarrow (P - M) \\
S \leftrightarrow P & \supset & (M - P) \leftrightarrow (M - S) \\
S \leftrightarrow P & \supset & (S \ominus M) \leftrightarrow (P \ominus M) \\
S \leftrightarrow P & \supset & (M \ominus P) \leftrightarrow (M \ominus S)
\end{array}
$$

4.3.3 Compound sets

To apply the set-theoretic operators defined above to the sets defined in NAL-2, we get the following definitions:

Definition 30 *If* $t_1, \cdots, t_n$ $(n \geq 2)$ *are different terms, a* compound extensional set $\{t_1, \cdots, t_n\}$ *is defined as* $(\cup \{t_1\} \cdots \{t_n\})$*; a* compound intensional set $[t_1, \cdots, t_n]$ *is defined as* $(\cap [t_1] \cdots [t_n])$.

In this way, extensional sets and intensional sets can both have multiple components. Intuitively, the former defines a term by enumerating its instances, and the latter by enumerating its properties. Again, the order of the components does not matter.

Now we can see that a set is a kind of compound term, whose internal structure fully specifies its extension (for an extensional set) or intension (for an intensional set), in the sense of identical terms. For example, for "$\{x\} \rightarrow \{t_1, t_2, t_3\}$" to be true, x must be identical to t_1, t_2, or t_3.

These compound sets satisfy the following identity relations (where t_1, t_2, t_3 are different terms).

Theorem 27

$$
\begin{array}{rcl}
(\{t_1, t_2\} \cup \{t_2, t_3\}) & \leftrightarrow & \{t_1, t_2, t_3\} \\
(\{t_1, t_2\} \cap \{t_2, t_3\}) & \leftrightarrow & \{t_2\} \\
(\{t_1, t_2\} - \{t_2, t_3\}) & \leftrightarrow & \{t_1\} \\
([t_1, t_2] \cap [t_2, t_3]) & \leftrightarrow & [t_1, t_2, t_3] \\
([t_1, t_2] \cup [t_2, t_3]) & \leftrightarrow & [t_2] \\
([t_1, t_2] \ominus [t_2, t_3]) & \leftrightarrow & [t_1]
\end{array}
$$

These relations can be extended to sets with more (or less) than two components. These results are similar to the ones in set theory, though in NAL not all terms can be treated as sets.

4.3.4 NAL-3 summary

The additional grammar rules of Narsese-3 are listed in Table 4.4.

$<term>$	::=	$\{<term>^+\} \mid [<term>^+]$
		$\mid (\cap <term><term>^+) \mid (\cup <term><term>^+)$
		$\mid (- <term><term>) \mid (\ominus <term><term>)$

Table 4.4: The New Grammar Rules of Narsese-3

The previous grammar rule for extensional set and intensional set becomes a special case of the new rule. For sets with multiple components, "," can be used to separate them. For an intersection or difference term, an "infix" format can also be used, as mentioned in the previous section.

Related to the new compound terms, the most important inference rules introduced in NAL-3 are those that combine two inheritance statements into a new one with a compound term as subject or predicate. These rules are listed in Table 4.5, which are applied only when T_1 and T_2 are different, and do not have each other as component.

$J_2 \setminus J_1$	$M \rightarrow T_1$	$T_1 \rightarrow M$
$T_2 \rightarrow M$		$(T_1 \cup T_2) \rightarrow M \ \ <F_{int}>$ $(T_1 \cap T_2) \rightarrow M \ \ <F_{uni}>$ $(T_1 \ominus T_2) \rightarrow M \ \ <F_{dif}>$
$M \rightarrow T_2$	$M \rightarrow (T_1 \cap T_2) \ \ <F_{int}>$ $M \rightarrow (T_1 \cup T_2) \ \ <F_{uni}>$ $M \rightarrow (T_1 - T_2) \ \ <F_{dif}>$	

Table 4.5: The Composition Rules of NAL-3

The truth-value functions in Table 4.5 are defined as the following:

$$
\begin{aligned}
F_{int}: \ & f = and(f_1, f_2) \\
& c = or(and(not(f_1), c_1), and(not(f_2), c_2)) + and(f_1, f_2, c_1, c_2) \\
F_{uni}: \ & f = or(f_1, f_2) \\
& c = or(and(f_1, c_1), and(f_2, c_2)) + and(not(f_1), not(f_2), c_1, c_2) \\
F_{dif}: \ & f = and(f_1, not(f_2)) \\
& c = or(and(not(f_1), c_1), and(f_2, c_2)) + and(f_1, not(f_2), c_1, c_2)
\end{aligned}
$$

To understand the above functions, let us look at the extensional cases, where the two premises have a common subject.

The frequency functions directly come from the definitions of the compounds: "$M \rightarrow (T_1 \cap T_2)$" if "$M \rightarrow T_1$" *and* "$M \rightarrow T_2$"; "$M \rightarrow (T_1 \cup T_2)$" if "$M \rightarrow T_1$" *or* "$M \rightarrow T_2$"; "$M \rightarrow (T_1 - T_2)$" if "$M \rightarrow T_1$" *but not* "$M \rightarrow T_2$".

The confidence functions are determined by listing the various cases where the conclusion gets full evidence (in idealized situations). For example, for "$M \rightarrow (T_1 \cap T_2)$," it happens when one premise is absolutely false, or when both premises have full evidence. The other confidence functions can be obtained similarly.

The intensional cases are completely symmetric to the extensional ones, so the same set of truth functions is used, though for the (extensional and intensional) intersections, the functions are switched.

As an example, given the following judgments,

$$(1)\ \{Tweety\} \rightarrow bird <1, 0.9>$$
$$(2)\ \{Tweety\} \rightarrow [yellow] <1, 0.9>$$
$$(3)\ \{Tweety\} \rightarrow canary <1, 0.9>$$

from (1) and (2) the system composes a compound term "$([yellow] \cap bird)$" ("yellow bird"), with "$\{Tweety\}$" in its extension:

$$(4)\ \{Tweety\} \rightarrow ([yellow] \cap bird) <1, 0.81>$$

Then, from (3) and (4) by induction the following judgment is derived:

$$(5)\ canary \rightarrow ([yellow] \cap bird) <1, 0.41>$$

In the last step, the compound terms involved are treated just like atomic terms.

4.4 NAL-4: products, images, and ordinary relations

One common criticism to Aristotle's Syllogism, or to term logic in general, is that it cannot represent and process a relation that is not a copula [Bocheński, 1970]. In NAL-4, these relations are handled with the help of certain types of compound terms, using an idea borrowed from set theory.

4.4.1 Ordinary relations and products

In NAL-4, "ordinary relations" indicates the relations among terms that are not the inheritance relation or its variants (such as similarity, instance, property, and instance-property). These relations may be not reflexive, not transitive, not merely reflexive and transitive, or not defined on all terms. The copulas are defined in the meta-language of NAL, with fixed (built-in) meaning to the system. In contrary, the ordinary relations are described in Narsese, with experience-grounded meaning.

Definition 31 *For two terms* T_1 *and* T_2*, their* product $(T_1 \times T_2)$ *is a compound term defined by*

$$((S_1 \times S_2) \rightarrow (P_1 \times P_2)) \equiv ((S_1 \rightarrow P_1) \wedge (S_2 \rightarrow P_2)).$$

This definition can be extended as before to allow more than two components in a product. Also, the "prefix" format can be used for products.

Unlike the term operators introduced in NAL-3, the product operator allows the components to be the same. That is, $(T \times T)$ is a valid compound term. $(T_1 \times T_2)$ and $(T_2 \times T_1)$ are usually different, and so are $(T_1 \times (T_2 \times T_3))$ and $((T_1 \times T_2) \times T_3)$.

Theorem 28

$$(S \rightarrow P) \equiv ((M \times S) \rightarrow (M \times P)) \equiv ((S \times M) \rightarrow (P \times M))$$
$$(S \leftrightarrow P) \equiv ((M \times S) \leftrightarrow (M \times P)) \equiv ((S \times M) \leftrightarrow (P \times M))$$

Theorem 29

$$((S_1 \times S_2) \leftrightarrow (P_1 \times P_2)) \equiv ((S_1 \leftrightarrow P_1) \wedge (S_2 \leftrightarrow P_2))$$

That is, two products are identical if and only if and only if their corresponding components are identical.

Theorem 30

$$\{(x \times y) \,|\, x \in T_1^E, y \in T_2^E\} \subseteq (T_1 \times T_2)^E$$
$$\{(x \times y) \,|\, x \in T_1^I, y \in T_2^I\} \subseteq (T_1 \times T_2)^I$$

The "$\subseteq$" cannot be replaced by "$=$" in the above theorem, because $(T_1 \times T_2)^E$ and $(T_1 \times T_2)^I$ may contain other terms that are not products.

Definition 32 *A* relation *is a term* R *such that there is a product* P *satisfying "*$P \rightarrow R$*" or "*$R \rightarrow P$*".*

Therefore in NAL a relation is not a set, because it is not only defined extensionally. A product is a relation (because "$(T_1 \times T_2) \rightarrow (T_1 \times T_2)$"), but a relation is not necessarily a product. In NAL, a relation can be an atomic term.

For example, "Acid and base neutralize each other" can be represented as "$(acid \times base) \rightarrow neutralization$," and "Neutralization happens between acid and base" can be represented as "$neutralization \rightarrow (acid \times base)$."

4.4.2 Images

Given one component, the "image" operator identifies the other one in the extension or intension of a given relation with two components.

Definition 33 *For a relation R and a product $(\times\ T_1\ T_2)$, the* extensional image *operator, "$\bot$," and* intensional image *operator, "$\top$," of the relation on the product are defined as the following, respectively:*

$$((\times\ T_1\ T_2) \rightarrow R) \equiv (T_1 \rightarrow (\bot\ R\ \diamond\ T_2)) \equiv (T_2 \rightarrow (\bot\ R\ T_1\ \diamond))$$

$$(R \rightarrow (\times\ T_1\ T_2)) \equiv ((\top\ R\ \diamond\ T_2)) \rightarrow T_1) \equiv ((\top\ R\ T_1\ \diamond)) \rightarrow T_2)$$

where "$\diamond$" is a special symbol indicating the location of T_1 or T_2 in the product, and it can appear in any place, except the first (which is the relation), in the component list. When it appears at the second place, the image can also be written in infix format as $(R \bot T_2)$ or $(R \top T_2)$ (in other cases, only the prefix format is used).

For example, "Acid corrodes metal" can be equivalently represented as "$(\times\ acid\ metal) \rightarrow corrosion$," "$acid \rightarrow (\bot\ corrosion\ \diamond\ metal)$," and "$metal \rightarrow (\bot\ corrosion\ acid\ \diamond)$."

The above definition can be extended to include products with more than two components, where the image can only be written in the prefix format.

In general, $(R \bot T_2)$ and $(R \top T_2)$ are different, but there are situations where they are the same.

Theorem 31

$$T_1 \leftrightarrow ((T_1 \times T_2) \bot T_2)$$

$$T_1 \leftrightarrow ((T_1 \times T_2) \top T_2)$$

Intuitively, the above theorem shows that starting from a term, a "product" followed by an "image" will go back to the same term. However, if the order of the two operators is switched, the result is different:

Theorem 32

$$((R \perp T) \times T) \rightarrow R$$

$$R \rightarrow ((R \top T) \times T)$$

The "$\rightarrow$" in the above theorem cannot be replaced by the "$\leftrightarrow$."

An image operator can be applied to both sides of an inheritance relation, though in the result, the subject and predicate may be switched:

Theorem 33

$$\begin{array}{lll} S \rightarrow P & \supset & (S \perp M) \rightarrow (P \perp M) \\ S \rightarrow P & \supset & (S \top M) \rightarrow (P \top M) \\ S \rightarrow P & \supset & (M \perp P) \rightarrow (M \perp S) \\ S \rightarrow P & \supset & (M \top P) \rightarrow (M \top S) \end{array}$$

4.4.3 NAL-4 summary

In summary, NAL-4 introduces the new grammar rules in Table 4.6.

$<term>$	$::=$	$(\times <term><term>^{+})$
		$\mid (\perp <term><term>^{*} \diamond <term>^{*})$
		$\mid (\top <term><term>^{*} \diamond <term>^{*})$

Table 4.6: The New Grammar Rules of Narsese-4

There is no new inference rule directly defined in NAL-4, except the equivalence rules in Table 4.7, given by the previous definitions. The table does not include all variants of a rule obtained by changing the component list of the compound.

The following example shows the capability of NAL-4. Let's start with three judgments:

(1) $vinegar \rightarrow acid <1, 0.9>$
(2) $baking\text{-}soda \rightarrow base <1, 0.9>$
(3) $(\times\ vinegar\ baking\text{-}soda) \rightarrow neutralization <1, 0.9>$

$S \rightarrow P$	$(S \times M) \rightarrow (P \times M)$
$S \leftrightarrow P$	$(S \times M) \leftrightarrow (P \times M)$
$(\times\ T_1\ T_2) \rightarrow R$	$T_1 \rightarrow (\bot\ R\ \diamond\ T_2)$
$R \rightarrow (\times\ T_1\ T_2)$	$(\top\ R\ \diamond\ T_2) \rightarrow T_1)$

Table 4.7: The Equivalence Rules of NAL-4

Then the following judgment can be derived as an equivalent form of (3):

(4) $vinegar \rightarrow (\bot\ neutralization\ \diamond\ baking\text{-}soda) <1, 0.9>$

From (1) and (4), by induction the system gets

(5) $acid \rightarrow (\bot\ neutralization\ \diamond\ baking\text{-}soda) <1, 0.45>$

which can be equivalently transformed into

(6) $baking\text{-}soda \rightarrow (\bot\ neutralization\ acid\ \diamond) <1, 0.45>$

From (2) and (6), by induction again:

(7) $base \rightarrow (\bot\ neutralization\ acid\ \diamond) <1, 0.29>$

which can be rewritten as

(8) $(\times\ acid\ base) \rightarrow neutralization <1, 0.29>$

This example shows that though the inference rules of NAL are defined on the inheritance relation and its variants, other relations (like "neutralization") can also be represented and processed properly.

Chapter 5

Higher-Order Inference

The NAL built so far is "first-order," in the sense that statements are relations among terms, but a statement cannot be treated as a term. In "higher-order inference," a statement can be treated as a term, therefore there are *statements on statements*, as well as inference on this kind of higher-order statements. In NAL, though it is possible to further divide higher-order statements into second-order, third-order, and so on, such a distinction is not practically useful. Therefore, they will be covered together under the same notion of "higher-order statements." In this chapter, NAL will be extended, step by step, to include various types of higher-order statements, as well as the inference on them.

5.1 NAL-5: statements as terms

In NAL-5, new grammar and inference rules are introduced, so that the system can treat a statement as a term.

5.1.1 Higher-order statements

First, the new grammar rules of Narsese-5 are listed in Table 5.1. In NAL-5 a statement can be used as a term (so it can be involved in various inheritance and ordinary relations). Some ordinary relations, such as "believe," "say," "know," and so on, take a statement as an argument. For example, "John knows that whale is a kind of mammal" is

$$
\begin{array}{rcl}
<term> & ::= & (<statement>) \\
<statement> & ::= & <term> \\
 & & \mid (\neg <statement>) \\
 & & \mid (\wedge <statement><statement>^{+}) \\
 & & \mid (\vee <statement><statement>^{+}) \\
<copula> & ::= & \Rightarrow \mid \Leftrightarrow
\end{array}
$$

Table 5.1: The New Grammar Rules of Narsese-5

represented in Narsese as "$(\{John\} \times \{whale \rightarrow mammal\}) \rightarrow know$," where the subject term is a product containing a statement.

On the other hand, a term can also be used as a statement (that corresponds to the name of a statement, such as "Newton's first law"). However, it does not mean that there is no difference between *term* and *statement*. In NAL, a statement has both meaning and truth value, but a non-statement term only has meaning (without truth value).

Compound statements can be formed using statement operators *negation* ("$\neg$"), *conjunction* ("$\wedge$"), and *disjunction* ("$\vee$"), whose meanings are intuitively similar to those in propositional logic.[1] As usual, conjunction and disjunction can be used both in the prefix format and the infix format.

Finally, two new copulas, *implication* ("$\Rightarrow$") and *equivalence* ("$\Leftrightarrow$"), are defined between statements. Intuitively, they correspond to "if" and "if-and-only-if," respectively. They are "higher-order relations" because they are only defined between two statements.

Please note that in general "$\Rightarrow$" and "$\Leftrightarrow$" are different from "$\supset$" and "$\equiv$," though their intuitive meanings are similar. The former two belong to the object language (Narsese), while the latter two belong to the meta-language of Narsese (propositional calculus). The two new copulas in NAL have different status from the three statement operators

[1]In this book, though the same symbols are used for the statement operators in Narsese and the logical connectives in the meta-language of NAL (as defined in propositional calculus), they should be distinguishable by context.

mentioned above, whereas in propositional calculus the five corresponding notions have the same status as truth-value operators.[2]

5.1.2 Implication and inheritance

First, the implication relation is defined by valid inference in NAL.

Definition 34 *If S_1 and S_2 are statements, "$S_1 \Rightarrow S_2$" is true (i.e., has truth value $<1, 1>$) if and only if from "$S_1 <1, 1>$" alone NAL can derive "$S_2 <1, 1>$."*

The derivation in the above definition can consists of any (finite) number (0, 1, 2, ..., n) of inference steps.

Theorem 34 *The implication relation, "$\Rightarrow$," is a reflexive and transitive relation from one statement to another statement.*

Since the above theorem of implication is parallel to the definition of inheritance (in NAL-0), higher-order inference in NAL can be defined as partially isomorphic to first-order inference. The correspondences are listed in Table 5.2.

first-order inference	**higher-order inference**
inheritance	implication
similarity	equivalence
subject	antecedent
predicate	consequent
extension	sufficient condition
intension	necessary condition
extensional intersection	conjunction
intensional intersection	disjunction

Table 5.2: The Isomorphism Between First-Order and Higher-Order

The definitions of the new notions in Table 5.2 are in the following.

Definition 35 *An* implication statement *consists of two statements related by the implication relation. In implication statement "$A \Rightarrow C$," A is the* antecedent, *and C is the* consequent.

[2]This difference will be discussed in detail in subsection 9.4.1.

Definition 36 *Given experience K, the* sufficient conditions *of a statement T is the set of statements $T^S = \{x \mid x \in V_K \wedge x \Rightarrow T\}$; the* necessary conditions *of T is the set of statements $T^N = \{x \mid x \in V_K \wedge T \Rightarrow x\}$.*

Definition 37 *For an implication statement "$A \Rightarrow C$," its* evidence *are statements in A^S and C^N. Among them, statements in $(A^S \cap C^S)$ and $(C^N \cap A^N)$ are* positive evidence, *while statements in $(A^S - C^S)$ and $(C^N - A^N)$ are* negative evidence.

Definition 38 *The* equivalence *relation is a symmetric implication relation. That is, "$A \Leftrightarrow B$" is defined to mean "$(A \Rightarrow B) \wedge (B \Rightarrow A)$."*

The amounts of evidence and the truth value for a higher-order statement are defined in the same way from evidence as for first-order statements.

Now the meaning of a statement includes not only its extension and intension, but also its sufficient and necessary conditions.

Please note that the truth value of an implication (or equivalent) statement do not depend on all its inheritance (or similarity) relations with other terms. Two statements can be fully equivalent (i.e., "$P \Leftrightarrow Q$" has a frequency $= 1$), but still have different meanings (i.e., "$P \leftrightarrow Q$" has a frequency < 1). On the other hand, if two statements have the same meaning, they should also have the same truth value.

Definition 39 *When S_1 and S_2 are different statements, their* conjunction, *$(S_1 \wedge S_2)$, is a compound statement defined by*

$$(\forall x)((x \Rightarrow (S_1 \wedge S_2)) \equiv ((x \Rightarrow S_1) \wedge (x \Rightarrow S_2))).$$

Their disjunction, *$(S_1 \vee S_2)$, is a compound statement defined by*

$$(\forall x)(((S_1 \vee S_2) \Rightarrow x) \equiv ((S_1 \Rightarrow x) \wedge (S_2 \Rightarrow x))).$$

The above two statement operators are symmetric, and can be extended to take more than two arguments.

Because of this isomorphism, there is an isomorphic inference rule in NAL-5 for the following rules defined previously:

- The NAL-1 rules for choice, revision, deduction, abduction, induction, exemplification, and conversion.
- The NAL-2 rules for revision, comparison, analogy, conversion, and deduction.
- The NAL-3 rules for the processing of intersections.

The term operators for (extensional/intensional) set, product, and (extensional/intensional) image are not involved in the isomorphism between first-order and higher-order terms. They treat higher-order terms just like first-order terms, and there is no special rule added.

To treat conditional statements (implications) and categorical statements (inheritances) in a similar way is not a new idea. The following opinion can be traced back to Leibniz: "Conditionals are categorical by virtue of the fact that the relationship between an antecedent and a consequent is exactly that the relationship between a subject and a predicate, namely, containment" [Englebretsen, 1981]. In predicate calculus, categorical statements are translated into conditional statements. What makes NAL different is that it does not treat the two as *the same*, but as *isomorphic to each other*. Consequently, the corresponding inference rules have different forms and meanings, though using the same truth-value function.

5.1.3 Implication as conditional statement

Besides the above isomorphism, higher-order inference and first-order inference in NAL can be directly related to each other, by extending the identity between an implication statement ($S_1 \Rightarrow S_2$) and an inference process (from S_1 to S_2) to actual judgments.

By definition, in NAL a judgment "$S < f,\ c >$" indicates that "The degree of belief the system has on statement S, according to available evidence, is measured by truth value $< f,\ c >$." Now if we assume that the available evidence currently used on the evaluation of S can be written as a compound statement E, then the same meaning can be represented by "$E \Rightarrow S < f,\ c >$," that is, "The degree of belief the system has on statement 'If E is true, then S is true' is measured by truth value $< f,\ c >$." In this way, a statement S is equivalently transformed into an implication statement "$E \Rightarrow S$" ("If the available evidence is true, then S is true"). This transformation is justified according to the semantics of NAL.

This transformation is a conceptual one, not an actual one in the sense that there is a statement used by NAL corresponding to the above E. This conceptual transformation is used to justify inference rules. We can add this implicit conditions into the premises, so as to change the premise combinations into the ones for which we already have inference rules. Finally, we remove the implicit condition from the conclusion. Table 5.3 contains several rules obtained in this way.

premises	**add condition**	**conclusion**	**drop condition**
$M \Rightarrow P,\ M$	$M \Rightarrow P,\ E \Rightarrow M$	$E \Rightarrow P$	$P <F_{ded}>$
$P \Rightarrow M,\ M$	$P \Rightarrow M,\ E \Rightarrow M$	$E \Rightarrow P$	$P <F_{abd}>$
$M \Leftrightarrow P,\ M$	$M \Leftrightarrow P,\ E \Rightarrow M$	$E \Rightarrow P$	$P <F'_{ana}>$

Table 5.3: The Conditional Syllogistic Rules (1)

Similarly, when the two premises can be seen as derived from the same evidence, it can be used as the common virtual condition of the two, and some conclusions can be derived accordingly, as in Table 5.4.

premises	**add condition**	**conclusion**	**drop condition**
$P,\ M$	$E \Rightarrow P,\ E \Rightarrow M$	$M \Rightarrow P$	$M \Rightarrow P <F_{ind}>$
$P,\ M$	$E \Rightarrow P,\ E \Rightarrow M$	$M \Leftrightarrow P$	$M \Leftrightarrow P <F_{com}>$
$P,\ M$	$E \Rightarrow P,\ E \Rightarrow M$	$E \Rightarrow (P \wedge M)$	$P \wedge M <F_{int}>$
$P,\ M$	$E \Rightarrow P,\ E \Rightarrow M$	$E \Rightarrow (P \vee M)$	$P \vee M <F_{uni}>$

Table 5.4: The Composition Rules of NAL-5

The truth-value functions in Table 5.3 and Table 5.4. are those defined in NAL-1 to NAL-3. For practical purpose, we can ignore the two columns in the middle, and treat the rules as directly go from the first column (premises) to the last column (conclusions).

Now we have three groups of syllogistic rules (deduction, abduction, and induction), one defined on the inheritance relation (in NAL-1), one on the implication relation (in the previous subsection), and one on a mixture of the two (above). In the three groups, each type of inference (deduction, abduction, or induction) has a different form, but uses the same truth-value function. The last group is similar to how the three types of inference are defined in extended propositional calculus [Flach and Kakas, 2000], except that in NAL the statements have truth values attached to indicate their evidential support.

Also according to the semantical interpretation of implication, we have

$$(M \Rightarrow ((\wedge\ A_1 \cdots A_m) \Rightarrow C)) \equiv ((\wedge\ M A_1 \cdots A_m) \Rightarrow C)$$

that is, a conditional statement of a conditional statement is equivalent to a conditional statement with a conjuncted condition. This is similar to what is called "exportation" in propositional logic [Copi, 1982]. This equivalence give us the rules in Table 5.5.

$\{(\wedge\ M A_1 \cdots A_m) \Rightarrow C,\ M\} \vdash (\wedge\ A_1 \cdots A_m) \Rightarrow C\ <F_{ded}>$
$\{(\wedge\ M A_1 \cdots A_m) \Rightarrow C,\ (\wedge\ A_1 \cdots A_m) \Rightarrow C\} \vdash M\ <F_{abd}>$
$\{(\wedge\ A_1 \cdots A_m) \Rightarrow C,\ M\} \vdash (\wedge\ M A_1 \cdots A_m) \Rightarrow C\ <F_{ind}>$

Table 5.5: The Conditional Syllogistic Rules (2)

The truth values of the premises are omitted in the rules. As before, the induction rule is applied only when the two premises are based on the same evidence.

These rules can be seen as generalizations of the previous table where $m = 0$.[3] Table 5.6 gives further extension of these rules.

In each group of the syllogistic rules, abduction and induction can be obtained from deduction by switching a (different) premise and the conclusion, so they are "reversed deduction" in a sense.

With the help of the isomorphism and the implicit condition technique, we also get the following implications in NAL-5 among statements.

[3]So C is $(\wedge\ A_1 \cdots A_m) \Rightarrow C$, and $M \Rightarrow C$ is $(\wedge\ M A_1 \cdots A_m) \Rightarrow C$.

$\{(\wedge\ M A_1 \cdots A_m) \Rightarrow C,\ A_0 \Rightarrow M\} \vdash (\wedge\ A_0 A_1 \cdots A_m) \Rightarrow C \ < F_{ded} >$
$\{(\wedge\ M A_1 \cdots A_m) \Rightarrow C,\ (\wedge\ A_0 A_1 \cdots A_m) \Rightarrow C\} \vdash A_0 \Rightarrow M \ < F_{abd} >$
$\{(\wedge\ A_0 A_1 \cdots A_m) \Rightarrow C,\ A_0 \Rightarrow M\} \vdash (\wedge\ M A_1 \cdots A_m) \Rightarrow C \ < F_{ind} >$

Table 5.6: The Conditional Syllogistic Rules (3)

Theorem 35

$$\begin{array}{rcl}(S_1 \wedge S_2) & \supset & S_1 \\ S_1 & \supset & (S_1 \vee S_2)\end{array}$$

Now we can also turn the (meta-level) implications in NAL theorems into inference rules. Since a meta-level implication theorem "$A \supset C$" corresponds to inference from statement A to statement C, it can be treated as an implication statement in Narsese "$A \Rightarrow C < 1, 1 >$," which can be used with "$A < f, c >$" to get "$C < f, fc >$," using the deduction rule. Similarly, each meta-level implication theorem "$A \supset C$" can also be used with "$C < f, c >$" to get "$A < f, c/(c+k) >$" as a special kind of abduction. Isomorphically, theorems in the form of inheritance statements can also be used in this way.

The above inference has some restrictions. For example, "$S_1 \Rightarrow (S_1 \vee S_2) < 1, 1 >$" is a theorem, and it is fine to use it and a judgment of S_1 to derives a judgment of $(S_1 \vee S_2)$ by deduction, where S_2 can be any term. However, this result should not be used with the theorem again to derive a judgment of S_2 by abduction, otherwise for any two arbitrary statements, from the truth value of one, the truth value of the other can be derived (though with a low confidence).

The equivalence statements in theorems and definitions are easier to be converted into inference rules: "$S_1 \equiv S_2$" corresponds to rules $\{S_1\} \vdash S_2$ and $\{S_2\} \vdash S_1$, where the truth value of the conclusion is the same as the premise. The same result can be obtained by treating the inference as a special case of analogy with "$S_1 \Leftrightarrow S_2 < 1, 1 >$." Isomorphically, theorems in the form of similarity statements can also be used in this way.

5.1.4 Negation

Since the negation operator in NAL-5 takes one argument, it is not directly isomorphic to the (extensional/intensional) difference operators defined in NAL-3. Instead, it is defined as the following:

Definition 40 *If S is a statement, its* negation, $(\neg S)$, *is a compound statement defined by switching the positive and negative evidence of S.*

The definition also gives us the negation rule, defined in Table 5.7.

$\{S < f_0,\ c_0 >\} \vdash (\neg S) < F_{neg} >$

Table 5.7: The Negation Rule

The negation rule use the following truth-value function:

$$F_{neg}:\ f = 1 - f_0,\ c = c_0$$

We still have the following theorems as in propositional logic:

Theorem 36 $(\neg(\neg S)) \equiv S$

Theorem 37 $(S_1 \Leftrightarrow S_2) \equiv ((\neg S_1) \Leftrightarrow (\neg S_2))$

However, the *Law of Contrapositive* in propositional logic [Copi, 1982] (i.e., the equivalence between "$S_1 \Rightarrow S_2$" and "$(\neg S_2) \Rightarrow (\neg S_1)$") is no longer true in NAL.

When the truth values of S_1 and S_2 are determined according to certain evidence E, the induction rule can be used to calculate the truth value of "$S_1 \Rightarrow S_2$." As a result, when both S_1 and S_2 are true, it is positive evidence for "$S_1 \Rightarrow S_2$"; when S_1 is true but S_2 is false, it is negative evidence (when S_1 is false, it is not evidence).

Similarly, when both S_1 and S_2 are false, it is positive evidence for "$(\neg S_2) \Rightarrow (\neg S_1)$" (because both $\neg S_1$ and $\neg S_2$ are true); when S_2 is false but S_1 is true, it is negative evidence (when S_2 is true, it is not evidence).

By comparing the above two cases, we can see that "$S_1 \Rightarrow S_2$" and "$(\neg S_2) \Rightarrow (\neg S_1)$" have the same negative evidence (S_1 true, S_2 false), but completely different positive evidence ("both true" for the former, and "both false" for the latter). Therefore, to derive one statement from the other, in NAL we use the *contraposition rule* defined in Table 5.8.

$\{S_1 \Rightarrow S_2 < f_0, c_0 >\} \vdash (\neg S_2) \Rightarrow (\neg S_1) < F_{cnt} >$

Table 5.8: The Contraposition Rules of NAL-5

Since only negative evidence is passed from the premise to the conclusion, and the premise at most is counted as evidence with unit amount, we have $w = (1 - f_0)c_0$, and therefore the truth-value function is:

$$F_{cnt}: \ f = 0, \ c = (1 - f_0)c_0/[(1 - f_0)c_0 + k]$$

To summarize, related to the traditional study of "conversion," "obversion," and "contraposition" [Copi, 1982], in NAL the corresponding relations with evidence are in Table 5.9.

$S_1 \Rightarrow S_2$ $\{S_1 \wedge S_2; S_1 \wedge (\neg S_2)\}$	$S_2 \Rightarrow S_1$ $\{S_1 \wedge S_2; (\neg S_1) \wedge S_2\}$
$(\neg S_1) \Rightarrow (\neg S_2)$ $\{(\neg S_1) \wedge (\neg S_2); (\neg S_1) \wedge S_2\}$	$(\neg S_2) \Rightarrow (\neg S_1)$ $\{(\neg S_1) \wedge (\neg S_2); S_1 \wedge (\neg S_2)\}$

Table 5.9: Negation and Evidence

In Table 5.9, there are four statements, each with its positive evidence and negative evidence listed under it. There are three relations among them:

conversion. The two statements in the same row become each other by exchanging the subject and the predicate. They have the same positive evidence, but different negative evidence.

obversion. The two statements in the same column become each other by negating the subject and the predicate. They have different positive and negative evidence.

contraposition. The two statements in the same diagonal line become each other by exchanging and negating the subject and the predicate. They have the same negative evidence, but different positive evidence.

Given the above relations, in NAL there are rules for conversion and contraposition, but not for obversion, since no evidence can be passed from the premise to the conclusion. For the former two, only one type of evidence is passed (positive in conversion and negative in contraposition), so the frequency value of the conclusion is a constant (1 in conversion and 0 in contraposition). In these two rules, at most a single unit of evidence can be provided, so the conclusion has a lower confidence value than the premise — this is the case, because in NAL, the evidence of a premise is never directly taken as evidence (of the same quality and quantity) of the conclusion, except in the revision rule, where the premises and the conclusion are all about the same statement.

By definition, the evidence of $(\neg(S_1 \Rightarrow S_2))$ is obtained by switching the positive and negative evidence of $(S_1 \Rightarrow S_2)$, so it is $\{S_1 \wedge (\neg S_2); S_1 \wedge S_2\}$. Since it is nothing but the evidence of $(S_1 \Rightarrow (\neg S_2))$, the two statements are equivalent. This relation is not true if "$\Rightarrow$" is interpreted as the material implication in propositional logic. It is also important to notice that in NAL, unlike in propositional calculus, "$A \Rightarrow C$" is usually different from "$(\neg A) \vee C$" in truth value.[4] Similarly, for the equivalence relation, there are two other equivalence rule. These equivalence rules are listed in Table 5.10.

Other negation-related inference rules include the ones obtained from the following implications (which are also true in propositional logic):

---

[4]This issue is discussed in detail in subsection 9.4.1.

$\neg(S_1 \Rightarrow S_2)$	$S_1 \Rightarrow (\neg S_2)$
$\neg(S_1 \Leftrightarrow S_2)$	$S_1 \Leftrightarrow (\neg S_2)$
$\neg(S_1 \Leftrightarrow S_2)$	$(\neg S_1) \Leftrightarrow S_2$

Table 5.10: The Equivalence Rules of NAL-5

Theorem 38

$$\begin{aligned}(\neg A) &\supset (\neg(A \wedge B))\\ (\neg(A \vee B)) &\supset (\neg A)\\ (A \wedge (\neg(A \wedge B))) &\supset (\neg B)\\ ((\neg A) \wedge (A \vee B)) &\supset B\end{aligned}$$

So far, we have seen three types of (statement level) negation. For a statement "$S \rightarrow P$" in NAL-0, its negation is implicitly represented, which is not a statement in Narsese-0, though we can talk about "$\neg(S \rightarrow P)$" in its meta-language, where "$\neg$" is used as in propositional logic. In NAL-5, a statement S is multi-valued (with truth value $<f, c>$), and so is $\neg S$ (with truth value $<1 - f, c>$), where "$\neg$" is a statement operator defined in Narsese-5, whose meaning is not exactly the same as the one in predicate logic.

Finally, as described previously, "$S \rightarrow P$" in NAL-0 becomes "$S \rightarrow P < 1, 1 >$" in NAL-1 (as well as in all of its extensions, including NAL-5). However, when it is not true, its truth value $<f, c>$ can be anything except $< 1, 1 >$, and it is not necessarily the case specified by the negation of the statement, which has truth value $<0, 1>$. This issue is very important when encoding knowledge in Narsese. Usually, by "S is not P," we mean that "There is negative evidence for $S \rightarrow P$" (i.e., "$S \rightarrow P <0, c>$," where $c < 1$), but not that "All evidence for $S \rightarrow P$ is negative" (i.e., "$S \rightarrow P <0, 1>$").

5.2 NAL-6: statements with variables

5.2.1 Variable terms

The terms we introduced so far are *constant* terms, in the sense that at any given time, each of them is unique in the system, and has a determined meaning; a *variable* term, on the other hand, may appears in more than one statements in the system, each of them with its own meaning.

In NARS, the meaning of a constant term is determined by its experienced relations with the other term *in the whole system*, while the meaning of a variable term is locally determined by its relations with other terms *within the same statement*. That means, if there are two variable terms with the same name but not in the same statement, they are not necessarily related to each other in meaning. On the contrary, occurrences of a constant term in different statements are always bounded together.[5]

There are two types of variable terms in NAL: *independent variables* (similar to the *universal variables* in first-order predicate logic) and *dependent variables* (similar to the *existential variables* and *Skolem functions* in first-order predicate logic). The latter is a function of the former. A dependent variable has a (maybe empty) independent variable list in it to indicate its dependence. In Narsese, a variable term (of either the above two types) is named by a word (or a number) preceded by '#'.

Given AIKR, no statement in Narsese is made about all terms. Whenever a statement is made about a group of terms, it is usually possible to put them into the extension or intension of a given term, then to make a statement about it. A dependent variable indicates a single term under the given condition; an independent variable indicates any term under the given condition.

---

[5]Since in NARS the meaning of a constant term may change over time, the distinction between constant terms and variable terms is not whether the meaning of a term changes or not. In the next chapter, we will see that a constant term in Narsese is the name of a specific *concept* in NARS, while a variable term indicates an unspecified concept.

In Narsese, variable terms can only appear in special positions. An independent variable always appears in both sides of an implication or equivalence relation, as extension or intension of two terms. A dependent variable always appear in two components of a conjunction, also as extension or intension of two terms. Therefore, the following are the simplest statements with variable terms.

$$(\#x \rightarrow S) \Rightarrow (\#x \rightarrow P) \qquad (\#x() \rightarrow S) \wedge (\#x() \rightarrow P)$$
$$(S \rightarrow \#x) \Rightarrow (P \rightarrow \#x) \qquad (S \rightarrow \#x()) \wedge (P \rightarrow \#x())$$

In this way, the scope of a variable is the statement in which it appears. The name of a variable is arbitrary, as far as it is unique in the statement. In statements with multiple variables, each of them uses a different name, therefore its scope does not need to be explicitly specified — it is the smallest statement that contains all occurrences of the variable.

As in predicate logic, the scope of a variable can be embedded in that of another one. For example, in Narsese the following situations can be represented:

- $((\#x \rightarrow key) \wedge (\#y \rightarrow lock)) \Rightarrow ((\#x \times \#y) \rightarrow open)$
  ["Every key can open every lock."]

- $(\#x() \rightarrow key) \wedge ((\#y \rightarrow lock) \Rightarrow ((\#x() \times \#y) \rightarrow open))$
  ["There is a key that can open every lock."]

- $(\#x \rightarrow key) \Rightarrow ((\#y(\#x) \rightarrow lock) \wedge ((\#x \times \#y(\#x)) \rightarrow open))$
  ["Every key can open a lock."]

- $(\#x() \rightarrow key) \wedge (\#y() \rightarrow lock) \wedge ((\#x() \times \#y()) \rightarrow open)$
  ["There is a key that can open a lock."]

Since an inheritance relation is identical to two subclass relations of the extension and intension of the two terms, independent variables becomes implicit in first-order NAL. Using the variable terms introduced above, we can see that "$S \rightarrow P$" is equivalent to "$((\#x \rightarrow S) \Rightarrow (\#x \rightarrow P)) \wedge ((P \rightarrow \#y) \Rightarrow (S \rightarrow \#y))$." It roughly means "For any term x in the extension of term S, it is also in the extension of term P; for any term y in the intension of term P, it is also in the intension of term S."

Variable terms become necessary when the extension or intension of a term needs to be specified separately, as well as when complicated relations among terms need to be described. For example, "$(\#x \rightarrow S) \Rightarrow (\#x \rightarrow P)$" and "$(P \rightarrow \#y) \Rightarrow (S \rightarrow \#y)$" have different evidence, and can be separately maintained to represent pure extensional or intensional relations between terms.

The independent variables in NAL are different from the universal variables in first-order predicate logic in that the former are restricted by their relations with other terms in the statement. For example, in "$(\#x \rightarrow S) \Rightarrow (\#x \rightarrow P)$," the independent variable '$\#x$' only represents terms in the extensions of S, but not the ones that are not there. On the contrary, in a predicate logic proposition "$(\forall x)(S(x) \supset P(x))$," the universal variable x can represent every individual in the domain. Roughly speaking, the former says "Every instance of S is also an instance of P," while the latter says "For everything in the domain, if it is an instance of S, then it is also an instance of P." The meaning of the two sentences are not exactly the same, though related.[6]

5.2.2 Inference with variable terms

Since an independent variable represents a unspecified term (under a certain condition), it can be replaced by a specific term (satisfying the condition) without changing the truth value of a judgment. Technically speaking, an independent variable can be substituted by another (variable or constant) term that has the same relation by *unification* (as defined in predicate logic), and then a conclusion can be derived. Some variable elimination rules are listed in Table 5.11.

$\{(\#x \rightarrow S) \Rightarrow (\#x \rightarrow P),\ M \rightarrow S\}$	$\vdash$	$M \rightarrow P\ <F_{ded}>$
$\{(\#x \rightarrow S) \Rightarrow (\#x \rightarrow P),\ M \rightarrow P\}$	$\vdash$	$M \rightarrow S\ <F_{abd}>$
$\{(\#x \rightarrow S) \Leftrightarrow (\#x \rightarrow P),\ M \rightarrow S\}$	$\vdash$	$M \rightarrow P\ <F'_{ana}>$

Table 5.11: Sample Independent-Variable Elimination Rules

[6] I will say much more about the difference between NAL and first-order predicate logic in Chapter 9 and 10.

Such a rule can be seen as a substitution (with $\#x$ replaced by M) followed by an inference defined in Table 5.3. The same technique can be applied to the other higher-order inference rules to use premises with variables.

The reverse of variable elimination rules introduces independent variables into conclusions, as listed in Table 5.12. These rules are justified in the same way as the rules in NAL-1 and NAL-2, except that here the "extensional inheritance" and "intensional inheritance" between S and P are separated, due to the using of a variable term.

$\{M \rightarrow P,\ M \rightarrow S\}$	$\vdash$	$(\#x \rightarrow S) \Rightarrow (\#x \rightarrow P)$	$<F_{ind}>$
$\{M \rightarrow P,\ M \rightarrow S\}$	$\vdash$	$(\#x \rightarrow S) \Leftrightarrow (\#x \rightarrow P)$	$<F_{com}>$

Table 5.12: Sample Independent-Variable Introduction Rules

The rule in Table 5.13 introduce dependent variables into conjunctions. Here the conclusion states the *existence* of an anonymous term that is in the extension of both S and P.

$\{M \rightarrow P,\ M \rightarrow S\}$	$\vdash$	$(\#x() \rightarrow P) \wedge (\#x() \rightarrow S)$	$<F_{exi}>$

Table 5.13: Sample Dependent-Variable Introduction Rule

This inference can only produce positive conclusion, and the conclusion reaches maximum confidence when both premises are absolutely true. Therefore, the truth-value function is the following:

$$F_{exi}: \quad f = 1, \quad c = and(f_1, c_1, f_2, c_2)$$

The reverse of the rule in Table 5.13 can be seen as a special type of unification to match a dependent variable with a constant, as given in Table 5.14. The truth-value function of the abduction rule is used here, because the conclusion gets evidence only when the first premise is positive, in which case term M is compared to the anonymous term

$\#x()$. Under the condition of $M \rightarrow S$, if $(\#x() \rightarrow P) \wedge (\#x() \rightarrow S)$, then $M \rightarrow P$ looks more likely, otherwise less likely.

$\{M \rightarrow S,\ (\#x() \rightarrow P) \wedge (\#x() \rightarrow S)\} \ \vdash \ M \rightarrow P \ <F_{abd}>$

Table 5.14: Sample Dependent-Variable Elimination Rule

The rules in Table 5.13 and Table 5.14 are only about the extension of S and P. Similarly, there are rules that only process the intension of the terms involved. As required before, in NAL a dependent variable is only introduced into a conjunction, and an independent variable into both sides of an implication or equivalence.

Variables can be introduced into statements where other variables exist. When an independent variable is introduced, the existing dependent variables become its function, until it is unified with a constant. The rules for multiple variables in Table 5.15 can be extended to handle more than two variables. Please note that here the four conclusions correspond to the four sentences in the previous "lock-key" example. It shows that NAL has rules to produce all meaningful combinations of variable terms.

The revision rule is also extended to unify variable terms. For example, statements $(\#x \rightarrow S) \Rightarrow (\#x \rightarrow P)$ and $(\#y \rightarrow S) \Rightarrow (\#y \rightarrow P)$ can be merged together.

Rule
$\{(\#x \rightarrow P) \Rightarrow (M \rightarrow (\perp R\ \#x\ \diamond)),\ M \rightarrow S\}$ $\vdash ((\#y \rightarrow S) \wedge (\#x \rightarrow P)) \Rightarrow (\#y \rightarrow (\perp R\ \#x\ \diamond)) \ <F_{ind}>$
$\{(\#x \rightarrow P) \Rightarrow (M \rightarrow (\perp R\ \#x\ \diamond)),\ M \rightarrow S\}$ $\vdash (\#y() \rightarrow S) \wedge ((\#x \rightarrow P) \Rightarrow (\#y() \rightarrow (\perp R\ \#x\ \diamond))) \ <F_{exi}>$
$\{(\#x() \rightarrow P) \wedge (M \rightarrow (\perp R\ \#x()\ \diamond)),\ M \rightarrow S\}$ $\vdash ((\#y \rightarrow S) \Rightarrow ((\#x(\#y) \rightarrow P) \wedge (\#y \rightarrow (\perp R\ \#x(\#y)\ \diamond))) \ <F_{ind}>$
$\{(\#x() \rightarrow P) \wedge (M \rightarrow (\perp R\ \#x()\ \diamond)),\ M \rightarrow S\}$ $\vdash (\#y() \rightarrow S) \wedge (\#x() \rightarrow P) \wedge (\#y() \rightarrow (\perp R\ \#x()\ \diamond)) \ <F_{exi}>$

Table 5.15: Sample Multi-Variable Introduction Rules

5.2.3 Hypothetical inference

As described above, a variable term is a special term which can be used as a *symbol* of other terms. By "symbol," it is meant that the meaning of a variable term is not fully grounded on the experience of the system, until it is "bounded" to a (non-variable) term. The same symbol can be used to stand for different terms in different statements. It is similar to pronouns in natural languages, such as "it," "that," or "everything."

Variable terms give NARS the capability of using the same term to indicate different concepts, or with different "interpretation." This is needed in hypothetical inference.

For example, conditional statement

$$(((\{\#x\} \times \{\#y(\#x,\ \#z)\}) \rightarrow parent) \wedge ((\{\#y(\#x,\ \#z)\} \\ \times\{\#z\}) \rightarrow brother)) \Rightarrow ((\{\#x\} \times \{\#z\}) \rightarrow uncle)$$

can be seen as a "rule" by which the system derives "$(\{Mary\} \times \{Tom\}) \rightarrow uncle$" ("Tom is the uncle of Mary") from "$(\{Mary\} \times \{Joe\}) \rightarrow parent$" ("Joe is the parent of Mary") and "$(\{Joe\} \times \{Tom\}) \rightarrow brother$" ("Tom is the brother of Joe"). However, accurately speaking, the inference rule used is the deduction rule, with the above "rule" (the conditional statement) as a premise.

In this way, what is usually called a "rule" in rule-based systems is formalized in NARS in two different levels. Conceptually, a rule indicates that a certain statement can be derived from certain other statement(s). In NARS, a small set of such rules are actually formalized as procedural inference rules, which are part of the logic. The other "rules" are formalized as implication and equivalence statements, which, when used with the above deduction rule, will effectively work as an inference rule. The former group includes rules defined on the built-in logical relations (*inheritance*, *similarity*, *implication*, and *equivalence*), while the latter group includes "rules" on the other acquired "ordinary" relations (such as *parent*, *brother*, and *neutralization*).

Consequently, NAL can be used as a meta-logic of an arbitrary logic, by representing the rules of that logic as NARS implication/equivalence statements. For example, we can "emulate" first-order predicate logic (FOPL) in NARS in this way. When implemented in a computer system, such an emulation will surely be more complicated and less efficient than a direct implementation of FOPL, but at the same time, it does have the benefit of allowing the system to reason "at the outside" of FOPL. For example, non-deductive inference (induction, abduction, analogy, ...) is invalid within FOPL, though it clearly plays an important role when a human mind uses a logic like FOPL. We often first reason outside the system to get a hypothesis, then inside the system to prove or disprove it.

The above description can be extended from FOPL to other formal or mathematical theory. NARS can embed them as subsystems, and do inference inside and outside, so that the system's empirical experience becomes relevant even when an abstract mathematical theory is thought about.

Another issue related to this is the idea of "local axiomization." As repeatedly mentioned before, the key assumption of NARS is that the system works with insufficient knowledge and resources, so that every judgment is only certain to a degree, and for empirical knowledge, confidence never reaches its maximum value 1, that is, it is always revisable. On the other hand, there are analytic statements, whose truth values are independent to the experience of the system, and is a matter of definition. A mathematical (or other formal) theory consists of a set of analytic statements as axioms, and a set of rules (which can be represented as NARS implications). Consequently, though NARS is non-axiomatic with respect to its empirical knowledge, it can contain axiomatic subsystems. When the system works "within" such a subsystem, no uncertainty is allowed.

How do these two parts interact with each other? It is similar to how the human mind uses mathematics and formal logics to solve practical problems. It contains several steps:

1. For a given task T, the relevant empirical knowledge is collected into a knowledge base K.

2. A formal theory F is selected to solve T given K.

3. An interpretation I is build, which maps T and K into T' and K', which are questions and statements in F.

4. A solution S' is found in F for T', according to K'.

5. A solution S is obtained for T from S', under the interpretation I.

In this process, usually only Step 4 is within an axiomatic system. As a result, to apply a mathematical theory to a practical problem does not give the conclusion the status of a mathematical theorem, and the process is highly biased by the experience of the system. The interpretation I in the above serves the same purpose as variable substitution, by which abstract terms in the formal theory are mapped into concrete terms in the domain. In this sense, all abstract terms are variable terms discussed in this section.

5.3 NAL-7: temporal statements

So far, the truth values of statements in NAL are determined by taking all past experience as relevant. However, sometimes we are only interested in the truth value of a statement at a given time. For these situations, NAL-7 introduces *time* into statements.

5.3.1 Time and events

In NAL, an "event" is defined as a statement whose truth value holds in a certain period of time. As a result, its truth value is time dependent, and the system can describe its temporal relation with other events.

Accurately speaking, almost all empirical statements are time dependent, and few statements are about relations holding forever. However, for practical purposes, it is not always necessary to take the time attribute of a statement into consideration. Therefore, whether a *statement* should be treated as an *event* may change from context to context, and events are just statements whose time attributes are specified. On the contrary, the time interval of a "non-event" statement is unspecified, except that it includes the current moment.

In NAL, time is represented indirectly, through events and their temporal relations. Intuitively, an event happens in a time interval, and temporal inference rules can be defined on these intervals [Allen, 1984]. However, in NAL each event is represented by a term, whose corresponding time interval is not necessarily specified. In this way, NAL assumes less information. When the duration of an event is irrelevant, it can also be treated as a point in the stream of time.

In NAL, the temporal order between two events E_1 and E_2 can be one of the following three cases:

- E_1 happens before E_2 happens,
- E_1 happens after E_2 happens,
- E_1 happens when E_2 happens.

This design is consistent with the observation that among human languages, there are three universal temporal primitives: "before," "after," and "when" [Wierzbicka, 1996]. Obviously, the first two cases correspond to the opposite directions of the same temporal relation. Therefore, the primitive temporal relations in NAL are:

"before": which is irreflexive, antisymmetric, and transitive;

"when": which is reflexive, symmetric, and transitive.

They correspond to the *before* and *equal* relation discussed in [Allen, 1984], respectively.

If the temporal relation between two events is more complicated than these three cases, it is always possible to divide an event into subevents, then describe their temporal relations in detail. For example, we can treat "when E_1 starts" and "when E_1 ends" as separate events. This representation covers all kinds of temporal relations between two events, including the ones discussed in the previous studies of temporal inference: *meets*, *overlap*, *during*, *starts*, and *finishes* [Allen, 1984]. In NAL, these temporal relations (and others) are represented not as logic constants, but as "ordinary relations" (discussed in NAL-4). Similarly, we can introduce a term "duration." If the system knows that "$(t_1 \times t_2) \rightarrow duration$", as well as that "$(t_1 \times t_2)$" and event E happen at the same time, it understands that E begins at time t_1, and ends at time t_2. Unlike in interval-based temporal logics, in NAL terms like t_1 and t_2 are events themselves (though their durations are usually omitted), not accurate measurement of absolute time.

If an absolute time is used to represent the temporal property of an event, then that time can be treated as a special event, and these two events are described as happening at the same time.

5.3.2 Temporal operators and relations

Instead of directly using the above two temporal relations between events by themselves, in NAL they are used in combination with certain other logic constants.

First, "E_1 happens before E_2 happens" and "E_1 happens when E_2 happens" both assumes "E_1 and E_2 happen (at some time)," which is "$E_1 \wedge E_2$" plus temporal information. Therefore, we can treat the two temporal relations as variants of the statement operator "conjunction" ("$\wedge$") — "sequential conjunction" (",") and "parallel conjunction" (";"). Consequently, "(E_1, E_2)" means "E_1 happens before E_2 happens," and "$(E_1; E_2)$" means "E_1 happens when E_2 happens." Obviously, "$(E_2; E_1)$" is the same as "$(E_1; E_2)$," but "(E_1, E_2)" and "(E_2, E_1)" are usually different. As before, these operators can take more than two arguments. These two operators allow Narsese to represent complicated events by dividing them into sub-events recursively, then specifying temporal relations among them.

Compared to the two "temporal conjunctions," the original conjunction "$E_1 \wedge E_2$" can be seen as with a default temporal information that both E_1 and E_2 hold at the same time, that is, "$(E_1; E_2)$," therefore the operator ";" is redundant. The other two statement operators, "$\vee$" and "$\neg$," have no temporal variant.

On the other hand, there are the temporal variants of implication and equivalence. For an implication statement "$S \Rightarrow T$" between events S and T, three different temporal relations can be distinguished:

1. If S happens *before* T happens, the statement is called "predictive implication," and is rewritten as "$S\ /\!\!\Rightarrow T$," where S is called a *sufficient precondition* of T, and T a *necessary postcondition* of S.

2. If S happens *after* T happens, the statement is called "retrospective implication," and is rewritten as "$S \setminus\!\!\Rightarrow T$," where S is called a *sufficient postcondition* of T, and T a *necessary precondition* of S.

3. If S happens *when* T happens, the statement is called "concurrent implication," and is rewritten as "$S \mid\!\Rightarrow T$," where S is called a *sufficient co-condition* of T, and T a *necessary co-condition* of S.

Similarly, three "temporal equivalence" (predictive, retrospective, and concurrent) relations are defined. "$S \not\Leftrightarrow T$" (or equivalently, "$T \backslash\!\!\!\Leftrightarrow S$") means that S is an *equivalent precondition* of T, and T an *equivalent postcondition* of S. "$S \,|\!\!\!\Leftrightarrow T$" means that S and T are *equivalent co-conditions* of each other. To simplify the language, "$T \backslash\!\!\!\Leftrightarrow S$" is always represented as "$S \not\Leftrightarrow T$," so the copula "$\backslash\!\!\!\Leftrightarrow$" is not actually included in the grammar of Narsese. Furthermore, since by default "$\Rightarrow$" and "$\Leftrightarrow$" assumes the pair of events connected by them happen at the same time (i.e., the current time), the relations "$|\!\!\!\Rightarrow$" and "$|\!\!\!\Leftrightarrow$" are redundant, and need not be actually implemented.

Adjectives like "past," "current," and "future" indicate temporal relations between events and "now," taken as a special event. When "now" is omitted in a statement, the temporal relations become temporal operators called "tense." The "current" tense is the implicit default, while the "past" and "future" tenses must be explicitly specified. For a statement S, "S happened" and "S will happen" are represented in Narsese as "$\backslash\!\!\!\Rightarrow S$" and "$\not\Rightarrow S$," respectively.

5.3.3 Temporal inference

The inference rules introduced in NAL-7 are variants of the rules defined in NAL-5 and NAL-6. The only additional function of these rules is to keep the available temporal information.

As an example, the following is a deduction rule introduced in NAL-5,

$$\{(\wedge\ MA_1 \cdots A_m) \Rightarrow C,\ A_0 \Rightarrow M\} \vdash (\wedge\ A_0A_1 \cdots A_m) \Rightarrow C \ \ <F_{ded}>$$

Now it has a variant in NAL-7, as listed in Table 5.16.

Since the logical factor and the temporal factor are independent of each other in the rules, these variants can be obtained by considering the two factors separately, then combining them in the conclusion.

$\{(M, A_1, \cdots, A_m) \not\Rightarrow C,\ A_0 \not\Rightarrow M\} \vdash (A_0, A_1, \cdots, A_m) \not\Rightarrow C \ \ <F_{ded}>$

Table 5.16: Sample Temporal Inference Rule

Before temporal information is introduced, in NARS whenever there are two judgments containing the same statement, it will be taken to mean different evidence about the same relation, and revision will be attempted. With temporal information, however, there is another possible interpretation, that is, as *update*, that is, as caused by a *change* in the environment. In this operation, the new judgment does not merge with the old one, but "pushes it into the past," by adding a "past tense" on it, so as to turn it into a different statement. Update usually causes chain reactions, that is, other updates in derived conclusions, as revision does.

In general, inference rules on tense can be derived from ordinary temporal inference rules, by first adding the term "now" into the premises (so as to turn the tense operators into temporal relations), and finally removing the "now" from the conclusions (so as to turn the temporal relations back into tenses). This procedure is similar to the usage of "virtual condition" (which turns an arbitrary statement into an implication) in NAL-5.

Clearly, inference on temporal implication relations can be used to predict the future and to explain the past. However, they are not the same as what we call "causal relation." We can think of the latter as a special case of the former, with additional requirements attached to the concept of "cause," which are highly context-dependent. That is why in NARS the above temporal implication/equivalence relations are defined as part of the logic, and with fixed meaning, while "cause" is left to be an empirically built relation, which will be learned and changed by the system according to experience and context. Even though, the basic capability of what we usually call "causal inference" is already in the system.[7]

5.4 NAL-8: procedural statements

In NAL-8 procedural interpretation is applied to events to represent operations of the system itself. Consequently, declarative knowledge and procedural knowledge are unified in NAL.

[7]This issue will be discussed in detail in subsection 9.4.2.

5.4.1 Operations and procedural inference

In NARS, *operation* is a special kind of event, which can be carried out by the system itself. Therefore it is system dependent: the operations of a system will be observed as events by other systems.

While relations and events are *declarative* knowledge, operations are *procedural* knowledge, in the sense that the meaning of an operation is not only revealed by what it *says*, but also by what it *does*.

Statement "$(\times\ \{A\}\ \{B\}\ \{C\}) \rightarrow R$" intuitively corresponds to "There is a relation R among (individuals) A, B, and C." If R is an event, it becomes "An event R happens among A, B, and C." If R is an operation, it becomes "To carry out R among A, B, and C." To be consistent with the other part of NAL, for the last case in the following we can also use the format "(R, A, B, C)," so that it is just a compound term defined in NAL, with an operator followed by an argument list.

An operation usually distinguishes input and output among its arguments. When an operation is described abstractly, its input arguments are typically independent variables, and its output are dependent variables. For instance, operation "$(plus, \#x, \#y, \#z(\#x, \#y))$" represents "$x$ plus y is z," where x and y are independent variables (input), and z is a dependent variable (output, as a function of x and y).

The knowledge about operations are usually represented as (temporal or not) implication/equivalence statements, which indicates the conditions, causes, and effects of each operation.

For the above example, if statement "$(\times\ \{A\}\ \{B\}\ \{C\}) \rightarrow sum$" represents "The sum of A and B is C," then the following statements describe the necessary postconditions (i.e., effects) of the operations *plus* and *minus*:

$$(plus, \#x, \#y, \#z(\#x, \#y)) \mathrel{/\!\!\!\Rightarrow} ((\times\ \{\#x\}\ \{\#y\}\ \{\#z(\#x, \#y)\}) \rightarrow sum)$$

$$(minus, \#x, \#y, \#z(\#x, \#y)) \mathrel{/\!\!\!\Rightarrow} ((\times\ \{\#y\}\ \{\#z(\#x, \#y)\}\ \{\#x\}) \rightarrow sum)$$

Here we see that the same logical relation (such as "*sum*") may correspond to multiple operations (such as "*plus*" and "*minus*") with different input/output partition among its arguments.

The (sufficient or necessary) preconditions of operations can be similarly specified. For example, to carry out $(divide, \#x, \#y, \#z(\#x, \#y))$, $\#y$ cannot be zero, and this necessary precondition can be represented as

$$(divide, \#x, \#y, \#z(\#x, \#y)) \mathrel{\backslash\!\!\!\Rightarrow} \neg(\#y \leftrightarrow 0)$$

More examples of operational knowledge are in the following list:

- "$((\{\#x\} \times \{\#y\}) \rightarrow r_1) \mathrel{/\!\!\!\Rightarrow} (op_1, \#x, \#y)$" indicates that a sufficient precondition for the operation "op_1" to be performed in two arguments is that there is a relation "r_1" between the two.
- "$(op_1, \#x, \#y) \mathrel{\backslash\!\!\!\Rightarrow} ((\{\#x\} \times \{\#y\}) \rightarrow r_2)$" indicates that a necessary precondition for the operation "op_1" to be performed in two arguments is that there is a relation "r_2" between the two.
- "$(op_1, \#x, \#y) \mathrel{/\!\!\!\Rightarrow} ((\{\#x\} \times \{\#y\}) \rightarrow r_3)$" indicates that a necessary postcondition of the operation "op_1" is that there will be a relation "r_3" between its two arguments.
- "$(((\{\#x\} \times \{\#y\}) \rightarrow r_4), (op_2, \#y, \#z)) \mathrel{/\!\!\!\Rightarrow} ((\{\#x\} \times \{\#z\}) \rightarrow r_5)$" indicates that after event "$(\{\#x\} \times \{\#y\}) \rightarrow r_4$" happens, executing "$(op_2, \#y, \#z)$" will cause the event "$(\{\#x\} \times \{\#z\}) \rightarrow r_5$" to happen.
- "$((op_3, \#x, \#y), (op_4, \#x, \#y), (op_4, \#x, \#y)) \Leftrightarrow (op_6, \#x, \#y)$" indicates that executing "op_6" on two arguments is equivalent to the sequential execution of "op_3," "op_4," and "op_5" on them.

What is expressed in these examples is quite similar to what is achieved by logic programming, except that in NARS the logic is fundamentally different from first-order predicate logic, and each statement has a truth value attached to indicate its uncertainty.

Under AIKR, in NARS the preconditions (restrictions) and postconditions (consequences) of an operation are never exhaustively specified. Instead, the system's beliefs on operations only reflect its (limited) experience on them. It is quite possible that certain conditions or consequences, though they exist, never become known to the system. Designed in this way, NARS takes a unusual position toward the well-known Frame Problem [McCarthy and Hayes, 1969], by accepting it as an inevitable consequence of AIKR.[8]

NARS will be implemented with certain primitive operations (both internally-oriented and externally-oriented) exposed to the inference engine, and they can be used as components to build compound operations. Both primitive and compound operations can be called from the inference engine. The system has beliefs, either built-in initially or acquired through experience, about these operations.

Not all operations in such a system are involved in reasoning in this way. NARS has a mixture of *deliberative* and *automatic* processes. The former consists of the system operations visible to the inference engine, but the latter is invisible. To make an operation exposed to the inference engine in this way usually makes its execution more flexible, but at the same time decreases the efficiency, and introduces various kinds of risks caused by the uncertainty in inference.

If an operation is accessible to the inference engine, it will be named by a Narsese term, and so will its (input and output) arguments (some of them may be variable terms). The system's beliefs about the operation will be represented by (temporal) implication and equivalence statements, as shown previously. Furthermore, inference on these statements will incrementally reveal its preconditions and postconditions, as well as its relation with other operations.

Since operations are just events under a procedural interpretation, the inference rules of NAL-8 are the same as NAL-7. The compound operations formed in the inference process correspond to "skills" (procedures) learned by the system. For example, initially the system may only have beliefs about operations "$op1$," "$op2$," and "$op3$" in isolation. By inference, the system will form beliefs on what may happen if the three are executed in a sequence — that may achieve a more

[8]Of course, it does not mean to do nothing at all about it. For a given operation, in NARS there are usually many judgments about its conditions and consequences. In the long run, however, only some of them will be kept, while most of the others will be forgot. For example, if a judgment omits an important condition, it will lead to many failures, and become useless; if a judgment mentions too many conditions, which are usually true, it will become unnecessarily complicated, and lost in resource competition to the simpler-but-equally-effective ones. The situation about consequence is similar. This issue will become more clear in Chapter 6, 11, and 12.

complicated consequence. Of course, due to the insufficiency of knowledge and resources, these derived beliefs may conflict with new evidence and get revised. Even so, after a while, the system will learn various skills, which are compound operations for which the system has beliefs. This process is similar to the "chunking" process in Soar [Newell, 1990], though in NARS it follows a very different logic.

5.4.2 Goals and desire values

In NAL-1, two types of sentences are defined: *judgment* and *question*, where the former is a statement with a truth value, and the latter is a statement whose truth value should be determined by the system. In NAL-8, the third type of sentence, *goal*, is introduced.

In Narsese, a goal has a format similar to a question, that is, it is a statement without a truth value attached, and may contain variables to be instantiated. However, their semantics are different. While a question asks the system to evaluate the truth value of the statement (and maybe find constant terms for the variables) by *inference* only, a goal asks the system to carry out some *operations* to make the statement true — of course, given the inevitable uncertainty in the consequence, it actually means "to make it as close to truth as possible."

NARS usually has multiple goals, and they may conflict with one another, in the sense that the achieving of a goal makes another one harder to be achieved. Therefore, the system must from time to time make decisions about whether to pursue various goals or whether to take various operations.

Since the conflicting goals may not be directly related to each other, we cannot expect the system to have explicit knowledge about how to handle every possible conflict. Instead, a common solution in the study of decision making is to define a "degree of desire," or *desire value*, on goals, so that every pair of goals is comparable.

Under AIKR, in NARS we cannot take the desire value of a goal as a known constant, as in most decision-making theories. Instead, it is something that the system needs to find out by inference. Since the desire value of a goal G is determined according to the system's experience, the truth value of statement is used to represent this degree of desire. To do that, a virtual statement D is introduced for the "desired state." Like the virtual statement E used for "all evidence" in NAL-5 or the virtual event "*now*" for tense definition in NAL-7, D does not appear within the system, but is used in the meta-theory to design related rules.

With the help of D, the desire value of a goal G is defined as the truth value of statement "$G \Rightarrow D$," that is, the degree that the desired state is implied by the achieving of this goal.

In this way, the desire-value functions can be derived from the truth-value functions. For example, if goal G has desire value d, and the system believes that G can be achieved by operation A (with truth value t), then the system has premises "$G \Rightarrow D <d>$" and "$A \Rightarrow G \ <t>$." From them, by deduction the system gets "$A \Rightarrow D < F_{ded} >$," which means that A becomes a derived goal, with a desire value obtained by using the deduction truth function with d and t as arguments. After dropping the virtual statement D, in the system we get a rule that derives a potential goal A (with a desire value) from a goal G (with a desire value) and a belief "$A \Rightarrow G$" (with a truth value).[9]

Now we can attach a desire value to every statement in the system, because it may become a goal in the future, if it is not already a goal. This value shows the system's "attitude" about the situation in which the statement is true.[10]

5.4.3 Goal-related inference

When there is no operation that can directly achieve a given goal, the system will do inference to indirectly achieve it. In NAL-1, we have seen how backward inference is used to derive questions. For all the rules introduced in NAL-2 to NAL-7, the same isomorphism holds between forward and backward inference, so all the rules defined in them can also be used to derive questions. A similar situation happens to goals.

[9]This treatment is directly related to the previous philosophical discussion on "desire as belief" [Lewis, 1988]. However, in NARS, a desire is not reduced into a belief. In the system, a belief on statement S and a desire for S to be true are clearly distinguished from, though also closely related to, each other.

[10]This desire value will eventually be attached to every term, to represent the system's "feeling" about it. If the term is not a statement, its desire value will be determined by the beliefs in which it appears.

When a goal G and a judgment J are taken as premises in an inference step, the judgment may provide a direct solution to the goal, if its truth value indicates that the goal has already been somehow realized (so nothing needs to be done). Otherwise, a derived goal G' can be produced, if and only if G can be derived from J and G'.

The backward inference on goals is related to *planning*. For a given goal, the inference engine can find a group of operations, organized by the "," and ";" operators defined in NAL-7, that achieve the goal (i.e., to make it true as the consequence of the execution of the operations). By executing the plan, and adjusting it when necessary, the internal or external environment is changed to turn the goal into reality. When repeatedly appearing groups of operations are memorized, repeated planning is avoided, and the system learns a new skill.

The planning process in NARS is different from what is usually called "planning" in AI, where the process starts with a goal and a set of primitive operations. In NARS planning incrementally builds complicated plans (i.e., procedural statements), and accumulates knowledge about them. It is just in rare cases that the system starts from the primitive operations, and goes all the way to achieve a complicated plan.

If an operation A will contribute to the achieving of goal G_1 but makes goal G_2 less likely to be achieved, the system will contains judgments "$A \Rightarrow G_1 < f_1, c_1 >$" and "$A \Rightarrow G_2 < f_2, c_2 >$," where f_1 is near 1, and f_2 is near 0. When both G_1 and G_2 are desired (to different degrees), the system will get two "$A \Rightarrow D$," that is, the goal A gets two desire values, obtained from different sources. Clearly, they should be merged using the revision rule.

Since the above process may be repeated when other goals are taken into consideration, after a while the desire value of a goal is usually influenced by many other goals. The final decision of executing an operation is made when the expectation of the desire value of the goal is above a certain threshold, m ($m > 0.5$), which is a system parameter (like the k defined in NAL-1). That is, the system only pursues goals whose overall expected desire value is sufficiently high.

As we have seen, the decision making procedure in NARS is specified differently from the conventional decision-making research (such as [Savage, 1954, Jeffrey, 1965]), in the following aspects:

- A goal is not a state, but a statement — under AIKR, in NARS no statement (within a system) can completely describe a state (of the environment).
- The desire value of a statement may change over time when new information is taken into consideration.
- The likelihood of an operation to achieve a goal is not specified as a probability, but as a truth value defined in NARS. In a truth value, there are two independent factors: the frequency value is similar to probability, and the confidence value is similar to the "third dimension" in decision making (beside desirability and likelihood) suggested by Ellsberg, that is, "a quality depending on the amount, type, reliability and 'unanimity' of information, and giving rise to one's degree of 'confidence' in an estimate of relative likelihoods" [Ellsberg, 1961].
- The decision is on whether to pursue a goal, but not on which goal is the best in a complete set of mutually-exclusive goals (or, as a special case, operations) — under AIKR, such a set cannot be obtained.

There are still similarities between this approach and traditional decision theory. Both approaches let decisions be based on quantitative comparison between alternatives, and in the measurement both the degree of desire of a goal and the likelihood that the goal will be achieved by an operation are involved.

Of course, the above discussion only provides the fundamentals of decision making, and there are advanced issues to be addressed in the future, such as the timing of operation execution and observation of the result of operation.

An important point about NAL-8 is that since some of its statements are actually system operations, the results of the inference process are no longer fully expressible in Narsese. That is, in the previous layers of NAL, no matter what inference rules are used, both input and output of the process are Narsese sentences, that is, the system does nothing else but generate new sentences from given sentences. After NAL-8 is implemented, this is no longer the case. Operations may have "side effects" that go beyond Narsese. As a simple example, if an operation converts its sole argument from a Narsese term into an ASCII string, then prints it out in the default printer, then it is something that the inference engine can control, but some of its effects happen in the physical world (e.g., as ink marks on a piece of paper). Similarly, some operations may be triggered by physical signals, and as a result, convert the signals into Narsese judgments. In this way, beside other things, NAL-8 provides an interface between NARS (as an inference engine) and various sensorimotor mechanisms.

5.4.4 NAL summary

Since NAL-8 is the last extension of the NAL system, now is the time to summarize NAL, as defined from NAL-1 to NAL-8.

The complete grammar rules of Narsese are listed in Table 5.17.

There are some additional notes about the Narsese grammar:

- Confidence values 0 and 1 are used in the meta-language of Narsese only, and cannot appear in actual sentences in the system.

- Many term operators can also be used in the "infix" form, that is, between components.

The semantics of Narsese has been introduced for Narsese-1. The later extensions of Narsese do not change the principle by which truth value and meaning are defined for the language, but allow other types of copula (beside inheritance) to be involved when determining the truth values of statements, as well as when determining the meanings of terms. Furthermore, the meanings of terms are also determined by the compositional relations between compounds and components.

The inference rules of NAL has been introduced in the inference tables in the previous chapters. However, these tables are by no means complete, for the following reasons:

- Many compound-related rules have variants, by changing the number and location of the involved components. Only certain representative cases of these rules are listed in the tables.

$<\mathit{sentence}>$	$::=$	$<\mathit{judgment}> \mid <\mathit{question}> \mid <\mathit{goal}>$
$<\mathit{judgment}>$	$::=$	$<\mathit{statement}><\mathit{truth\text{-}value}>$
$<\mathit{goal}>$	$::=$	$<\mathit{statement}><\mathit{desire\text{-}value}>$
$<\mathit{question}>$	$::=$	$<\mathit{statement}>?$
		$\mid\ ?<\mathit{copula}><\mathit{term}> \mid <\mathit{term}><\mathit{copula}>?$
$<\mathit{statement}>$	$::=$	$<\mathit{term}> \mid (<\mathit{term}><\mathit{copula}><\mathit{term}>)$
		$\mid (\neg <\mathit{statement}>)$
		$\mid (\wedge <\mathit{statement}><\mathit{statement}>^{+})$
		$\mid (\vee <\mathit{statement}><\mathit{statement}>^{+})$
		$\mid (\ ,\ <\mathit{statement}><\mathit{statement}>^{+})$
		$\mid (\ <\mathit{word}>\ <\mathit{term}>^{+})$
		$\mid (\ <\mathit{tense}>\ <\mathit{statement}>)$
$<\mathit{term}>$	$::=$	$<\mathit{word}> \mid (<\mathit{statement}>) \mid <\mathit{variable}>$
		$\mid \{<\mathit{term}>^{+}\} \mid [<\mathit{term}>^{+}]$
		$\mid (\cap <\mathit{term}><\mathit{term}>^{+}) \mid (\cup <\mathit{term}><\mathit{term}>^{+})$
		$\mid (- <\mathit{term}><\mathit{term}>) \mid (\ominus <\mathit{term}><\mathit{term}>)$
		$\mid (\times <\mathit{term}><\mathit{term}>^{+})$
		$\mid (\perp <\mathit{term}><\mathit{term}>^{*} \diamond <\mathit{term}>^{*})$
		$\mid (\top <\mathit{term}><\mathit{term}>^{*} \diamond <\mathit{term}>^{*})$
$<\mathit{variable}>$	$::=$	$\# <\mathit{word}> \mid <\mathit{variable}>\ (<\mathit{variable}>^{*})$
$<\mathit{copula}>$	$::=$	$\rightarrow \mid \leftrightarrow \mid \circ\!\!\rightarrow \mid \rightarrow\!\!\circ \mid \circ\!\!\rightarrow\!\!\circ$
		$\mid \Rightarrow \mid \Leftrightarrow \mid /\!\!\!\!\Rightarrow \mid \backslash\!\!\!\!\Rightarrow \mid /\!\!\!\!\Leftrightarrow$
$<\mathit{tense}>$	$::=$	$/\!\!\!\!\Rightarrow \mid \backslash\!\!\!\!\Rightarrow$
$<\mathit{truth\text{-}value}>$	:	a pair of real number in $[0, 1] \times (0, 1)$
$<\mathit{desire\text{-}value}>$	:	the same as $<\mathit{truth\text{-}value}>$
$<\mathit{word}>$	:	a string in a given alphabet

Table 5.17: The Complete Grammar of Narsese

- As explained in NAL-5, the meta-level implication relations between statements can be used as inference rules, which are not listed in the inference tables.
- The inference rules of NAL-7 and NAL-8 have not been finalized.

Even so, these tables are quite comprehensive in representing the inference rules of NAL.

As mentioned previously, a "logic" consists of a language, a semantics, and a set of inference rules. Now we have seen these parts of NAL, though we haven't fully discussed the nature of this logic, or how to implement it in a computer system. That is what the following chapters will do.

Chapter 6

Inference Control

The previous three chapters specify a Non-Axiomatic Logic, consisting of a formal language (Narsese), an (experience-grounded) semantics, and a set of (extended syllogistic) inference rules. Given them, we now know what a sentence in Narsese looks like, what it means to the system, and how to derive new sentences from existing ones in a single step.

However, this is not the whole NARS. Beside the above logical part, there is a control part, which consists of the following major components:

Memory structure: How does the system store its judgments, questions, and goals?

Control strategy: How does the system decide what rule to use and on what sentences at a given time?

Carnap called some of these components "methodological," and distinguished them from the "logical" components of a system [Carnap, 1950]. A similar distinction is made by Kowalski in the formula "algorithm = logic + control" [Kowalski, 1979]. Generally speaking, the logical part of a reasoning system provides the possibility for a conclusion to be obtained, and the control part selectively realize some of these possibilities. With insufficient knowledge and resources, not all possibilities will be realized.

Since in this book I will not talk about the implementation details of NARS, the description on control mechanism will be given at a general level. In the following, only the basic principles of inference control will be covered.

6.1 Task management

6.1.1 Task and belief

In a system where NAL, NAL-1 to NAL-7, are implemented, the interaction between the system and its environment only happens in Narsese, that is, all input and output information is carried out by Narsese sentences. The implementation of NAL-8 will break this limitation (through operations), but the communication in Narsese still plays a central role.

If we focus on Narsese, the input to the system is a stream of Narsese sentences. Since NARS is an open system working in real time, any legal sentence in Narsese can enter the system at any time. Since the time between sentences matters (we will see why and how later), an accurate description of the life-long experience of the system, in Narsese, looks like the following:

$$S_1, N_1, S_2, N_2, \cdots, S_m, N_m$$

where S_i is a sentence in Narsese, and N_i is a number that measures the length of the time interval between the arriving of S_i and that of S_{i+1}. The unit of this measurement will be introduced later.

As described before, the input sentences have three types: judgments, questions, and goals, and each of them define a *task* for the system to process.

Judgment. An input judgment is a piece of new knowledge to be absorbed. To process such a task not only means to add it to the knowledge base (memory) of the system, but also means to use it and the existing knowledge to derive new knowledge, by forward inference.

Question. An input question is a user query to be answered. To process such a task means to find a judgment that answers the question as well as possible (as defined by the choice rule). Backward inference may be used to get answers through derived questions.

Goal. An input goal is a user command to be followed, or a statement to be realized. To process such a task means to check if the statement is already true, and if not, to execute some operation to make the statement true. Backward inference may also be used to generate derived goals.

Such a bidirectional (both forward and backward) flow of activity is critical to NARS. If it worked only backwards, the system could not answer questions or achieve goals (because truth-value functions are attached only to forward inference rules). On the other hand, the system does not possess enough resources to work forwards only (thus exhaustively generating all conclusions, then matching them to questions and goals), so it must use questions and goals to guide its inference process. Backward inference is just the reverse of forward inference, and it works by "waking up" the related judgments in order to answer questions or to achieve goals.

A Narsese judgment that is already part of the system's knowledge base is called a *belief* of the system. The relation between forward and backward rules means that in NARS each inference step always takes a task and a belief as premises, and generates one or more new tasks as conclusions, where a task can be a judgment, a question, or a goal, but a belief can only be a judgment. Furthermore, the rule(s) that can be applied is determined by the combination of the task and belief (such as the position of the shared term, the relation types of the two, whether there is variable involved, and so on). Consequently, the "control strategy" is reduced to the selection of task and belief in each inference step.

On the output port, the system's behaviors, that is, its responses to the environment, is also a stream of the form

$$S_1, N_1, S_2, N_2, \cdots, S_m, N_m$$

where the sentences include answers for questions, conclusions to be shared, questions to be answered, and goals to be achieved by the other system (i.e., commands). In this way, the format of input and output are exactly the same in NARS. It also means the two sides on the communication channel can be two NARS systems, each of which has its own memory, though using the same logic.

According to the above description, we see that in NARS to process a task means to repeatedly let it interact with the beliefs of the system. Due to insufficient resources, the system cannot use all beliefs for each task; due to insufficient knowledge, the system cannot optimally select the beliefs that lead to the best solution for each task.

As mentioned in Chapter 2, NARS is said to work "with insufficient time resources," or "under time pressure," because

1. new (input and derived) tasks appear from time to time;

2. each task has a time requirement attached;

3. the system's task-processing ability has an upper bound.

For this reason, part of NARS looks like an operating system, because it is responsible for managing the system's (time and space) resources. The control mechanism needs to decide the amount of resource to be spent on each task, as well as which beliefs to use within the given budget. The memory structure should be designed in such a way that makes the resource management easy and efficient.

Clearly, not all control strategies are equally good (or equally bad) in this situation. As in the case of logic, where we need a new standard for the validity of inference rules, here we need a new standard for the validity of control mechanisms. If the system's resource is always in short supply, the best thing it can do is to use it to satisfy the goals *as much as possible.*

Intuitively speaking, if every task t_i in the system has a measurement on the extent to with the goal has been achieved, called "the degree of satisfaction," s_i, and each goal has a relative utility value, u_i, attached, then the optimum control mechanism should maximize the total degree of satisfaction $S = \sum u_i s_i$.

Now the problem becomes to decide how to estimate the effect of each decision on this S. As an adaptive system with insufficient knowledge and resources, NARS can neither accurately predict the future, nor can it try all possibilities before a decision. Instead, it has to use its experience as the only guidance for this kind of decision.

6.1.2 Priority and durability

NARS' aim is not to process every task to a predetermined quality, but for the whole system to work as efficiently as possible when resources are in short supply; for this reason, NARS *distributes* its resources among many tasks. Consequently, the time resource given to a task is not determined by an absolute amount, but by a relative "share," which depends both on the request from the external environment and on the internal situation of the system.

In general, there are two ways to distribute time among tasks: sequential and parallel. "Sequential" means to process the tasks one by one, and "parallel" means to have more than one task being processed at the same time. NARS processes its tasks in parallel, because it is a more flexible way to distribute resources. It should be stressed that for this purpose we do not need parallel processing at the hardware level (i.e., multiple processors) — such an implementation is possible, but is not necessary for the model.

What should a system do if it is occupied by one task when another one shows up? Since for NARS new tasks do not come from a predetermined set, usually the system cannot tell at the beginning what kind of solution can be found, or how much resources it will cost. In such a situation, it is usually undesired either to let the new task wait for an unlimited period, or to let the new task interrupt the processing of the current task for a unlimited period.

NARS distributes its resources among the tasks in a time-sharing manner, meaning that the processor time is cut into fixed-size slices, and in each slice a single task is processed. Because NARS is a reasoning system, its processing of a task divides naturally into inference steps, one per time-slice.

In each inference step, a task is chosen probabilistically, and the probability for a task to be chosen is proportional to its "priority." As a result, priority determines the (expected) processing speed of a task. The priority value of a task is a real number in [0, 1]. At any given instant, if the priority of task t_1 is u_1 and the priority of task t_2 is u_2, then the amounts of time resources the two tasks will receive in the near future keep the ratio $u_1 : u_2$. Priority is therefore a relative rather than an absolute quantity. Knowing that $u_1 = 0.4$ tells us nothing about when task t_1 will be finished or how much time the system will spend on it. If t_1 is the only task in the system, it will get all of the processing time. If there is another task t_2 with $u_2 = 0.8$, the system will spend twice as much time on t_2 as on t_1.

If the priority values of all tasks remain constant, then a task that arises later will get less time than a task that arises earlier, even if the two have the same priority value. A natural solution to this problem is to introduce an "aging" factor for the priority of tasks, so that all priority values gradually *decay*. In NARS, a real number in (0, 1), called *durability*, is attached to each priority value. If at a given moment a task has priority value u and durability factor d, then after a certain amount of time has passed, the priority of the task will be du. Therefore, durability is a relative measurement, too. If at a certain moment $d_1 = 0.4$, $d_2 = 0.8$, and $u_1 = u_2 = 1$, we know that at this moment the two tasks will get the same amount of time resources, but when u_1 has decreased to 0.4, u_2 will only have decreased to 0.8, so the latter will then be receiving twice as much processing time as the former.

If the durability value of a task remain constant, the corresponding priority will become a function of time, $u = d_0^{ct} u_0$, where u_0 and d_0 are the initial values of priority and durability at instant 0, respectively, and c is a constant. Taking the integration of the function, we get the *relative time-budget* of a task (at instant 0):

$$T = \int_0^\infty u dt = -\frac{u_0}{\ln d_0}$$

As a relative measurement, the constant is omitted in the result.

By assigning different priority and durability values to tasks, the user can put various types of time pressure on the system. For example, we can inform the system that some tasks need to be processed immediately but that they have little long-term importance (by giving them high priority and low durability), and that some other tasks are not urgent, but should be processed for a longer time (by giving them low priority and high durability).

Aging is not the only factor that changes the priority distributions among the tasks. The amount of time spent on a task is determined not only by requirements of the environment (user), but also by the current result(s) the system is getting for the task. For example, if the system has found a good answer to a question, the question should become less urgent, and its durability factor should also be decreased (so it will get less time).

For these reasons, each time a task is processed (i.e., during each inference step), the system re-evaluates the task's priority and durability values, to reflect the current situation. As a result, NARS maintains a *dynamic* distribution of priority in its task pool.

6.1.3 Anytime processing

If a task is a question asked by the user, when to report an answer? In a traditional computing system, an answer gets reported only at a "final state," where the system has completed its processing of the question. However, when time is treated as a limited resource and no answer is final, it is better to let the system provide some sort of answer as soon as possible.

NARS keeps a record of the best answer it has found for each question, and whenever a new candidate is found, it is compared with the current best. If the new one is better, it is reported to the user, and the record is updated. After that, the question remains active, but with a lower priority, so the system will continue to spend time on it to find better answers. Therefore, it is possible for the system to report more than one answer for a question — it can "change its mind" when more beliefs are taken into consideration.

When should the system stop processing a task? Ideally, as in conventional computer systems, this should happen when the task has been "finished." For a new judgment, this would mean that the system had generated all its implications. For a question (or goal), it would mean that the system had found the best possible solution, given its beliefs. In both cases, the system would need to access all relevant beliefs. However, under the assumption of insufficient resources, such exhaustive use of beliefs is not possible for NARS. When time is in short supply, some beliefs have to be ignored, even if they may be relevant.

The most common method for dealing with insufficient resources is to retreat to a stance of accepting *satisficing* solutions instead of *complete* solutions [Simon, 1996]. For example, we can limit the maximum number of steps of forward inference for new knowledge, or set a threshold for the confidence or expectation of the answers. These types of methods are widely used in heuristic-search systems, but they are too inflexible for the purpose of NARS. Because the supply and demand of resources are constantly changing in NARS, such thresholds are sometimes too high (therefore the system still cannot satisfy them) and sometimes too low (therefore the system makes no attempt to get better results even though resources are still available).

The solution used in NARS is: when a task is removed from the task pool, it is not because the processing of the task has met some predetermined goal, but because the task has lost too much ground in the competition for resources. Thus, when the task pool is exceeded (it has a constant capacity), tasks at the low end of the priority spectrum are removed. It is possible for a question to be removed from the task pool before even a single answer is found for it.

Resource allocation in NARS is context-dependent. Even if we give the system the same task, with the same priority and durability values, it may be processed differently: when the system is busy (that is, there are many other tasks with higher priority values), the task will be processed briefly, and only "shallow" implications or answers will be found; when the system is relatively idle (that is, there are few other tasks), the task will be processed more thoroughly, and "deep" results can be obtained.

Generally speaking, in NARS a task can be processed for any number of steps, as in "anytime algorithms" [Boddy and Dean, 1994]. The actual number of steps to be carried out for a task is determined both by its initial priority and durability values, and by the resources competing in the system at that time.

6.1.4 Task derivation

NARS constantly generates derived tasks with its inference rules. Each such task is assigned priority and durability values by the system (according to the type of the inference and the priority and durability of its parents), and then put into the task pool. After that, it is treated just like a task provided by the user. Even if a "parent" task has been removed (by losing out in the competition for resources), "children" tasks derived from it may still be processed, provided that they have sufficiently high priority values.

For example, when answering a question Q, NARS may generate two subquestions, Q_1 and Q_2, as alternative ways to answer Q. Later, it finds an answer to Q_1, which leads to an answer to Q. At this point, the priority values of Q and Q_1 are decreased more rapidly than that of Q_2, and it is possible for Q_2 to be processed even after Q has been removed from the system's task pool.

If the purpose of a system were solely to answer questions coming from the user, the above strategy would seem pointless, because Q_2 is merely a means to solve Q, hence should go away if Q goes away. However, the purpose of NARS is to adapt to its environment, which means that Q_2, as a derived question, has value for its own sake, even in a situation where the question that engendered it has utterly vanished. The system will benefit from the processing of Q_2 when similar questions appear thereafter.

As a result, after running for a while, there will be tasks in the system that are only remotely related to the tasks provided by the user. Furthermore, since new tasks are derived from existing tasks according to the current beliefs, it is quite possible for the beliefs involved to be rejected by new evidence obtained latter. In this case, the derived task may have little relevance to its "parent" task, or, even worse, may conflict with its parent.

For example, the system at a certain time has a belief "$P \Rightarrow Q$," with a certain truth value. As a result, a goal Q with this belief will produce a derived goal P. However, it turns out later that the realizing of P makes the realizing of Q more difficult or even impossible.

Under AIKR, the historical relation and the factual relation among tasks are not necessarily consistent. If one task is derived from another in one or more than one step, it does not necessarily mean the satisfaction of the former will indeed satisfy the latter, even to a degree.

Though this "task alienation" often has negative effect to the system's performance, it is also an important source of creativity, originality, and autonomy. No matter whether the effect is good or bad, such a phenomenon is inevitable for a system with insufficient knowledge and resources.

What was described in this section is the "controlled concurrency" mechanism of NARS. Its main features, summarized, are these: the environment provides the system tasks from time to time, giving each task a priority value and a durability factor. In each inference step, the system picks a task according to the current priority distribution, generates new tasks as results of the interaction between that task and a relevant belief, then adjusts the priority of the task according to its durability value and the result of the inference. An answer is reported to the environment as long as it is the best the system has so far found for an input question. Tasks with low priority are removed as a result of competition for resources.

This control mechanism is different from that of ordinary time-sharing, because here the tasks work on a common knowledge base, the processing of a task does not follow a determined procedure, and it is not guaranteed that all tasks will be processed all the way to their final conclusions. Consequently, the interaction among tasks in NARS is much stronger and more competitive than that among the processes in a conventional time-sharing system.

6.2 Memory structure

Now let us see how the memory of NARS is organized to support controlled concurrency.

6.2.1 Probabilistic priority queue

In each inference step, the system chooses a task, according to the priority distribution, then interacts it with a belief, to get results. However, how is the belief chosen?

Here the problem is very similar to the problem discussed previously. With insufficient resources, the system cannot consult all relevant beliefs for a task. On the other hand, beliefs cannot be used indiscriminately — some beliefs are more useful than the others.

To solve this problem, the above idea of "controlled concurrency" is generalized. Let us say that a system has some items to process in a certain way. Because new items may arrive at any time, and because the time requirements of the items would exceed the system's capacity, it is impossible for the system to do the processing exhaustively. It has to distribute its time resources among the items, and to truncate the processing of an item before reaching its "final destination." Furthermore, items are not treated equally, though the system does not have all the information at the beginning to make an optimal plan in advance. Instead, the system has to evaluate the relative priority of each item as a function of several factors, and adjusts its evaluation when the situation changes. In addition, the system's storage capacity, which is a constant, is also in short supply. When an overflow happens, items with high priority values should be kept.

Because this phenomenon pervades the discussion of systems with insufficient resources, it will be useful to design a special data structure for it. In NARS, there is such a data structure called "bag."

A *bag* is a kind of *probabilistic priority queue* with a constant capacity, and it can contain some *items*. Each item has a *priority value*, which is a positive real number.

There are two major operations defined on a bag:

put in. The operation takes an item as an input argument, and puts it into the bag. If the bag is already full, the item with the lowest priority is first removed from it, and then the new item takes its place in the bag.

take out. The operation has no input argument, and returns one item when the bag is not empty. The probability for a given item to be chosen is approximately proportional to its priority.

There are other common operations defined on bag, such as finding an item in it by key.

It is possible to implement these operations in such a way that each of them takes a small constant time to finish, independent of the number of items currently in the bag. The implementation detail will not be discussed here.

6.2.2 Concepts

As mentioned previously, NARS implements a "term logic." This kind of logic is characterized by the use of categorical sentences and syllogistic inference rules. A property of term logic is that almost all inference rules use two premises sharing a term. This convenient property of term logic naturally localizes the choice range of beliefs.

For example, if the statement of a task (which may be a judgment, a question, or a goal) is "$raven \rightarrow bird$," we know that the beliefs that can directly interact with it must have "*raven*" or "*bird*" in it as subject or predicate.[1]

In NARS, all tasks and beliefs that share a common term are collected into a *concept*, which is named by the shared term. As a result, any valid inference step necessarily happens within a single concept. Therefore, a concept becomes a unit for resource allocation, which is larger than a task or a belief. The system distributes its resources firstly among the concepts, and then secondly, within each concept, among the tasks and beliefs.

The result is a two-level memory structure. On both levels, the notion of "bag" applies. We can describe the memory of NARS as a bag of concepts, with, within each concept, a bag of tasks and a bag of beliefs.[2]

For instance, if the statement of a new task is "$raven \rightarrow bird$," then the task will be put into the task bag of the concept "C_{raven}" and that of the concept "C_{bird}." If the task is a judgment, it will add a corresponding belief into the belief bag of the two concepts, too. If there is no such concept (i.e., the term is novel to the system), it will be created as a result of the acceptance of the task.

Now we can see the distinction (in NARS) between *task* and *belief* more clearly. Within the system, every judgment is a belief. Every question and goal is a task. New judgments also serve as tasks for a short time. If a belief provides an answer for a question, it will be "activated," that is, its corresponding task will be generated.

---

[1]There are exceptions, but they can be handled with some special treatments.

[2]By introducing concepts as a unit of inference and resources management, NARS can also be described as a model of categorization. This topic will be discussed in detail in Chapter 11.

Because of this distinction, the system has, at any given moment:

- a small number of tasks, which are active, remembered for a short time, and highly relevant to the current situation;

- a much larger number of beliefs, which are passive, remembered for a long time, and mostly irrelevant to the current situation.

6.2.3 Dynamics in memory

The memory of NARS changes its content and structure dynamically when the system is running.

As the result of inference steps, new tasks, beliefs, and concepts may be put into the corresponding bags, old items may be removed when the bags are full, and structure parameters (priority and durability values of the related items) may be adjusted according to the immediate feedback.

For a given item (task, belief, or concept), its priority will be increased by the "activating" process, and decreased by the "forgetting" process.

Two types of "forgetting" happen in NARS. The first type, "relative forgetting," is caused by the insufficiency of *time* resource — the priority values of items decay over time, and items with low priority are seldom accessed by the system, though they are there all the time. The second type, "absolute forgetting," is caused by the insufficiency of *space* resource — items with the lowest priority are removed from overloaded bags.

It follows from the assumption of insufficient resources that in NARS the results of task processing are usually only derived from part of the system's beliefs, and which part of the beliefs is used depends on the context at the run time, since the items (concept, task, and belief) processed in each step depends on the priority distributions among the items in the memory. Consequently, NARS is no longer "logically omniscient" [Fagin and Halpern, 1988] — it cannot recall every belief in its memory, not to mention being aware of all their implications.

6.3 Inference processes

In a reasoning system, an "inference process" consists of a sequence of inference steps.

6.3.1 Life cycle and execution cycle

In conventional computing systems, the processing of a (user provided) problem has a well-defined starting point and ending point. For a problem to be accepted, it must correspond to an algorithm already implemented for this type of problem. The problem is accepted when the algorithm is waiting to be executed. After an execution, the working space is cleared, the relevant states are reset, and the system waits for the next problem.

NARS does not work in this way. When a NARS system starts to run for the first time, its memory may either be empty, or comes with certain "innate" beliefs. Roughly speaking (i.e., with some implementation details omitted), the system runs by repeating the following "execution cycle":

1. To check the input buffer. If there are new (input or derived) tasks, add them into the corresponding concepts (after some simple preprocessing).

2. To take out a concept from the concept bag.

3. To take out a task from the task bag of the concept.

4. To take out a belief from the belief bag of the concept.

5. To apply inference rules on the task and belief. Which rule is applicable is determined by the syntax of the task and the belief.

6. To adjust the priority and durability values of the involved task, belief, and concept, according to the quality of the results.

7. To return the involved task, belief, and concept are returned to the corresponding bags.

8. To put the results generated in this step into the input buffer as new tasks. If a result happens to be a best-so-far answer of a question asked by the user, it is reported to the user.

It should be noticed that each execution cycle may produce new tasks (as well as related beliefs and concepts), without removing any old items. In every inference step, the promises used are but back into the memory afterwards. This is true even for revision, where a conclusion is drawn by pooling together the evidence provided by the premises, but after it, the premises are still valid (with respect to their own evidence), and will be kept in the memory, though may be with lower priority values.

Clearly, each execution cycle carries out a single step of inference. It is possible to implement the above cycle in such a way that it takes roughly constant time, no matter how large are the involved bags. Such a step is like an "atomic operation" of the problem-solving processes in NARS. The user can interrupt the process between execution cycles, then resume the running at a later time.

Unlike a conventional computing system, in NARS the processing of tasks are interwoven, even when the tasks are not directly related to each other in contents. Also, the starting and ending point of a task processing are not clearly defined, because the system never waits for new tasks in a special state, and it never reports a final answer, then stops working on a task right after it. What it does to a task is strongly influenced by the existence of other tasks.

If the user stops the program and resets the memory, the system will start another life cycle, with no memory about its "previous life." Unless its experience accurately repeats itself, the system may develop different beliefs, and process tasks differently, even though its innate capability (inference rules and control mechanism) remain unchanged.

The system is "task-driven" in the sense that every execution cycle is an attempt to process some task. If we concentrate on the processing of a single task in NARS, we can see that the process consists of a sequence of execution cycles, each of them is like an instruction in a programming language, in the sense that it carries out some simple work. However, the whole process does not act like a typical program, where the order of instructions are accurately determined in advance, and the process can be accurately repeated as far as the same task is given to the system. In NARS, for a given task, the execution cycles form a sequence only at the run time, and the order is unpredictable unless from the complete history of the system, including its experience that is not directly related to the given task.

6.3.2 Operation execution

One omitted process in the previous description of the running of NARS is the execution of operations that match current goals.

In NAL-8, after introducing the procedural interpretation of Narsese statements, operations and goals are defined. If an operation O happens to achieve a goal G, the system would like to execute it when G is selected as a task in an executing cycle.

In the first glance, the situation here looks like logic programming. When a Prolog program runs, whenever a goal is nothing but a "built-in predicate" (like an operation in NARS), it is executed, and the goal is achieved. Under AIKR, however, the situation becomes much more complicated. Because of insufficient knowledge, there is always uncertainty in the conditions and effects of an operation, therefore "The system currently believes that O can achieve G" does not guarantee the achieving of G by executing O. Furthermore, since at the same time there are usually many goals to be achieved, the executing of O may make other goals harder to be achieved, even if it indeed realizes G.

As a first approximation, in NARS the execution of operations will be handled by the following mechanism.

Whenever the executing of an operation is proposed by the processing of a goal, usually the action will not be taken immediately. Instead, the operation is put into the task base of the corresponding concept, with its desire value (as defined in Section 5.4.2). While the system continues to run, other goals also "cast their votes" on the execution of the operation, by either increasing or deceasing its desire value, depending on whether its execution contributes to their satisfaction.

The contribution of a goal to the desire value of an operation is determined by several factors, including

importance of the goal. Of course, if the goal is important to the system, the operation has a higher desire value.

certainty of the belief. If the belief that links the goal and the operation is strong (that is, has a high expectation value), it will contribute more to the desire value.

At a given time, the system only executes one operation (though it can be a compound operation with internal structure), and the selection is based on both the desire value and the priority value of the operations (as goals). When the system is implemented in a programming language like Java, the execution of operations may be carried out by another thread, so this process will go in parallel with the reasoning process that runs the execution cycles, and be monitored and controlled by the latter.

There is an important difference between the executing of operations and the processing of tasks, in that the former is a binary decision, while the latter is a matter of degree. Since the processing of a task typically consists of many steps, the system can carry out many such processes in parallel, at different speeds. On the contrary, whether to execute an operation is usually a matter of "yes/no," so these processes must be carried out sequentially, one at a time.[3]

6.4 Budget assessment

What has not been discussed is the subtlest part of the control mechanism: how to assess the priority and durability values of concepts, tasks, and beliefs, both initially and after each inference step. To this end, we need to design a set of numerical functions.

6.4.1 Relevant factors

In each execution cycle, only the directly involved concepts, tasks, and beliefs will have their priority and durability values adjusted. Thus, no

[3]Of course, the situation will be different if NARS is implemented in such a way that parallel processing can also happen at the hardware level. For example, some operations may need to operate multiple devices at the same time.

matter how big the memory is, this set of adjustments takes roughly constant time. This is exactly what one would wish for. On the other hand, these adjustments have global effects. Since priority and durability are *relative* values, the increase of the priority of any particular task means that all competing tasks will have less chance. Though such effects are small at each step, they accumulate in the long run, and result in a dynamic allocation of resources.

Since all the quantities involved are real numbers in [0, 1], these functions have been designed in a manner reminiscent of the design of the truth-value functions. First, the boundary condition constraining the desired function is determined, and then a function satisfying this condition is formed with the help of T-norm, T-conorm, and other general-purpose operators (such as difference, average, and so on). Functions obtained in this way are by no means optimal. They only reflect my current analysis of the situation, and my ideas are still being revised from time to time. In the following, therefore, they will only be discussed qualitatively.

Generally speaking, the priority of an item reflects the amount of resources the system plans to spend on it in the near future. Generally, it is determined by the following considerations:

1. The *origin* of the item, according to its source of generation. This factor is available initially, and remains constant.

2. The *quality* of the item, according to its attribute values. This factor is available initially, and may or may not remain constant.

3. The *usefulness* of the item, according to its contribution in the past. This factor is acquired gradually, and indicate an individualized evaluation.

4. The *relevance* of the item, according to its relation to the current situation. This factor is re-evaluated from time to time.

The first three factors mainly determine the long-term aspect in priority, while the last one determines the short-term aspect. The current priority value is a function of these two aspects.

For different kinds of item, the above factors are determined in different ways.

6.4.2 Assessing tasks

The priority and durability values of an input task are assigned by the environment (a user or another computer system), and reflect the time constraint placed by the environment on the task. Of course, the system can provide default values.

The priority and durability values of a *derived* task are determined by several factors: generally, it inherits them from its parents. If both the task and belief that derive the new task have high priority (or durability) values, the new task will get a high priority (or durability) value. However, the values will also be affected by the "quality" of the inference result.

- In revision, a new task gets a high priority value if the two premises have significantly different expectations (a belief conflict).

- In forward inference, a new task gets high priority and durability values if it is a confident conclusion.

- In backward inference, if the answer to the derived task can (with its parent belief) generate a good answer to the parent task, it gets high priority and durability values. For instance, if the related forward rule is deduction, it is better than the case of induction, for this purpose.

- If the given task is a question and the given belief provides an answer for it, the priority and durability values of the new task (which is a copy of the belief) depends on how good the answer is.

After the inference, the priority value of the parent task is decreased (to indicate the cost of the inference). In addition to the decrease due to the durability factor described earlier, there may be further decrease, depending on the quality of the result. If an answer to a question is found and is quite good, then the priority of the question is decreased more than if the answer is poor.

6.4.3 Assessing beliefs

Initially, a belief is generated by copying a (judgment) task. It is given the highest priority value (that is, 1), and its durability value is determined by the priority and durability values of that task. In this way, a new belief gets some initial attention from the system, but usually decays rapidly unless it corresponds to a task with high priority and durability.

After a belief is used, its priority and durability values are adjusted. In contrast to the case of tasks, the priority value of a belief can be increased. In each adjustment, there are two competing forces pulling the priority value in opposite directions. One force is the universal aging force, which decreases the priority value by multiplying the durability value (a number less than 1). On the other hand, the belief gets rewarded each time (by increasing its priority value) depending on how good the result is in this step. Therefore, in the long run, useful beliefs (which has provided good answers and implications in the past) get high priority values, which makes it more accessible in the future.

The durability value of any given belief is increased each time. As a result, the longer a belief stays in the memory, the more slowly its priority value will decrease.

6.4.4 Assessing concepts

The priority value of a concept increases when a task is put into it, with the size of the increase depending on the priority value of the task. At the same time, the concept's durability value is also adjusted. If the concept already has high priority before the task is inserted into it, it will be given a high durability value, which ensures that the current high priority will be kept longer. On the other hand, if the concept's priority took a big, upwards jump, then the durability value will be set low, so the concept will lose its priority soon.

The priority value of a concept decreases whenever any task is removed from it. In addition to the decrease due to the durability factor, the size of the decrease also depends on the priority of the task that was just removed. Specifically, if the task had a high priority value, the decrease will be larger.

Another factor that influences the resources allocated to a concept is its "quality." Roughly speaking, the system will give more resources to a concept whose extension and intension are balanced, and have relatively sharp boundary (i.e., between elements and non-elements), because such a concept tends to produce better results.[4]

[4]We will return to this topic in Section 11.1.6.

Part III

Comparison and Discussion

Chapter 7

Semantics

Generally speaking, *semantics* is the study of the relation between a *language* and the *environment* in which the language is used. The language can be either *artificial* or *natural*. The former usually has well-defined grammar rules followed by the users of the language (so they are often also called "formal languages" or "symbolic languages"), while the latter is usually formed in history, described by some loose grammar rules summarized from its common usage, which may change from time to time, from place to place, and often with exceptions.

A reasoning system normally uses a language for communicating with other systems, as well as for knowledge representation within the system. For each of the two functions, the semantic theory of the language plays an important role:

- Outside the system, semantics specifies how the language should be understood (such as how to be translated into other languages) in communication, so that other (human or computer) systems know how to "talk" with the system. For this purpose, the semantic theory specifies how the *meaning* of a word or a sentence in the language is determined, by relating it to the outside of the language.

- Within the system, semantics provides justification for the inference rules, that is, to explain why these rules and not others are proper to be used to carry out inference on the language. For this purpose, the semantic theory specifies how the *truth values* of declarative sentences of the language are determined, so that the rules can be validated as preserving truth in the inference process.

In this chapter, the Experience-Grounded Semantics (EGS) of NARS (first introduced in Chapter 3) is compared with other schools of semantics, and several important issues are discussed.

7.1 Experience vs. model

7.1.1 Model-theoretic semantics

The "native language" of NARS is Narsese. It is a formal language, in the sense that its grammar is formally defined (see Table 5.17 for the complete Narsese grammar). Since the dominant semantics for formal languages is Model-Theoretic Semantics (MTS), the first issue to be discussed is why Narsese does not use MTS, but uses a new semantic theory, EGS.

The basic of MTS can be roughly described as the following. For a formal language **L**, a *model* **M** consists of descriptions about objects and their factual relations in a domain. The descriptions are written in a *meta-language* **Lm**, which can either be a natural language, like English, or another formal language. An *interpretation* **I** maps the words in **L** onto the objects and relations in **M**.

According to this theory, the *meaning* of a word in **L** is defined as its image in **M** under **I**, and whether a statement in **L** is *true* is determined by whether it is mapped by **I** onto a fact in **M**.

The study of formal languages was started as part of the study about the foundation of mathematics by Frege, Russell, Hilbert, and others. A major motivation of using formal languages was to avoid the ambiguity in natural language, so that objective and accurate artificial languages were created. MTS was founded by Tarski's work. Although Tarski's primary target was formal language, he also hoped that the ideas could be applied to reform natural language [Tarski, 1944].

To directly use this kind of semantics in a reasoning system (such as NARS) means to understand the meaning of a word in Narsese according to the object or relation it refers to (under a given interpretation), and to choose inference rules that are truth-preserving under all possible interpretations. According to this viewpoint, "semantics is a discipline which deals with certain relations between expressions of a language and the objects 'referred to' by those expressions." [Tarski, 1944].

According to MTS, for any formal language **L**, the necessary and sufficient condition for its terms to have meaning and for its statements to have truth value is the existence of a model. In different models, the meaning of a term and the truth values of a statement may change; however, these changes are not caused by *using* the language. A reasoning system **R** that processes sentences in **L** does not depend on the semantics of **L** when the system runs. That means, on the one hand, that **R** needs no *access* to the meanings of terms and truth values of statements — it can distinguish terms only by their forms, and derive statements from other statements only according to its (syntactically defined) inference rules, but it puts little constraint on how the language can be interpreted [Putnam, 1981]. On the other hand, what beliefs **R** has and what operations **R** performs have *no influence* on the meanings and truth values of the terms and sentences involved.

Such a treatment is desired in mathematics. As Russell put it, "*If* our hypothesis is about *anything*, and not about some one or more particular things, then our deductions constitute mathematics. Thus mathematics may be defined as the subject in which we never know what we are talking about, nor whether what we are saying is true." [Russell, 1901] In mathematical logic, abstract patterns of inference are studied, and the patterns can be applied to different domains by constructing different models. Here we do enjoy the freedom provided by the separation of "syntactic processing" and "semantic interpretation." The study of semantics has contributed significantly to the development of meta-mathematics. As Tarski said, "As regards the applicability of semantics to mathematical science and their methodology, i.e., to meta-mathematics, we are in a much more favorable position than in the case of empirical sciences." [Tarski, 1944].

As all normative theories, MTS is based on certain assumptions, and it should be applied to a problem only when the assumptions are satisfied. In asserting the existence of a model **M**, the theory presumes that there is, at least in principle, a consistent, complete, accurate, and static description of (the relevant part of) the environment in a language **Lm**, and that such a description, a "state of affairs" is at least partially known, so that the truth value of some statements in **L** can be determined accordingly. These statements then can be used as premises for the following inference activities. It is also required that all valid inference rules must be truth-preserving, which implies that only true conclusions are desired. After the truth value of a statement is determined, it will not be influenced by the system's inference activity.

Such conditions hold only when a system has *sufficient knowledge and resources* with respect to the problems to be solved. "Sufficient knowledge" means that the desired results can be obtained by deduction from initially available knowledge alone, so no additional knowledge will be necessary; "sufficient resources" means that the system can afford the time–space expense of the inference, so no approximation will be necessary. These are exactly the assumptions we usually accept when working within a mathematical theory. Therefore, it is no surprise that model-theoretic semantics works fine there.

Of course, what we just described is merely the basic form of MTS. Many variations and extensions of MTS have been proposed for various purposes, such as possible worlds, multi-valued propositions, situational calculus, and so on [Barwise and Perry, 1983, Carnap, 1950, Halpern, 1990, Kyburg, 1992, Zadeh, 1986b]. However, these approaches still share the same fundamental framework: for a reasoning system **R** working in an environment **E** with a language **L** (for knowledge representation and communication), the semantics of **L** is provided by descriptions of **E** in another language **Lm** and a mapping between items in **L** and **Lm**.

No matter how the details are specified, this kind of semantics treats the semantics of **L** as *independent* of the two processes in which **R** is involved (and where the language **L** plays a central role): first, the communication between **R** and **E**, and second, the internal reasoning activity of **R**. According to MTS, these processes are purely syntactic, in the sense that only the form of the words and the structure of the sentences are needed. Since the above two processes can be referred to as the "external experience" and "internal experience" of the system, we say that MTS is "experience-independent," and it does not even need to assume the existence of a reasoning system **R** that actually uses the language.

7.1.2 Why NARS does not use MTS

Though MTS can be applied to Narsese, it provides little help for the design and use of NARS. If we give Narsese a model, it tells us what the words mean to *us*, but says nothing about what they mean to *the system*, which does not necessarily have access to our model. Similarly, the model tells the truth value of statements to *us*, but not to *the system*.

By "to the system," I mean that to solve the semantic problems in NARS (that is, to understand the language and to justify the rules), we need to explain why the system treats each term and statement as different from other terms and statements, and such an explanation should be based on the relation between the language and the environment, not only on the syntactic natures defined within the language. Since the relation between NARS (the system in which the language is used) and its environment (which is the "world" to the system) is indicated by the *experience* of the system, the semantic features of a term or a statement has to be determined according to its role in the experience of the system, because in NARS there is no other way to talk about the outside world.

If we still define truth as "agreement with reality," in the sense that truth values cannot be threatened by the acquisition of new knowledge or the operation of the system, then no statement can ever be assigned a truth value by the system under AIKR, because by the very definition of *open* system, all beliefs can be challenged by future evidence.[1]

Moreover, since non-deductive inferences (which are absolutely necessary when knowledge is insufficient) are not truth-preserving in the model-theoretic sense, they are hardly justifiable in the usual MTS way. MTS also prohibits the system from using the same term to mean different things in different moments (which is often inevitable when resources are in short supply, to be discussed later), because meaning is defined as independent of the system's activity.

However, it is not true that in such a situation semantic notions like "truth" and "meaning" are meaningless. If that were the case,

[1]One proposed solution of this problem is to treat "truth value" and "degree of belief" as different to each other. This issue will be discussed in detail in Section 7.4.2.

then we could not talk about truth and meaning in any realm except mathematics, because our mind faces exactly the same situation.

For an intelligent system likes NARS (or for adaptive systems in general), the concept of "truth" still makes sense, because the system *believes* certain statements, but not other statements, in the sense that the system chooses its actions according to the expectation that the former, not the latter, will be confirmed by future experience; the concept of "meaning" still makes sense, because the system uses the terms in Narsese in different ways, not because they have different shapes, but because they correspond to different experiences.

For these reasons, in NARS we need an *experience-grounded* semantics, in which truth and meaning are defined according to the experience of the system. Such a theory is fundamentally different from MTS, but it still qualifies to be a "semantics," in a broad (and original) sense of the notion.

The idea that truth and meaning can be defined in terms of experience is not a new one. For example, it is obviously related to the theory of pragmatism of Peirce, James, and Dewey. In recent years, related philosophical ideas and discussions can be found in the work of Putnam and many others [Dummett, 1978, Field, 2001, Fodor, 1987, Lynch, 1998, Putnam, 1981, Segal, 2000, Wright, 1992]. In linguistics and psychology, similar opinions can be found in [Barsalou, 1999, Ellis, 1993, Kitchener, 1994, Lakoff, 1988, Palmer, 1981].

In AI research, the situation is different. Unlike in philosophy, linguistics, and psychology, where MTS (with the related theories, such as realism, the correspondence theory of truth, and the reference theory of meaning) is seen as one of several candidate approaches in semantics (by both sides of the debates), in AI not only is MTS accepted explicitly by the "logical AI" school [McCarthy, 1988, Nilsson, 1991], and implicitly by the "symbolic AI" school in general [Newell, 1990], but also it is taken to be the only possible semantics, both by its proponents and its critics. As McDermott put it, according to the logicist opinion, "The notation we use must be understandable to those using it and reading it; so it must have a semantics; so it must have a Tarskian semantics, because there is no other candidate." [McDermott, 1987] When people do not like this semantics, they usually abandon it together with the idea of formal language and inference rules, and turn to neural networks, robots, dynamic systems, and so on, with the hope that they can generate meaning and truth from perception and action [Birnbaum, 1991, Brooks, 1991, Harnad, 1990, Smolensky, 1988, van Gelder, 1997].

Therefore, though the philosophical foundation of MTS is under debate, and its suitability for a natural language is doubtable [Haack, 1978], few people have doubt about its suitability for a formal language. We have not seen a *formal semantics* that is not *model-theoretic*, and such concept may even sound self-contradictory to some people.

7.1.3 EGS vs. MTS

The EGS theory used in NARS has been formally defined in Chapter 3. Briefly speaking, an EGS first defines the form of experience a system can have, then defines truth value and meaning as functions of given experience.

Though both are descriptions of an environment (or "world"), "model" and "experience" are different in the following aspects:

- A model is static, whereas experience stretches out over time.
- A model is a complete description of (the relevant part of) an environment, whereas experience is only a partial description of it, in the sense that novel terms may appear that were not known previously.
- A model must be consistent, whereas judgments in experience may conflict with one another.
- A model of a language **L** is described in another language **Lm**, whereas experience is represented in **L** itself.
- The existence of a model **M** of **L** is independent of the existence of a system **R** using **L**. Even when both **M** and **R** exist, they are not necessarily related to each other in any way. On the contrary, experience must happen in a system.

These two types of descriptions serve different purposes. A reasoning system assuming sufficient knowledge and resources makes no attempt to answer questions beyond the scope of available knowledge and resources — when such a question is provided, the system simply replies "I don't know," "Invalid question," or gives no reply at all. For such a system, it is fine to describe its environment as a model. On the contrary, a system designed under AIKR always attempts to answer a question with available knowledge and resources, which means that the system may revise its beliefs from time to time. For such a system, it is better to describe its environment by its experience.

Generally speaking, the human mind works with insufficient knowledge and resources [Medin and Ross, 1992]. However, for certain relatively mature and stable beliefs, it is more efficient to treat them assuming the sufficiency of knowledge and resources. This is exactly the role played by mathematics. In such a theory, we do not talk about concrete objects and properties. Instead, we talk about abstract ones, which are fully specified by postulates and conventions. After we figure out the implications of these postulates and conventions, we can apply such a theory into many situations, because as far as the postulates and conventions can be "instantiated" by substituting the abstract concepts with the concrete ones, all the ready-made implications follow. This is the picture provided by MTS.

On the other hand, if the beliefs embedded in a reasoning system are not mathematical, but empirical, then what we have is a system where the concepts are no longer abstract and can be interpreted freely — no matter how an external observer interprets them, for the system their meaning and truth come from experience, and an EGS should be used.

Some researchers suggest that the reasoning system itself (human or computer), rather than the world it deals with, should be used as the "domain" of the language the system uses. Thus, one could posit that the meaning of a particular term is a particular "concept" that the system has, and the truth value of a statement is the system's "degree of belief" in that statement. This idea sounds reasonable, but it does not answer the original question: how are "concepts" and "degrees of belief" dependent upon the outside world? Without an answer to that question, such a solution "simply pushes the problem of external significance from expressions to ideas" [Barwise and Perry, 1983], that is, it turns the problem of *word* meaning into the problem of *concept* meaning.

The meaning of a concept is not simpler than the meaning of a word at all. It often changes from time to time and from place to place, and such changes cannot always be attributed to the changes in the world. People in different cultures and with different languages usually have different opinions on what "objects" are there even if they are in the same environment [Whorf, 1956]. People often use concepts metaphorically [Lakoff, 1987] or with great "fluidity" [Hofstadter and FARG, 19995]. These issues are hard to handle in MTS.

What if we take a concept as a Platonic entity that never changes, and treat the changes in the meaning of a word as mappings to different concepts? Now the problem becomes to explain why a certain new concept, rather than many others, becomes the new meaning of a word. We still need a way to link the change of meanings of words, concepts, ideas, or whatever we call them, to the experience of the system.

In NARS, since a term is the name of a concept, and a statement is the name of a conceptual relation, the meaning/truth defined for the language and the meaning/truth defined for the concepts system are defined similarly.

Though overall NARS uses EGS, there are still places where MTS is used. One example is the symbols in the inference rule. As mentioned in Chapter 3, the induction rule of NARS is

$$\{M \rightarrow P <f_1, c_1>,\ M \rightarrow S <f_2, c_2>\} \vdash S \rightarrow P <f, c>$$

Written in this way, the symbols S, M, and P have no experience-related meaning until they are instantiated by constant terms "*bird*," "*raven*," and "[*black*]," respectively, and then the meanings of the symbols are determined by the meanings of the constant terms.

Similarly, when mathematical knowledge is provided to NARS, it is used with MTS. This kind of knowledge always needs an interpretation step when being applied to a practical situation. Roughly speaking, MTS is usually used to base one language in another language, while EGS is usually used to base a language in the experience of a system using the language.

According to the above discussion, we see that MTS and EGS are designed under different assumptions, and therefore should be used for different purposes. In this sense, EGS is not proposed as a competitor of MTS. However, since now MTS is used everywhere, including in situations where EGS (or its variations) should be used, EGS indeed competes with MTS as candidates of application, especially in AI and cognitive sciences (but not in meta-mathematics).

It is very often implicitly assumed that the semantics of a formal language has to be model-theoretic. Such an inductive conclusion seems warranted by our experience — almost all formal languages have traditionally been assigned their semantics in this way. As a result, people who do not like the semantics usually abandon the language at the same time.

However, a language can be "formal" in two different senses. In a *syntactic* sense, "formal" means merely that the language is artificial, and is defined by a formal and symbolic grammar; in a *semantic* sense, "formal" means that the language is used in conjunction with an MTS. Narsese is "formal" in the syntactic sense only. From a technical point of view, it would be easy to give the language a model-theoretic semantics, but with such a semantics, the language would no longer be suitable for our current purposes.

Logicians, in distinguishing themselves from other scholars (such as psychologists), tend to stress the *normative* nature of logical theory. As a result, in their study of semantics, the goal is often that of looking for *the* real, objective meanings of terms or truth values of sentences. Even if such an opinion has some degree of justifiability when one's purpose is to study the logic of mathematics, that justifiability goes away when one turns to the study of the "logic" of empirical science and common sense. For the purposes of AI, what we need is another kind of normative model, in which meanings and truth values are founded on the system's experience.

Since NARS is a *normative* theory of intelligent reasoning, not a *descriptive* theory of it, the semantics proposed here is about how truth and meaning *should be* used in a system, not how they *are* actually used in the human mind. I do not present NARS as a psychological or linguistic model of truth and meaning. However, since the human mind is basically an adaptive system evolved in an environment where its knowledge and resources are generally insufficient with respect to the problems to be solved, I do believe that in general this model is closer to a descriptive model than MTS is. Though it is not the major goal of the current research on EGS, it will be interesting to explore the implications of this theory in philosophy, linguistics, and psychology.

7.2 Extension and intension

Traditionally, *extension* and *intension* refer to two different aspects of the meaning of a term: roughly speaking, its *instances* and its *properties*.

In previous theories, a term's extension is usually defined as a set of *objects* in a "physical world" that are denoted by the given term; a term's intension is usually defined as a *concept*, or a set of *attributes*, in a "Platonic world" which denotes or describes the given term [Bocheński, 1970, Copi, 1982]. In spite of minor differences among the exact ways the two words are used by different authors, they always indicate relations between a term in a language and something *outside* the language.

By contrast, in NAL (as defined in Chapter 3) a term's extension and intension are sets of term linked to the given term by an inheritance relation (in the opposite directions). Here extension and intension are defined *within* the language, and become symmetric to each other. Yet even so, the definition retains the intuitive feature that "extension" refers to instances, and "intension" refers to properties.

Such a departure from tradition has important reasons and implications. One of them is already addressed previously in the "experience vs. model" discussion, that is, NAL does not assume a "physical world" or a "Platonic world" that is described in another language, and everything that is semantically relevant must be based on the experience of the system, described by the same language. In the following, we focus on the symmetry, or duality, of extension and intension in NAL, which lead to a unified treatment of the two.

7.2.1 The need for a unification

One feature that distinguishes NAL from other logical systems is its unified representation and processing of extension and intension.

In semantics, this unification happens in two places. For each term, its meaning consists of its extension and intension. For each judgment, its truth value is usually determined by both extensional and intensional factors.

It needs to be stressed that in the terminology of set theory, what is being counted in NAL as a piece of evidence in extension is not an "element," but a "subset," because of the use of inheritance relation. For example, if the system's experience is $\{A \circ\!\!\rightarrow B,\ B \rightarrow C,\ C \rightarrow D\}$, then the extension of D is $\{\{A\},\ B,\ C\}$, not just $\{A\}$. Similarly, what is being counted in intension is a "superset," not a "property."

In other theories, following the tradition of set theory, a concept is often treated as a set, with its instances as elements. Consequently, the logic developed on these concepts is a kind of extensional logic. If intension is as important as extension, how can the traditional logic have been used for such a long time without being challenged on this issue?

This is the case because traditional formal logic has been developed and mainly used in mathematics, where given the extension of a concept (i.e., what instances it has), its intension (i.e., what properties it has) is uniquely determined; and given its intension, its extension is uniquely determined, too. We have seen this in the discussion of NAL-0 in Section 3.1. In such a situation, to process both extension and intension becomes unnecessary, and even confusing. When concepts are explicitly defined and processed according to their extension, their intension is implicitly defined and processed.

This is no longer the case for a system like NARS, which is built under AIKR. In this situation, not only that extension and intension do not fully determine each other, even known extension (intension) cannot determine future extension (intension).

For example, in set theory, $S \subseteq P$ means that the members of S are also members of P. This extensional statement implies the following:

- if M is an element of S, it is also an element of P;
- if M is a property shared by the elements of P, it is also shared by the elements of S.

On the contrary, in NAL if all known instances of S are also instances of P, the system is not necessarily certain to the same extent (i.e., indicated by the same truth value) about the implications listed above. What it means is that in this situation, we cannot only process extension of concepts, and expect intension to follow automatically.

For practical purposes, we may prefer to treat concepts as purely defined by extension (or intension), so that we can use various mathematical tools on them. By doing that, however, we are assuming certain beliefs to be "axioms" that won't be challenged, and doing inference accordingly. In that situation, it is fine to concentrate on extension alone. For example, we can define a "subset" relation among concepts (that are treated as sets), and do inference on the relation accordingly. The pure extensional logic used in this case is not part of NAL, but a system that can be called by NAL as a tool to solve specific problems. NAL is not designed as a logic that *includes* all other useful logics, but one that allows the others to be used by it. To a system using NARS as its "intelligent core," it works like an operating system, which use various logics and algorithms as application programs to solve various concrete problems.

However, no matter what logic is used, the conclusions obtained are only as good as the assumptions, and the price of ignoring intensional information will be paid anyway. If certain concepts should not be treated as pure extensional, but the system chooses to do that anyway, the conclusions will be less confident than the ones obtained by considering both extensional and intensional beliefs, because the former is based on less evidence.

It is possible to develop extensional or intensional term logics separately, as shown in [Wang, 1994b]. As stated previously, "$S \rightarrow P$" means, when it is understood *extensionally*, that P inherits S's instances; but when it is understood *intensionally*, the same relation means that S inherits P's properties. Therefore, if "$S \rightarrow M$" is completely false and "$M \rightarrow P$" is completely true, what can be derived from them is different in the two logics. In the extensional logic, the premises are understood as "S and M have no common instances, and all instances of M are also instances of P." From these two relations, we cannot decide whether S and P have common instances. On the other hand, in the intensional logic, the premises are understood as "S and M have no common properties, and all properties of P are also properties of M," which implies that "S and P have no common properties."

Symmetrically, if "$S \to M$" is completely true and "$M \to P$" is completely false, the extensional implication is "S and P have no common instances," and there is no intensional implication.

Though the extensional logic and the intensional logic, defined in this way, are different, formally they are isomorphic to each other, much as union and intersection are isomorphic to each other in set theory. This isomorphism is described in [Wang, 1994b], and it comes directly from the "dual" definitions of extension and intension in NARS.

The dual definitions of extension and intension make it possible for NARS to treat them *uniformly*, as, for instance, in the definition of "amount of evidence." We need systems to deal with them together, because the coordination of the extensions and intensions of concepts is an important principle in the development of human cognition [Inhelder and Piaget, 1969], and when evidence is used to judge a conceptual relation, whether the evidence is extensional or intensional is often irrelevant or unimportant. We often determine the extension (instances) of a concept according to its intension (properties), or the other way around, and seldom judge a relation between concepts by considering the extensional or intensional factor *only*, especially when the system has insufficient knowledge and resources.[2]

Therefore, though a pure extensional (or intensional) logic is easier to define, it is less interesting from the viewpoint of AI, so NARS is not designed in that way. On the other hand, if really necessary, it is possible for NARS to express pure extensional or pure intensional relations using more complicated methods (such as with variable terms defined in NAL-6), and to support an extensional (or intensional) logic as a subsystem.

7.2.2 Unification in meaning

As discussed above, the meaning of a term has two aspects, its extension and intension, related to the "reference" and "sense" of Frege, respectively [Copi, 1982].

A common practice in AI and cognitive science is to take a term as the name (or label, symbol, and so on) of an object, or a set of objects, in the world. This intuition is the foundation of the MTS.

---

[2]We will come back to this issue in Chapter 11 when discussing categorization.

The problem of this approach is the assumption that there is an objective way to describe the world as objects with relations among them, and that terms have one-to-one mapping to objects or sets of object. This assumption, though sounds natural, conflicts with the assumption that the system may have all possible kinds of future experience.

As usual, this idea is not completely wrong. The meaning of a term does partially depend on its relation to its instances (extension). The difference is that in NARS the extension of a term consists of other *terms*, not *objects*. By following the extensional relations, the system will eventually reach terms that cannot be further specified, though actually they still exist within the system, for instance, as the "mental image" that formed by perception. For example, we can say that "Mars" is in the extension of "planet," but here "Mars" is something in our mind, not something that exists independent of our mind.

According to EGS, the meaning of a term is determined by its (experienced) relations with other terms. These relations can either be extensional or intensional. If "$S \rightarrow P$" is a new belief, then it contributes to the meaning of both S (by indicating part of its intension) and P (by indicating part of its extension). In a sense, S and P are partially defining each other, and no one is "more primitive" than the other semantically.

As special cases, there are terms whose meaning is mostly determined by its extension, as well as terms whose meaning is mostly determined by its intension. Only in these situations, a pure extensional or pure intensional theory of categorization may work approximately. Since NARS determines the meaning of a term according to *available* relations, it can handle these special situations, as well as the general situation where both extensional and intensional factors should be considered.

One consequence of using a unified approach is that that system tends to keep the extension and intension of a concept in coherence. In an extensional logic, since properties are derived from given instances, the system makes no attempts to use them to evaluate membership. It is possible to define a concept by a set of instances, where one instance is very different from the others (in terms of its properties), and yet is a perfect instance. In NARS, however, the result is different. The membership (i.e., the frequency) of the special instance will be decreased, because of its difference with the other instances. In this way, there is a "feedback loop" between extension and intension.

7.2.3 Unification in truth value

According to the definitions introduced in Chapter 3, in the truth value of a judgment, the extensional factor and the intensional factor are merged together.

This is a controversial issue in NARS. Intuitively, even if both extensional and intensional factors need to be considered, it is more informative to represent them separately, and to process them in parallel. For example, for "$S \to P$," why do not we measure the evidence collected from the extension and intention of S and P with different numbers? As shown in [Wang, 1994b], it can be done in binary logic, and the same idea could had been applied to NAL.

NARS does not follow that path because, once again, AIKR. Here this assumption means "to make judgment according to whatever knowledge is available," no matter where it comes from.

Let's take medical diagnosis as an example. Suppose a doctor wants to determine whether a patient P is suffering from disease D, that is, to evaluate statement "$\{P\} \to D$." For this task, at least two types of information can be taken into account: (1) whether P has D's symptoms S (that is, to derive the conclusion from "$D \to S$" and "$\{P\} \to S$" by abduction), and (2) whether D is a common illness among reference class C to which P belongs (that is, to derive the conclusion from "$C \to D$" and "$\{P\} \to C$" by deduction). With respect to the term D, the inference is intensional in (1), and extensional in (2). To get a summarized conclusion means to merge the two conclusions by the revision rule, and the result is neither pure extensional nor pure intensional for D. Such a merging also means that different types of evidence (extensional and intensional) can be balanced against each other, or be accumulated together.

After the truth value of "P is suffering from D" is evaluated, it can be combined with the truth value of "T is a proper treatment to D" (which is usually a statistic statement, too, therefore extensional) to get the truth value for "T should be applied to P." In such a situation (which is the usual case, rather than an exception), even if extensional and intensional evidence can be distinguished in the premises, they are mixed in the intermediate and final conclusions. If the system insists in separating extensional and intensional truth values, the above inference cannot be carried out.

Here we get to this conclusion once again, though along a different path: technically, it is possible to build a pure extensional or intensional logic, but when they are used with insufficient knowledge and resources, there are many situations where they cannot use the available knowledge like NAL does. Compared to NAL, they are not "wrong," but "weak."

If extensional and intensional evidence are collected in different ways, is it valid to merge them into a single truth value? It is valid, because though they are from different sources, the evidence contributes in the same way to the truth value. In the design of truth-value functions, no assumption is made on whether the evidence of the premises are extensional or intensional, so the system is consistent on this issue.

Sometimes we do hope to distinguish extensional and intensional factors in truth value for the purpose of explanation, such as in answering the question "Why do you believe that S is a special kind of P?." In this case, "Because S has the properties of P" and "Because P has the instances of S" are obviously different. However, we do not keep this information in the truth value of the belief "$S \rightarrow P$," because the truth value is used to summarize evidence, not to keep detailed information about evidence.

With insufficient knowledge and resources, the system makes no attempt to keep all the information about how a conclusion is obtained in the truth value of a statement. If we really need to separate the extensional factor and the intensional factor of "$S \rightarrow P$," it can be done by instead talking about "$(\#x \rightarrow S) \Rightarrow (\#x \rightarrow P)$" and "$(P \rightarrow \#x) \Rightarrow (S \rightarrow \#x)$," as mentioned previously.

7.3 Meaning of term

7.3.1 Meaning in NARS

The definition of meaning in EGS has the following implications:

1. The meaning of a term is its experienced relations with other terms.

2. The meaning of a term consists of its extension and intension.

3. Each time a term is used in an inference process, only part of its meaning is involved.

4. Meaning changes with time and context.

5. Meaning is subjective, but not arbitrary.

As said previously, a human observer can still interpret the terms appearing in NARS freely by identifying them with words in a natural language or human concepts, but that is their meaning *to the interpreter*, and has little to do with the system itself. For example, if the term "*bird*" never appears in the system's experience, it is meaningless to the system (though meaningful to English speakers). When "$bird \rightarrow animal < 1, 0.8 >$" appears in the system's input stream, the term "*bird*" begins to have meaning to the system, revealed by its inheritance relation with "*animal*." As the system knows more about "*bird*," its meaning becomes richer and more complicated. The term "*bird*" may never mean the same to NARS as to a human (because we cannot expect a computer system to have human experience), but we cannot say that "*bird*" is meaningless to the system for this (human chauvinistic) reason. This is just like that a child often uses a word in a different way, compared to its common usage. We can say that the usage is "different," or even "wrong," but we cannot say that the word is "meaningless" to the child. As long as a term has experienced relations with other terms, it becomes meaningful to the system, no matter how poor its meaning is.

An adaptive system should never processes a term only according to its shape without considering its position in the system's experience. The shape of a term may be more or less arbitrary, but its experienced relations with other terms are not.

This conclusion to an extent agrees with Wittgenstein's claim that the meaning of a word is its use in the language [Wittgenstein, 1999]. For NARS, the meaning of a term, such as "*game*," is not determined by a definition or a set of "things" in the world, but by how the term is related to the other terms according to the system's experience. As a result, there may be no common property shared by all instances of "*game*." Instead, there is only a "family resemblance" among them, indicated by the overlapping properties here or there (without a definitive property for all of them).[3]

7.3.2 Symbol grounding

By saying the above, I do not mean that in the human mind a word in a natural language gets its meaning *only* by its relation with other words in the language, because *human experience* is not limited to a language channel, but closely related to sensation, perception, and action [Barsalou, 1999, Harnad, 1990]. However, the general principle is still applicable here, that is, a word gets its meaning by its experienced relations with the system's other *experiential components*, which may be words, perceptive images, motor sequences, and so on. In a system like this, the meaning of a word is much more complex than in a system whose experience is limited to a language channel, but it does not rule out the latter case as a possible way for words (terms, symbols) to be meaningful. For example, a software agent can get all of its experience in this manner, and we cannot deny that it is genuine experience.

For a symbolic system built according to an EGS, the symbols in the system are already *grounded* — in the system's experience, of course. The crucial point here is that for a symbol to be meaningful (or grounded), it must be related somehow to the environment. However, such a relation is not necessarily via a sensorimotor mechanism. The experience of a system can be *symbolic*, as in the case of NARS. This type of experience is much simpler and "coarse-grained" than sensorimotor experience, but it *is* real experience, so it can ground the symbols which appear in it, just as words in natural language are grounded in human experience. In the future, when NARS can accept visual input, an image will be related to the concept of "Mona Lisa," so it does not merely mean "a painting by Leonardo da Vinci." This additional link changes the meaning of the concept, but it does not change the semantic principle of the system: the meaning of the concept is not completely determined by the "object in the world referred to by it," but by its experienced relations with other things in the system's experience.

---

[3]In this way, the semantics of NARS also implies a new theory of categorization, which will be discussed in Section 11.1.

The definition of meaning in EGS is similar to that of conceptual role semantics and semantic network [Harman, 1982, Kitchener, 1994, Quillian, 1968], where the meaning of a concept (or word) is defined by the role it plays in a conceptual system (or a natural language). The difference between EGS and these theories are:

- In NARS, the relations among terms are not definitional or conventional, but are *experienced* through the interaction between a system and its environment. Therefore, they are dynamic and subjective in nature.
- In NARS, the relations between a term and others are concretely specified by its extension and intension, consisting of inheritance judgments, whose meaning and properties are formally specified.
- In NARS, whenever a term is used, only part of its meaning is involved. In other words, the "current meaning" of a term is not exactly its "general meaning" in the long run.

Similar ideas are called "dictionary-go-round" by Harnad — he hopes that meaning of symbols can "be grounded in something other than just more meaningless symbols" [Harnad, 1990]. Here we should notice a subtle difference: in EGS, the meaning of a term is not *reduced into (or grounded on) the meaning of other terms* (that will indeed lead to circular definition in a finite language), but *defined by its relations with other terms*. These relations are formed during the interaction between a system and its environment, and are not arbitrary at all.

Another relevant factor is that in NARS, the copulas, term operators, and other syntactic markers are logical constants in Narsese. Their meaning is innate to the system, because they are directly recognized and processed by the inference rules and control mechanism in predetermined ways. Even when all the terms in an input statement are novel, the inheritance relation is known. Therefore, NARS is not "getting meaning out of the meaningless."

For these reasons, though NARS has a language defined by a formal grammar and used by a set of formally defined inference rules, it is not a "symbol system" discussed by Harnad, where symbols get their meanings "as standing for objects, as describing states of affairs" [Harnad, 1990].

7.3.3 Chinese room

This leads us to Searle's "Chinese room" argument [Searle, 1980], by which he claimed that a system using syntactic rules cannot have meaning.

Searle's argument is based on the assumption that a symbol can get meaning only from a model, by an interpretation. If one accepts the idea of an EGS, this is an untenable argument. As said above, as soon as a term has experienced relations with other terms, it becomes meaningful to the system, no matter how impoverished or diluted its meaning is. An adaptive system like NARS never processes a term solely on the basis of its shape, without considering its relations with other terms in the system's experience.

The feeling of meaninglessness in Searle's "Chinese room" comes from his deliberate cutting-off of his experience in Chinese from his sensorimotor experience and his experience represented in his native language. If we put an intelligent computer system into the same situation, there are two possible cases. If the computer system already had profound sensorimotor experience and/or a "native language," it might also consider the Chinese characters to be meaningless, because it could not relate them to its previous experience.

However, if the system entered the room with no previous experience, Chinese would become its "native language" — that is, the system would build up meanings for the characters on the basis of how they are related to one another, and would not attempt to ground them on some "more fundamental" stuff, nor would it complain about "meaningless squiggles and squoggles" when it failed in doing so. If the system also had sensorimotor capacities and communicated with other similar computer systems in Chinese, we might find that the meanings of Chinese words, to these systems, were as rich and as complex as they are to human Chinese speakers, though it is possible that they might occasionally have different opinions about the "correct" meaning of a given word.

Please note that my response to Searle's problem is not "the robot reply" he rejected in his paper [Searle, 1980]. I agree with him that even if a sensorimotor mechanism is introduced into the system, it does not directly bring meanings to symbols. According to EGS, a term is meaningful if it has experienced relations with other components of the system. Here, what matters is whether a symbol has (recognizable) relations in the system's experience, not whether the experience comes from a sensorimotor mechanism.

7.3.4 Subjectivity

As mentioned in Chapter 6, due to insufficient resources, the system cannot consult all known beliefs associated with a term each time the term is used. Instead, in NARS a priority distribution is maintained among the beliefs, which determines the chance for a certain belief to be taken into consideration at the current time. The distribution is adjusted by the system according to the feedback of each inference step (to make the more useful beliefs more accessible), as well as according to the current context (to make the more relevant beliefs more accessible).

Consequently, the meaning of a term becomes context-dependent — it does not only depend on what the system knows about the term, but also depends on the system's current tasks and how the relevant beliefs are ranked in terms of their priority. When the system gets new beliefs, or turns to another task, the meaning of the involved terms may change (more or less). Again, these changes are anything but arbitrary, and the meaning of some terms may remain relatively stable during a certain period. Without such a restriction, a "relational" theory of meaning cannot be practically used, because in a sufficiently complicated system, a concept may (in principle) be related to other concepts in infinite number of relations, and to take all of them into account is impossible.

Since the meaning of a term is determined by the system's experience, it is fundamentally subjective. However, as soon as the term is used in the communication with another system, the two systems begin to have common experience, and they begin to know how the term is used by the other. In the long run, the meaning of such terms gradually become "objective" in the sense that it reflects the common usage of the term within the language community, and less biased by the idiosyncratic usage of a single system.

Therefore, we can still understand what NARS means by a certain term and agree with a belief of the system, because of the partial overlap of its conceptual system with ours. However, we cannot expect its conceptual system to be identical to that of a human being, due to the fundamental difference between its experience and our experience.

Accurately speaking, no two people have identical conceptual systems (so misunderstanding and disagreements happen all the time), but we can still communicate, and understand each other to various extents on various topics, because we co-exist in the same world and the same human society, therefore have shared (physical and social) experience.

7.4 Truth of statement

7.4.1 Truth in NARS

As defined in Chapter 3, in NARS "truth" corresponds to statements with truth value $< 1, 1 >$, and non-analytic truth can only be approached, but not reached, by actual beliefs in the system. In general, in NARS truth is a matter of degree, represented by a $<f, c>$ pair.

The definition of truth value in the semantics of NARS has the following implications:

1. Truth is a matter of degree, and, in the case of first-order statements, is determined by the extent to which the subject term and the predicate term of the statement can be substituted by each other in certain ways.

2. A truth value consists of a pair of real numbers, one for the relative amount of positive evidence (with respect to negative evidence), and the other for the relative amount of all available evidence (with respect to future evidence).

3. A truth value is assigned to a statement according to the *past* experience of the system. It does not indicate whether the statement will be consistent with *future* experience, though an adaptive system behaves according to it.

4. For a statement, in the same time the system may has it in several beliefs, with different truth values, derived from different parts of the system's experience. Which one will be used at a given time depends on many factors, including the priority distribution

among the beliefs. As a result, the truth value of a statement seems to change from context to context.

5. Since truth comes from the experience of the system, it is subjective (i.e., different systems usually have different opinions on what is truth), but not arbitrary (i.e., such an opinion can be explained according to the experience of the system).

It is well known that when the evidential support of a scientific hypothesis is evaluated, we not only pay attention to the *amount* of (positive and negative) evidence, but also to its *diversity*, that is, we hope the evidence to be different in other aspects, while remain to be evidence for the hypothesis under evaluation. In NARS, the diversity of evidence is not directly measured in the amount of evidence (and therefore, in truth value), but works indirectly. When a hypothesis is supported by uniform evidence, the same evidence usually also support some other competing hypotheses, and therefore will eventually make the hypothesis unlikely to be chosen. On the contrary, diverse evidence means that the other factors show less regularity than the one captured in the hypothesis, therefore the competing hypothesis get less support, and the hypothesis under consideration is more likely to be chosen (even though its truth value may remain the same as the case of uniform evidence).

7.4.2 Truth value and degree of belief

EGS is similar to the "coherence" theories of truth [Haack, 1978], in that the truth value of a belief is partially determined by the truth values of other beliefs in the system, and the system will try its best to resolve inconsistency among its beliefs. However, in NARS some of the beliefs come from the experience of the system, so they are not necessarily consistent initially, and the system usually cannot achieve complete consistency among its beliefs, no matter it has tried. On the other hand, the system will not accept a set of consistent beliefs if it is not related to its experience.

On the contrary, MTS provides a "correspondence" theory of truth [Haack, 1978], where the truth value of a statement is determined by whether it corresponds to the state of affairs, as described in a metalanguage. According to MTS, "truth value" and "degree of belief" are fundamentally different — a system can strongly believe a false statement. This is from the viewpoint of an observer who knows the "objective truth" and can compare it with a system's belief. However, for the system itself, if it has insufficient knowledge and resource, the sole way to judge the truth of a statement is to consult experience. Here "experience" is used in the broad sense, not limited to personal perceptual experience only. In this situation, "truth value" and "degree of belief" are conceptually the same.

In everyday language, for a statement S, to say "S is true" is different from to say "I believe S," though their difference is not necessarily fundamental. To me, the former is like "S is not only believed by me, but also by everyone else (or that will be the case);" the latter is like "S sounds true to me, though may be not to the others." When we use the word "truth," we do imply certain *objectivity*, but it is more about "according to relatively complete and unbiased evidence" than about "as the world really is."

We can still say that in NARS "true" means "corresponds to reality," except that here reality is only revealed by the system's experience. When later we find a previous belief to be "false," it does not mean that we have had a chance to directly check the belief with reality (bypassing our experience), but that it conflicts with our updated belief based on more experience.

With such a semantics, we can still say "I strongly believe S, though it may be false," which means "I can imagine it to be rejected in the future." All of these differences cannot be used to argue that truth value cannot be the same as degree of belief.

Similarly, in NARS there is no fundamental difference among "hypothesis," "fact," "knowledge," "belief," and "guess." Instead, the difference is a matter of degree, and depends on usage convention of the words.

If we talk about such a system from an observer's point of view, then the situation is different. For example, if we have control over the experience of NARS, we may construct a situation in which the system strongly believes in a false statement. However, here "false" is from our point of view, and judged according to our knowledge about the system's future experience, which is not available to the system yet. Still, the general principle, that is, truth value is a function of experience, remains the same.

EGS challenges the traditional distinction between *ontology* and *epistemology*, that is, between *what is out there* and *what a system beliefs to be out there*. By accepting such a semantics, I do not reject the principle of naturalism — that is, the natural world exists independent of us, and it is the origin of all our knowledge [Kitchener, 1994]. What I stress here is that all *descriptions* of such an objective world in a system with insufficient knowledge and resources are intrinsically revisable. The interaction between the system and its environment is a process of assimilation and accommodation [Piaget, 1960], which usually does not maintain a one-to-one mapping between the terms/statements within the system and the objects/facts beyond the system. Actually, we cannot even talk about "objects" without assuming the cognitive capacity of some kind of system, which is what cuts reality into pieces. Since where to cut and how to cut depend on the nature and the experience of a system, there is no objective way to describe the world. What we call "ontology" is just a description of the world accepted by a community of observers, not a description of the world "as it is."

7.4.3 Validity of inference

A major motivation for the creation of EGS was to provide a justification for non-deductive inferences. As revealed by Hume's "induction problem," there is no sure way to get infallible predications about the future [Hume, 1748]. From limited past experience, we cannot get accurate descriptions of state of affairs, neither can we know how far our current belief is from such an objective description.

Based on this, Popper made the well-known conclusion that an inductive logic is impossible [Popper, 1959]. However, from the previous discussion, we can see that what is really pointed out by Hume and Popper is the impossibility of an inductive logic *with a MTS*. That is, inductive inference is invalid as far as validity is defined as "truth-preserving in all models."

If the conclusions derived in NARS are fallible, in what sense are they "better" than arbitrary guesses? This leads us to the concept of "rationality."

When infallible predictions cannot be obtained (due to insufficient knowledge and resources), beliefs based on past experience are better than arbitrary guesses, if the environment is *relatively stable* (as discussed in Section 2.1.2). To say a belief is only a summary of past experience (thus no future confirmation is guaranteed) does not make it equal to an arbitrary conclusion — it is what "adaptation" means.

Adaptation is the process in which a system changes its behaviors *as if* the future is similar to the past. It is a rational process, even though individual conclusions it produces are often wrong, and we know that the future cannot be identical to the past. For this reason, valid inference rules (deduction, induction, abduction, and so on) are the ones whose conclusions correctly (according to the semantics) summarize the evidence in the premises. They are "truth preserving" in this sense, not in the model-theoretic sense that they always generate conclusions which are immune from future revision.

7.4.4 Two types of truth

One important character of EGS is its dynamic and subjective nature. The truth value of a judgment may change from time to time in NARS, due to the arrival of new evidence. The system's inference activity also changes the truth values of judgments by combining evidence from different sections of the experience. Since truth values are based on the system's experience, they are intrinsically *subjective*. To be more precise, the system's beliefs are not objective descriptions of the world, but summaries of its own experience, so it is from the *system's point of view*. Even two systems in precisely the same environment may have different beliefs, obtained from their different individual experiences.

To say that truth values are dynamic and subjective does not mean that they are arbitrary. Different systems in the same environment can achieve a certain degree of "objectivity" by communicating to one another and thus sharing experience. However, here "objective" means "common" or "unbiased," not "observer independent." The common beliefs are still bounded by the experiences of the systems involved, though no longer by that of a single system.

The model-theoretic "truth" still has its place in NARS, though it plays a secondary role here. Whenever mathematical (or other conventional) statements are under consideration, their truth values are fixed, and are independent of the system's experience and the system's degrees of belief on them. We still do not know the truth value of Goldbach's Conjecture, though it has been confirmed in all the previous testing cases. Such a usage of the word "true" does not conflict with the fact that in the system this statement does have a truth value, calculated according to the system's experience.

From a philosophical point of view, this definition of truth is similar to Putnam's "rational acceptability" [Putnam, 1981]. In AI, a similar approach is discussed in Kowalski's paper "Logic without Model Theory," in which he defines "truth" as a relationship between sentences of the knowledge base and observational sentences [Kowalski, 1995]. However, the technical details of these approaches are quite different from NARS. For instance, Kowalski still uses predicate logic.

Chapter 8

Uncertainty

To reason with insufficient knowledge and resources, NARS needs to deal with various types of uncertainty. As described previously, a major component of the design is a mechanism for the representation and processing of uncertainty. In this chapter, I compare the NARS approach to the other approaches of uncertainty representation and processing, and discuss its special features.

8.1 The non-numerical approaches

The existing approaches for uncertainty management can be divided into two types: numerical and non-numerical. The former attaches one or several *numbers* to each statement to represent the degree of uncertainty, while the latter represents uncertainty qualitatively (not quantitatively) [Bhatnagar and Kanal, 1986, Bonissone and Decker, 1986].

As we have seen, NARS uses numerical values to represent the uncertainty of a statement. Though it seems to be a natural idea, there have been strong objections proposed in AI research against using numbers to express "degree of uncertainty." One major argument is the observation that in daily communications in natural languages, people rarely use numbers to express uncertainty. A numerical approach may have difficulty in interpreting what the number measures, as well as in actually collecting this kind of data [McCarthy and Hayes, 1969, Sullivan and Cohen, 1985, Wallsten et al., 1993]. Based on such an opinion, many people prefer a non-numerical approach to represent uncertainty in a reasoning system. In the following, NARS is compared with two non-numerical approaches.

8.1.1 Endorsement theory

Endorsement Theory uses verbal labels to represent "reasons to believe or disbelieve uncertain propositions" [Sullivan and Cohen, 1985].

Obviously, this representation is more natural because it is closer to how uncertainty is represented in our daily communication. Furthermore, verbal labels not only can make quantitative differences, but also can make qualitative differences, to indicate *why* to believe, as well as *how much* to believe, therefore more information is preserved, compared with numerical approaches.

This approach works for certain purposes, but cannot be used in a general-purpose reasoning system like NARS. In such a system, this approach encounters a dilemma. If the endorsements are interpreted as different degrees along the same semantic dimension, what we get is a "coarse numerical scale," which has finite different values to take. Beside its naturalness (because verbal labels are used), such an approach has few advantage over numerical approaches [Wallsten et al., 1993]. On the other hand, if the endorsements are interpreted as along different semantic dimensions (as in [Sullivan and Cohen, 1985]), there must be labels that cannot be compared or combined. In the situations where our main purpose is to *record* uncertainty, such an approach may be appropriate, but it is inapplicable for a reasoning system, where the system *must* set up a common representation for uncertainty from different sources, so that to carry out the operations on them.

8.1.2 Non-monotonic logics

There are several formal systems within the category of *non-monotonic logics*. Though built differently, they share some opinions about human common-sense reasoning: with incomplete knowledge, some *convention* or *default rules* can (and should) be used to get tentative conclusions, which can be rejected by later acquired facts [Reiter, 1987].

Though no numerical or verbal value is attached to statements, in a reasoning system using a non-monotonic logic, statements are (explicitly or implicitly) divided into three groups.

1. facts, such as "Tweety is a bird."

2. defaults, such as "Birds fly."

3. tentative conclusions, such as "Tweety flies."

When a tentative conclusion conflicts with a fact, the former is rejected.

Though non-monotonic logics can be used for various purposes, they cannot be used in a system like NARS, for several reasons.

- When two tentative conclusions conflict with each other, the system has to rely on domain-specific preference ranking to make the selection. For a system working with insufficient knowledge, it cannot be assumed that this kind of information is always available. This is the well-known "multiple extensions" problem.

- No matter what happens, in non-monotonic logic no fact or default will be revised or modified, and this is inconsistent with the assumption of insufficient knowledge. For example, given enough counter examples, we will expect "Birds fly" to become "Birds do not fly," but it will never happen in a common non-monotonic logic.

- In practical applications, there is not always an effective way to distinguish facts and defaults from tentative conclusions.

Reiter wrote that "Nonmonotonic reasoning is intimately connected to the notion of prototypes in psychology and natural kinds in philosophy" [Reiter, 1987]. However, for an open system, the notion of prototypes and natural kinds have to be defined and maintained according to quantitative information. Concretely, in human cognition, prototype or default are based on the "central tendency" [Rosch, 1973], but with constantly coming evidence, whether a tendency is "central" is usually a matter of degree. Another property of an open system is that every piece of knowledge is revisable, with different sensitivity or stability, which is also a matter of degree. It is possible to indicate these degrees by verbal labels, but such an approach will be less general and less efficient than a numerical one.

To treat default rules as convention is possible and even desired in many situations. In communication between systems (human or computer), these conventions are often intentionally followed, and if we already have lots of evidence, or if we only study a reasoning system for a short period, the influence of new evidence can be ignored. In these situations, a binary logic is preferred for its simplicity and clarity. However, such assumptions about environment are not always valid.

Another thing we need to keep in mind is: when treated as *conventions*, these rules become *a priori* to the system, and are fundamentally different from when they are treated as *generalized experience*. As conventions, their generation, acceptance, comparison, modification, and rejection are no longer determined, or even influenced, by the experience of the system. Though it is correct to say that "normality" or "typicality" should not be interpreted in a pure frequentist way as "in most cases," we still have reason to argue that for many purposes, it is better to see them as closely related to empirical evidences, and have different degrees [Kahneman and Miller, 1986, Rosch, 1973]. Therefore, it makes sense, and is often necessary, to *measure* the relations between a default and available evidence, which cannot be done in the framework of binary logic.

NARS has some similarity to non-monotonic logics: it can make guesses when the knowledge is insufficient, and the guesses are based on the "typical" or "normal" situations, according to the system's knowledge. When such guesses conflict with new evidence, they can be modified, or even rejected. As a numerical approach, NARS have more types of operations on uncertainty than non-monotonic logics, for example, NARS can generate hypotheses from evidences by induction and abduction, and all of its domain knowledge is revisable. For these reasons, NARS is not a non-monotonic logic, though the size of its "believed statements" (i.e., those with expectation values above a certain threshold) does change non-monotonically in time.

8.1.3 The necessity of a numerical measurement

Therefore, for a general-purpose system like NARS that is open to all possible new evidences, a numerical measurement of uncertainty is *necessary* for the *internal representation* of the system, not because it is accurate, but it is uniform and simple, especially because the system needs to have rules for induction, abduction, comparison, and so on. In these types of inference, uncertainty emerges even if all the premises are certain, and the amount of evidence is a dominant factor for the processing of the uncertainty. For this reason, even if verbal labels of uncertainty are used in the interface language for the sake of simplicity and naturalness, it is still desired to represent uncertainty numerically within the system.

However, the advocators of non-numerical approaches are correct in their claim that a numerical measure is *insufficient* for uncertainty management in the system, since some operations are sensitive to the source of uncertainty. To solve such problems, in NARS some other methods are used as *supplements*, rather than *replacements*, of the numerical approach. For example, a serial number mechanism is used in NARS (described in Section 3.3) to detect correlated evidence. Also, higher-level statements can be used to explicitly describe properties and relations of a particular statement (see Section 5.1).

Though a numerical approach is used in NARS for the internal representation of uncertainty, the system does not insist in using numbers in the input and output sentences. Rather, in the implemented system, it is allowed for the truth values to be omitted (and filled with default values). In the future, when a natural language interface is added to NARS, the system will use verbal expressions of uncertainty when high accuracy is unavailable or undesired. Such an idea has been briefly explained when discussing the "frequency interval" (Section 3.2).

Another important issue raised by the advocators of non-numerical approaches is that the existing numerical approaches often suffer from a lack of proper interpretation of the measurement of uncertainty they used. If the meaning of a "degree of uncertainty" is unclear, there will be little guidance in how the data can be collected or how it should be processed. Consequently, it will be hard to justify the validity of the approach. To solve this problem, in the semantics of Narsese (Section 3.2), I first define the concept of evidence (in terms of idealized experience), then introduce three equivalent forms (amount of evidence, frequency and confidence, and frequency interval) of the uncertainty of a statement. In this way, it is easier for a user to understand what the values mean, as well as for the designer to choose a proper truth-value function for each inference rule.

In summary, for NARS a numerical measurement is necessary for internal uncertainty representation, though it is not sufficient for all uncertainty related operations, not necessary for external communication, and not "natural" or "self-evident" in the sense that no interpretation is required.

8.2 The fuzzy approach

In this section, NAL is compared with Zadeh's fuzzy logic.

8.2.1 Grade of membership

The idea of "fuzzy set" was proposed by Zadeh to capture the phenomenon that whether an entity is a member of a category is often a matter of degree [Zadeh, 1965]. Though a lot of works have been done in this field, the theory is still suffering from an innate shortage: the grade of membership, or "fuzziness," has not been properly interpreted by the theory.

Here are Zadeh's opinions:

1. Fuzziness comes from the description of complex systems. He proposed the "Principle of Incompatibility," which states that as the complexity of a system increases, our ability to make precise and yet significant statements about its behavior diminishes [Zadeh, 1973].

2. A membership function usually maps a continuous numerical variable to a discrete linguistic variable, so that the information can be summarized approximately. For example, "John is young" is an approximative way to say "John is 28." Since the underlying numerical variable changes continuously, there is no well-justified

way to cut the boundary between the values of the linguistic variable. But, by introducing a membership function, we can describe the *compatibility* between a linguistic label and a numerical value [Zadeh, 1975].

3. Such a compatibility has no frequency interpretation, so it cannot be processed according to probability theory. Thus, stating that "The membership of John's age to *young* is 0.7," we do not mean that John's age is a random number, which takes the value of *young* 70% of the time, but that 0.7 is the degree to which *young* and 28 are compatible [Zadeh, 1978].

4. Membership functions of primary terms are subjective and context-dependent, so there is no general method to determine them. Their specification is a matter of definition, rather than objective experimentation or analysis. The task of fuzzy logic is to provide rules to compute the meaning of composite terms, once the meaning of the primary terms is specified in a given context [Zadeh, 1972, Zadeh, 1979].

As a result, various methods, chosen according to the designers' preference and experience, are used to get membership functions when fuzzy logic is applied to practical domains [Dubois and Prade, 1980, Turksen, 1991]. Some people are satisfied with this outcome, and even argue that it can increase the flexibility of fuzzy logic. However, it is easy to see that there are many negative consequences.

Without a clear interpretation, it is hard for a person to assign such a value or to understand a value, or for a computer system to generate the memberships automatically or to get them from sensory devices. By "hard," I mean that although some values can be easily assigned, they look quite arbitrary and unjustified. In such a case, when are the system's results, which are determined by these initial assignments, better than random choices?

It is well-known that memberships are context dependent, and may be influenced by new knowledge [Barsalou, 1987]. For example, "If 'Mary is young' is uttered in a kindergarten or in a retirement home situation, the effect on the expected age of Mary will be very different" [Cheeseman, 1986]. However, without a clear interpretation, there is no reasonable way to modify the memberships by new knowledge, so as to make them context-sensitive. On the other hand, it is unimaginable for the designer to provide a system with a membership function for every concept (for instance, *young*) in every possible context (kindergarten, elementary school, ..., retirement home, even basketball team or cabinet) that the system may meet.

The *max* and *min* operators (for set union and intersection, respectively), which are the most distinguished components of fuzzy theory, are not strongly supported by experimental evidence or theoretical consideration. They sometimes lead to counter-intuitive results [Oden, 1977b, Osherson and Smith, 1981, Smith and Osherson, 1984]. Though there are some works which show that the operations can be deduced from certain axioms [Bellman and Giertz, 1973, Gaines, 1978], it is still unclear if human cognition really follows these axioms or why we should follow them. Zadeh admitted that in some contexts the union/intersection operators should be *sum/product* rather than *max/min* [Zadeh, 1975], but he did not indicate *how* to determine which pair should be used when facing a new context. If what is measured by a membership function is unclear, how can we argue that some operations on it are better than others?

In summary then, fuzzy logic is not proposed as a pure formal system that only has some interesting mathematical properties, but as a formal model of fuzziness management that happens in human cognition, and as a tool of fuzziness management for practical purposes. Why should we accept such a claim? The popular arguments are: (1) there is fuzziness in human cognition, (2) no frequency interpretation of the fuzziness has been found, so probability theory cannot be applied, and (3) some practical problems have been solved successfully by fuzzy logic [McNeill and Freiberger, 1993, Zadeh, 1986a]. Without a clear analysis of fuzziness, these arguments are not enough for fuzzy logic to be accepted as a general cognitive model [Smets, 1991].

As described previously, in NARS the membership relationship is represented by inheritance (and its variants), and an inheritance statement is true to a degree (represented by the frequency–confidence pair). In the following, I will argue that this solution is better than fuzzy logic.

Some people may argue that in this way, we will lose one of the advantages of fuzzy logic, that is, the freedom for the system designer to determine the membership functions and operators. I disagree. On one hand, the lack of an interpretation is a disadvantage for a theory, since it provides less guidance for its users. On the other hand, with a detailed interpretation, a system (like NARS) can still be flexible. What the interpretation does is not to provide for each fuzzy set an "objective" membership function, but to indicate how a membership function can be established and modified by the system according to available evidence. In this sense, the interpretation takes certain kinds of "freedom" from a human designer, and gives them to the system itself.

8.2.2 Fuzziness from similarity

First, let us distinguish between two types of fuzziness: that which happens mainly with nouns and verbs, and that which happens mainly with adjectives and adverbs. In the following, these two classes are called "Type 1" and "Type 2" Fuzziness, respectively.

Fuzziness of Type 1 happens in concepts such as "animal," "furniture," "to play," "to cause," and so on. Psychologists have shown that people judge some instances to be better examples of a concept than some others, and can intuitively assign a numerical "degree" to a membership relation [Oden, 1977a, Rosch, 1973, Rosch and Mervis, 1975].

Several theories have been proposed by psychologists to explain the phenomenon. One explanation, *prototype theory*, suggests that from given members of a category, people abstract out the central tendency or *prototype* that becomes the summary mental representation for the category. Then, membership of a novel instance is measured by how similar it is to the prototype [Rosch, 1973, Rosch and Mervis, 1975]. Another explanation, *exemplar theory*, assumes that membership of a novel instance is evaluated by directly comparing it with given members of the category [Medin and Schaffer, 1978, Nosofsky, 1991].

Generally speaking, the basic cause of Type 1 Fuzziness is that the concept is not defined by sufficient/necessary conditions, but is exemplified by many objects/actions/events which share common properties.

These results are often quoted as evidence in favor of fuzzy logic (for example, [McNeill and Freiberger, 1993]). Speaking precisely, however, they only support the existence of fuzziness, rather than Zadeh's explanation and suggested operations on it. To psychologists, fuzziness, or grade of membership, is not a primary attribute of a concept that cannot be further analyzed. Rather, it is usually treated as a result determined by some other (more primary) factors, following certain rules [Dubois and Prade, 1980, Medin and Schaffer, 1978, Rosch, 1973].

More concretely, there is a consensus that grade of membership is strongly influenced by the degree of *similarity* between an instance to be judged and a prototype or an exemplar of the concept, so in the simplest cases, the membership measurement is reduced to a similarity measurement [Nosofsky, 1991, Tversky, 1977].

In psychological literatures, two kinds of similarities are distinguished: those that are asymmetric and those that are symmetric [Tversky, 1977]. They roughly correspond to the "inheritance relation" and "similarity relation" in NARS, respectively. As defined in NAL-1 and NAL-2, in both cases the frequency of a statement is measured as the proposition of positive evidence among all evidence, so is a special case of the *ratio model* of similarity [Tversky, 1977]. As a result, this kind of grade of membership is a special case of the truth value in NARS.

8.2.3 Fuzziness from relativity

Now we turn to "Type 2 Fuzziness." What makes it different from the previous type is this: though a concept of the Type 1 can be treated as a fuzzy set with a relatively stable membership function, the same is not true for a concept of the Type 2 — the fuzziness caused by an adjective or adverb usually depends on the described noun or verb.

For example, if we treat "*big*" as a fuzzy set, just like "*flea*" and "*animal*," then "*big flea*" can be represented as "$(big \cap flea)$." According to this interpretation, from "S is a big flea" and "Fleas are animals," "S is a big animal" would be derived, which is counter-intuitive. In other words, many adjectives are not "predicative" [Kamp, 1975].

This problem is usually explained by saying "the membership function of *big* (as well as *tall*, *young*, *far*, etc.) is context dependent." Of course it is, but *why* and *how*?

This phenomenon is not new at all to linguistics. As early as 1944, Sapir proposed the idea that although *wide*, *young*, and *big* linguistically precede *wider*, *younger*, and *bigger*, respectively, the comparative forms precede the simple forms *logically*. The simple forms are *implicitly graded antonyms*. They can only be understood in terms of the comparative forms, with respect to some *norm* set by the object being described [Sapir, 1944]. This opinion has been accepted by many other linguistics [Cruse, 1986, Palmer, 1981]. An interpretation of fuzziness follows naturally from this clarification about the meaning of "fuzzy adjectives and adverbs" (such as *big*, *young*, and *soon*).

Let us analyze and compare the following sentences:

1. "S is big."

2. "S is bigger than T."

3. "S is a big flea."

4. "S is a big animal."

According to the above semantic theory, if there is no default or assumption about the context, "S is big" provides no information about its size.

"S is bigger than T" does provide information about the size of S, but in a relative way. The "bigger than" relation may become uncertain, due to incomplete information or imprecise measurement, but usually there is no fuzziness, since "bigger than" is a well-defined binary relation, at least in principle.

In "S is a big flea" (or "S is a big animal"), what is the norm to which S is compared? Somebody may suggest that it is a "normal," "typical," or "average-sized" flea (or animal). However, if this is the case, then these sentences can be reduced to "S is bigger than T," and no fuzziness is present.

In my opinion, fuzziness, or graded membership, appears when the norm in the above sentences are the *class* of other fleas and other animals. When a well-defined binary relation between *two objects* is used between an object and a *class of objects*, uncertainty emerges because the relation is no longer well-defined, and the result is only "true to a certain extent," due to the internal diversity of the norm, the class.

Generally speaking, the Fuzziness of Type 2 appears in sentences with the pattern "S is a RC," where C is a class, S is a member of C, and R is an adjective (or adverb). The comparative form of R is "R-er than," indicated by R_t, is a binary relation on C, and is asymmetrical, transitive, and non-fuzzy. In such a case, "RC" is a fuzzy concept (such as "big flea," "young men," and so on), and its membership is a matter of degree, since when a member is compared with a class, the relation usually holds with some members of the class, but fails with others. The extent to which the relation holds in general depends on the member's relative ranking in the class, with respect to the relation.

Under this interpretation, in Narsese "S is a RC" is can be represented by statement "$(\{S\} \times C) \rightarrow R_t$." If we focus on the extensions of the terms, the frequency of this statement can be taken as the degree of membership for S to be in RC, which is

$$f_{RC}(S) = \frac{|(\{S\} \times C) \cap R_t|}{|C - \{S\}|}$$

In the case of "big flea," the degree of membership is the ratio

$$\frac{\text{the number of fleas that are smaller than } S}{\text{the number of fleas minus 1}}$$

The "minus 1" is there, because it is not necessary to compare S to itself.

Now $f_{RC}(S) = 1$ means that S is the biggest flea; $f_{RC}(S) = 0$ means that S is the smallest flea. Actually, this function identifies $f_{RC}(S)$ with the *percentage* of fleas that are smaller than S in class C.

If the probability distribution function for the size of fleas is given as $P(x)$, we can get a direct relation between the size of a flea y, $m(y)$, and its grade of membership to *big flea*, $f_{RC}(y)$:

$$f_{RC}(y) = \int_0^{m(y)} P(x)dx$$

This equation can be generalized to all fuzzy concepts of the Type 2 by considering $m(y) : C \rightarrow (-\infty, \infty)$ as a measurement corresponding to the relation "R-er than," and $P(x) : (-\infty, \infty) \rightarrow [0, 1]$ as the probability distribution of objects in C with respect to $m(y)$.

In this way, we get a function that calculates the membership of an object from a *fundamental argument*, as Zadeh did. However, there is a basic difference. According to Zadeh, "The label *young* may be regarded as a *linguistic value* of the variable *age*, with the understanding that it plays the same role as the numerical value 25 but is less precise and hence less informative" [Zadeh, 1975]. But in NARS, "*young*" is interpreted as an approximate way to tell someone's relative youthfulness, with respect to a reference class. Only with a corresponding probability distribution, can the relative measurement be mapped to the absolute measurement.

If "John is tall" is an approximate way to tell John's height, then it follows that this type of sentence is always less informative than a sentence like "John is 6 feet high." However, this is not always the case. For example, the sentence "In basketball, tall players usually have an advantage" cannot be rewritten by replacing "tall" by an accurate height, without losing its generality. The sentence makes the same sense in many contexts (from elementary school to college), where how "high" being mapped to height is drastically different.

To say "S is a big flea," what one needs to know is not the size of S, but how it compares with the other fleas. If S is the only known flea, we cannot say if it is a "big flea," even when we know its size exactly. On the contrary, if we always observe fleas through a magnifying glass whose magnifying power is unknown, then we may have little idea about the actual size of S, but "S is a big flea" still makes sense. Actually, the sentence makes the *same* sense, no matter how the sizes of fleas are distributed.

In communication, the context is often omitted in sentences. As a result, we only say "John is tall" or "S is big." Such omissions may cause understanding problems. If the default reference class of the speaker and that of the listener are different, misunderstandings will happen; if the listener is not sure of the speaker's intended reference class, a guess has to be made, perhaps according to the frequency of various possible contexts. Even when the reference class is presented explicitly in the sentence, as in "S is a big flea," it is still possible for the speaker and the listener to make different estimations about the size of S, because from personal experience they may have different objects in mind when "flea" is mentioned (so here it is also explained that why there is an interpersonal difference in membership judgment). The above factors cause *ambiguity*, which is closely related to fuzziness, but it should not be confused with fuzziness, because communication is not a precondition for fuzziness to appear.

As analyzed above, although membership functions in a system are subjective because they are related to idiographic experience of the system, fuzziness is not caused by interpersonal difference. "John is a young man" is a matter of degree, it is not because each person uses the word "young" differently (though that is true), but because John is younger than some men, but older than the others. Therefore, the interpretation of fuzziness proposed here is different from the opinion that grade of membership can be determined by polls [Dubois and Prade, 1980, Turksen, 1991]. According to this opinion, to say "The membership of John's age to *young* is 0.7" means that "70% of people will agree that John is young." It is assumed that each person has a determined standard for "young," and the concept is fuzzy because we accept different standards. However, to force people to answer "yes" or "no" to the question "Is John young?" contradicts to the very idea that "young" is a fuzzy concept, and it is not explained where the difference among people comes from. As discussed before, this opinion confuses the uncertainty caused by interpersonal difference (ambiguity) and the uncertainty caused by diversity in a reference class (fuzziness), so it fails to capture the essence of fuzziness that makes it different from other types of uncertainty.

8.2.4 A practical application

The above result has been used in the design of a "recommendation system," which returns a "top N list" from a database according to the preference of a user.

Assume that a user is accessing a flight reservation system, and is looking for *cheap* ticket for any flight that, in a given date, leaves city C_1 *around* 9 AM and arrive city C_2 *as early as possible.*

If the system is based on a relational database, the user typically has to get rid of the fuzziness in the request, and to force it into a query where each of the attributes of the returned entries must fall into an interval. For example, "cheap" becomes "below $200," "around 9 AM" becomes "between 8:30 and 9:30 AM," and "as early as possible" becomes "before noon."

The problem in this solution is that unless the user is familiar with the value distribution of the attributes, very often the number of items returned by the query is either too large or too small. This is the case because for a binary query, an entry in the database will either satisfy it or fail to satisfy it, and beside that, no additional information can be provided.

To solve this problem, we need conditions that different values satisfy to *different degrees*. Under this consideration, one possible solution is to apply fuzzy logic into database query, which leads to the idea of "fuzzy database" [Yang et al., 2001]. In such a solution, the user's preferences are specified using "linguistic variables." Each linguistic variable corresponds to a fuzzy set, with a membership function to calculate the score for each attribute value. The total score of a product is the minimum of all the individual scores, because in fuzzy logic the (default) function for conjunction of conditions is "*min*." Finally, a "top-N list" is reported to the user, based on the total scores of the candidates.

In this way, we can indeed get a recommendation system that no longer suffers from the above problem found in conventional database. Since the idea of fuzzy logic has been well known for many years, and this application is not that difficult, why have not we seen many such systems?

Among all reasons, a major issue is the design and maintenance of the membership functions. Fuzzy database is a good solution only when the membership functions can be designed once, and do not need further adjustment or revision. Unfortunately, for many problems this assumption is not true.

According to the previous discussion on the interpretation of fuzziness, we see that NARS can provide a more general and flexible solution for this problem. Actually, here we do not need the full power of NARS. Rather, a special program can be designed, according to the relevant part of NARS, as the following.

The database DB is a collection of data items, each of which is a vector $d_i = < d_{i1}, d_{i2}, \cdots, d_{it} >$, in which each d_{ij} is the value of d_i on an attribute A_j. In other words, DB is a matrix, where each row is a data item, and each column corresponds to an attribute.

A recommendation request r consists of two components, a constraint vector $c = < c_1, c_2, \cdots, c_t >$ and a preference vector $p = <p_1, p_2, \cdots, p_t>$. Each c_j has the form "$relation_j \ v_j$," where v_j is a constant value, and $relation_j$ is one of the following relations: $=$, $\neq$, $<$, $\leq$, $>$, $\geq$. The evaluation of c_j against a value d_{ij} should return 1 (*true*) or 0 (*false*). Each p_j has the form '$\approx v_j$" (for "as close to v_j as possible"), "$\ll$" (for "as small as possible"), or "$\gg$" (for "as large as possible"). The evaluation of p_j against a value d_{ij} should return a real number in [0, 1].

Unlike in fuzzy logic, where degree of membership is a subjective judgment that cannot be further analyzed, in my approach the "score" of each value for a given preference is the *proportion of positive evidence among all evidence*, that is, $s = w^+/(w^+ + w^-)$, where w^+ and w^- are the amount of *positive* and *negative* evidence, respectively, for the preference to be satisfied by the value.

How is evidence defined and measured? Let us start from a concrete example. If the price of a notebook computer is $1250, then to what extent does it belongs to the concept of "cheap notebook computers"? According to my previous interpretation, such a question cannot be answered without a "reference class," that is, it depends on the answer of another question: "Compared to what?." Adjectives like "cheap" get their meaning from relations like "cheaper than," though the object of the comparison is often omitted in the expression. In the recommendation process, it is assumed that the default objects of comparison are the other candidates in the database that satisfy the constraint vector. Therefore, "cheap" is interpreted here as "cheaper than the other candidates." Usually there are multiple candidates, and some of them may be cheaper, while others more expensive, than the product under consideration. Therefore, whether one candidate is "cheaper than the other candidates" is usually a matter of degree.

If there are M candidates that satisfy the constraints, then they are used to score one another for the preferences. To decide the score for a $1250 computer to be labeled as "cheap," the other $M - 1$ candidates are compared to it one by one in price, where more expensive ones are counted as positive evidence, and cheaper ones as negative evidence, for the labeling (candidates with the same price provide no evidence). The total amount of evidence is the sum of the amount of positive evidence and the amount of negative evidence. Therefore, among the M candidates if there are m_1 of them are more expensive than \$1250, and m_2 of them cheaper than \$1250, then the score for a \$1250 computer to be labeled as "cheap" can be simply taken as $m_1/(m_1+m_2)$. Especially, the cheapest candidate gets a score 1.0, and the most expensive one gets 0.0, for the given preference.

The above approach can be applied to a preference as far as the values of the corresponding attribute form a total order, even if the values are not numerical. The following is a general definition of evidence in the recommendation process:

- When a value d_{ij} is evaluated according to a preference p_j of the form "$\approx v$" ("similar to v"), if another value d_{kj} (of another candidate) is farther away from v than d_{ij} is, it is positive evidence; if d_{kj} is closer to v than x is, it is negative evidence.

- When a value d_{ij} is evaluated according to a preference p_j of the form "$\ll$" ("has a small value"), if another value d_{kj} (of another candidate) is larger than d_{ij}, it is positive evidence; if d_{kj} is smaller than d_{ij}, it is negative evidence.

- When a value d_{ij} is evaluated according to a preference p_j of the form "$\gg$" ("has a large value"), if another value d_{kj} (of another candidate) is smaller than d_{ij}, it is positive evidence; if d_{kj} is larger than d_{ij}, it is negative evidence.

After separating positive and negative evidence from non-evidence among the other candidates, their number can be used as the above m_1 and m_2, respectively, then from them the score of the given value can be calculated according to the previous formula.

For a given attribute, if its values are not only comparable, but also numerical, sometimes the distance between values should be taken into account when scores are calculated. For example, to evaluate the score for \$1250 to be labeled as "cheap," the existence of a \$750 and a \$1200, as the prices of other candidates, are very different. Though both are negative evidence, the former is clearly a much "stronger" one than the latter. For this situation, a more meaningful way to calculate the amount of evidence is to give each piece of evidence a "weight," which is the difference of that value and the given value. For the above case, the weights of the two pieces of evidence are $1250 - 750 = 500$ and $1250 - 1200 = 50$, respectively. Now the amount of (positive and negative) evidence m_1 and m_2 are weighted sum of pieces of evidence.

For a product d_i, after its attribute values get their scores as a vector $<s_{i1}, s_{i2}, \cdots, s_{it}>$ according to a given preference $p = <p_1, p_2, \cdots, p_t>$, the next step is to combine them into a total score s_i, which represents the membership for the product to be an instance of the concept C_p. Here each preference p_i is treated as an independent channel to collect (positive and negative) evidence for the membership relation. Therefore, evidence collected in each channel should be pooled together. As a default rule (assume all the preferences are equally weighted and each score is obtained from the same number of comparisons), the total score is simply the average of the individual scores, that is, $s_i = (\sum_{j=1}^{t} s_{ij})/t$.

Now the top N data items can be selected according to their total scores, and return them as the recommendation to the user. As options, the total score of each may be displayed, and they may be sorted according to their total scores.

As of today, most shopping websites still use conventional database query to carry out the selection process. As a result, non-expert users have to spend lots of time in fine tuning their query to get a desired number of candidates for the final comparison and selection. Compared to that process, the recommendation procedure introduced above has the following advantages:

- Both (binary) constraints and (fuzzy) preferences are allowed, where the former is expressed in absolute terms, while the latter in relative terms. This is a more natural way to set selection criteria for most users, especially for users who are not experts in the field, and what really matters is the relative value, not the absolute value, of a data item on an attribute.

- Trade-offs and compromises among multiple preferences are allowed and supported. Actually, all difficult selection problems happen in the cases where different criteria have to be balanced against each other. Using the concept of evidence (for membership relation), the above algorithm maps values of different attributes (with different units) into a common (unit-free) dimension.

- By presenting a top-N list to the user, the selection process is simplified without losing promising candidates. Still, the user can consider other factors (that are not in the recommendation request) in making the final decision.

Compared to the similar solution based on fuzzy logic, this recommendation procedure has the following advantages:

- The degree of membership for an instance to belong to a concept is no longer a subjective opinion, but a measurement justified according to a theory on cognition and intelligence.

- The scores are determined by the available data according to a domain-independent algorithm. Consequently, there is no need to manually design and maintain the membership functions. Instead, such functions are automatically learned from the data, and so are adaptive to the changes in data.

Because of the adaptive nature of the membership function, the system treats "cheap computer" and "cheap pencil" with different standards, simply because the available instances in these two categories have different price distributions. When a new product with the lowest price is added into the system, all other products in the same category automatically become "more expensive," as the result of comparing to it. This is closer to how the human mind works in these situations. Also, this "maintenance-free" solution is easier to be used in practical situations.

8.3 The Bayesian approach

There is an obvious relation between the truth value defined in NAL and probability. The frequency of a statement is defined as $f = w^+/w$, which intuitively looks like a probability value. On one hand, probability theory is a solid mathematical theory which is often used to measure the "degree of belief," and the Bayesian approach is dominating the research of reasoning under uncertainty; on the other hand, the design of truth-value function in NARS is not derived from any theory, and in several places the decisions look arbitrary. If this is the case, why bother to develop a new calculus rather than use the Bayesian approach for uncertainty processing in NARS?

8.3.1 An analysis of the Bayesian approach

Bayesianism comes in different shapes. Since this book is about AI, what I call "the Bayesian approach" is the most influential form of Bayesianism in AI, applied in reasoning systems. Here I will not attempt to evaluate all kinds of Bayesianism, or all interpretations and applications of Bayes Theorem.

In recent years, Bayesian networks have achieved great success in AI. It has been not only applied to various problems, but also taken by more and more people as a normative theory of reasoning, both for the human mind, and for AI systems [Cheeseman, 1985, Pearl, 1988, Spiegelhalter, 1989].

Though the Bayesian approach is indeed a powerful tool for many theoretical and practical problems, its limitation is often seriously underestimated, due to a conceptual and notational confusion. The problem was first addressed in [Wang, 1993a], but it has got little attention, and the confusion continues to spread.

According to [Pearl, 1990], traditional Bayesianism is defined by the following attributes:

- willingness to accept subjective belief as an expedient substitute for raw data,
- reliance on complete (i.e., coherent) probabilistic methods of belief,
- adherence to Bayes' conditionalization as the primary mechanism for updating belief in light of new information.

When probability theory is applied into a reasoning system, it usually starts by assuming a proposition space S, which contains all the propositions that the system can represent and process.[1] S is often gen-

[1]Of course, probability theory, including Bayes Theorem, is not limited to propositions. Very often, the subjects of discussion are random variables or events. However, since the current discussion is about the application of probability theory to a reasoning system, propositions will be taken as the subject matter.

erated from a set of atomic propositions, using logical operators "*not*" ($\neg$), "*and*" ($\wedge$), and "*or*" ($\vee$). A probability distribution P is defined on S, and for every proposition $h \in S$, its probability evaluation $P(h)$ is a real number in [0, 1], and satisfies the following axioms of probability theory [Kolmogorov, 1950]:

- $P(h \vee \neg h) = 1$.
- $P(h \vee h') = P(h) + P(h')$, if $h' \in S$ and $h \wedge h'$ is false.

For any h and e in S, the probability of h under the condition that e is true is a conditional probability evaluation $P(h|e) = P(h \wedge e)/P(e)$. From it we get the Bayes Theorem

$$P(h|e) = \frac{P(e|h)P(h)}{P(e)} \tag{8.1}$$

Though the above mathematical definitions and results are acknowledged by all people using probability theory for reasoning, the Bayesian approach interprets them differently. According to Bayesianism (as defined above), the probability of a proposition h in a system is the system's degree of belief on h, according to certain background knowledge k (or call it experience, data, evidence, and so on).

The system starts with a *prior* probability distribution P_0, determined by background knowledge k_0 at time t_0. At time t_1, when a piece of new knowledge e is collected, Bayes Theorem is applied to change P_0 into a *posterior* probability distribution P_1, where

$$P_1(h) = P_0(h|e) = \frac{P_0(e|h)P_0(h)}{P_0(e)} \tag{8.2}$$

Now P_1 is based on k_1, which includes both k_0 and e. By repeatedly applying Bayes Theorem, the system learns new knowledge, and adjusts its beliefs accordingly [Heckerman, 1999, Pearl, 2000]. This process is called "conditioning."

According to the above description, we see that under the Bayesian interpretation, a probabilistic evaluation $P(h)$ is always "conditional," in the sense that it is not an objective property of the proposition h, but a relation between h and background knowledge k. For this reason, the previous inference rule can be written as

$$P_{k_1}(h) = P_{k_0}(h|e) = \frac{P_{k_0}(e|h)P_{k_0}(h)}{P_{k_0}(e)} \tag{8.3}$$

where k_0 and k_1 are the background knowledge the system has at time t_0 and t_1, respectively. In the following, I will call them "implicit conditions" of the corresponding probability distribution function, because they are conditions of the probability functions, and they are usually implicitly assumed in the formula.

A common practice is to represent the dependency of a probability to an implicit condition as a conditional probability, that is, to represent the above rule as

$$P(h|k_1) = P(h|e \wedge k_0) = \frac{P(e|h \wedge k_0)P(h|k_0)}{P(e|k_0)} \tag{8.4}$$

This kind of treatment can be found in many publications on the Bayesian approach [Cheeseman, 1985, Heckerman, 1999, Pearl, 1988, Pearl, 2000].

Since in conditional probability the condition is explicitly represented, in the following, I will call them "explicit conditions." My argument is that, in general, it is improper to represent an implicit condition as an explicit condition, and that the difference between the two shows a serious limitation of the Bayesian approach, which is related to several previous debates on related topics.

Since Bayesian learning is carried out by Equation 8.2, the knowledge the system can learn, by applying Bayes Theorem, must be represented as explicit condition e. This means:

1. It is a (binary) proposition (otherwise it cannot be in S).
2. It is in S (otherwise its probability $P_0(e)$ is undefined).
3. $P_0(e) > 0$ (otherwise it cannot be used as a denominator in Bayes' Theorem).

These restrictions are not unknown (for example, a similar list is discussed in [Diaconis and Zabell, 1983, Pearl, 1990]). Since all learning methods have their restrictions, it is not a surprise that Bayesian conditioning cannot learn everything. However, here the problem is that the above restrictions are not applied to the implicit conditions of a probability distribution function:

1. An implicit condition may include statistical conclusions and subjective probabilistic estimates, which are not binary propositions.

2. An implicit condition only needs to be related to S, but not necessarily in S. For example, "Tweety is a bird and cannot fly" can be part of an implicit condition, even though S includes only "Birds fly," and does not include the name "Tweety" at all.

3. Even if a proposition is assigned a prior probability of zero according to one knowledge source, it is still possible for the proposition to be assigned a non-zero probability according to another knowledge source.

Now we can see that only certain types of implicit conditions can be represented as explicit conditions. It follows that if some knowledge is not available when the prior probability is determined, it is impossible to be put into the system through Bayesian conditioning. We cannot assume that we can always start with a "non-informative" prior probability distribution, and learn the relevant knowledge when it becomes available.

In fact, when S is finite, conditioning can only accept a finite amount of (different) new knowledge after its prior probability distribution is determined. To see it, we only need to remember that the new knowledge must be in S, and each time a proposition e is provided to the system as a piece of new knowledge, at least e and $\neg e$ (as well as $\neg e \wedge h$, and so on) cannot be used as new knowledge in the future.

If we insist that all implicit conditions must satisfy the same three restrictions, the prior probability distribution will degenerate into a consistent assignment of 0 or 1 to each proposition in S, and, after the assignment, the system will be unable to accept any new knowledge at all.

From a practical point of view, the three restrictions are not trivial, since they mean that although the background knowledge can be probabilistic, all new knowledge must be binary; no novel concept and proposition can appear in new knowledge; and if a proposition is given a probability 1 or 0, such a belief cannot be changed in the future, no matter what happens. We could build such a system, but unfortunately it would be a far cry from the everyday reasoning process of a human being.

Therefore, it is wrong to represent an implicit condition as an explicit condition, and the previous Equation 8.3 and Equation 8.4 are not equivalent. Though both formulas are correct, they have different meaning.

Some authors attempt to represent revision as "deriving $P(h|e_1 \wedge e_2)$ from $P(h|e_1)$ and $P(h|e_2)$" [Deutsch-McLeish, 1991]. According to the previous discussion, we can see that this treatment has the same problem, because it only considers explicit conditions, while in general we cannot assume that conflict beliefs come under the same implicit condition. Furthermore, revision happens when there are conflicting opinions based on different evidence, and the result should be a compromise of the opinions, but $P(h|e_1 \wedge e_2)$ is not necessarily between $P(h|e_1)$ and $P(h|e_2)$. Therefore, this is not a proper way to formalize revision.

Some people claim that the Bayesian approach is sufficient for reasoning with uncertainty, and many people treat Bayes Theorem as a generally applicable learning rule, because explicit conditions and implicit conditions of a probability evaluation are seldom clearly distinguished in related discussion. Without such a distinction, the illusion arises that all the knowledge supporting a probability distribution function can be represented by explicit conditions, and can therefore be learned by the system using Bayes Theorem. As a result, the capability of Bayes Theorem is overestimated.

Within the Bayesian tradition, there is a way to handle new evidence that is not a binary proposition. After a prior probability distribution P_0 is assigned to a proposition space S, some new evidence may show that "The probability of proposition e ($e \in S$) should be changed to p" (i.e., $P_1(e) = p$). In this situation, assuming the conditional probabilities that with e or $\neg e$ as explicit condition are unchanged (i.e., $P_1(x|e) = P_0(x|e)$), we can update the probability evaluation for every proposition x in S to get a new distribution function by using Jeffrey's rule [Jeffrey, 1965]:

$$P_1(x) = P_0(x|e) \times p + P_0(x|\neg e) \times (1 - p) \tag{8.5}$$

If we interpret "e happens" as "e's probability should be changed to 1," then learning by conditioning, as in Equation 8.2, becomes a special case of Jeffrey's rule, with $p = 1$.

A related method was suggested to process uncertain evidence e, where a "virtual proposition" v is introduced to represent the new knowledge as "a (unspecified) proposition v is true, and $P_0(e|v) = p$" [Jeffrey, 1965, Pearl, 1988]. Then a new conditional probability distribution can be calculated (after considering the new knowledge) for each proposition $x \in S$ in the following way:

$$P_1(x) = P_0(x|v) = P_0(x|e \wedge v) \times P_0(e|v) + P_0(x|\neg e \wedge v) \times P_0(\neg e|v) \tag{8.6}$$

Under the assumption that

$$P_0(x|e \wedge v) = P_0(x|e) \text{ and } P_0(x|\neg e \wedge v) = P_0(x|\neg e)$$

Equation 8.6 can be reduced into Equation 8.7:

$$P_1(x) = P_0(x|v) = P_0(x|e) \times p + P_0(x|\neg e) \times (1 - p) \tag{8.7}$$

Therefore we end up with Jeffrey's rule. The only difference is that here the prior probability is not *updated* directly, but is instead *conditionalized* by a virtual condition (the unspecified proposition v). However, no matter which procedure is followed and how the process is interpreted, the result is the same [Pearl, 1990].

Jeffrey's rule (and its equivalent forms) avoids the first restriction of Bayes Theorem, that is, the new evidence must be a binary proposition. Also, if conditional probability is directly defined by de Finetti's coherent conditional probability (that is, not as $P(x|y) = P(x \wedge y)/P(y)$), it is possible to do conditioning on an event which has prior probability 0 [Coletti et al., 1993]. Furthermore, Pearl defines "Neo-Bayesianism" by adding structural information (i.e., the topological structure of a Bayesian network) into tradition Bayesianism. With this kind of information, conditional probability can be introduced independent of the absolute probability values, and, therefore, the above limitations of conditioning is overcome [Pearl, 1990].

Though the above methods are well justified, they only covers a special case. In general, by "revision" (or "learning," "belief change," and so on), I mean the process by which a system changes the uncertainty value (no matter what it is called) of a statement h from $P(h) = p_1$ to $P(h) = p_2$, according to evidence e. By "updating," I mean a special case of the above process where e takes the form of "$P(h)$ should be p_2," and it is indeed the result of the process, no matter what p_1 is.

Though "updating" is a valid operation in uncertain reasoning, it is only a special case of "revision," because it does not cover the situation where the result is a compromise of conflicting beliefs/evidence.

In certain situations, it is proper to interpret belief changes as updating [Dubois and Prade, 1991], but revision is a more general and important operation. When there are conflicts among beliefs, it is not always the case that one piece of evidence should be *completely* suppressed by another piece of evidence, unless the beliefs are about an event which the system is sure that has changed its status.

Concrete speaking, revision happens when the system's current belief on h is $P_{k_0}(h)$, the new knowledge is $P_{k_0'}(h)$, and the result is $P_{k_1}(h)$, where k_1 summarized the knowledge in k_0 and k_0'. We cannot do it in the Bayesian approach, because $P_{k_0'}(h)$ contains information that cannot be derived from P_{k_0}, nor can the operation be treated as updating, where $P_{k_0}(e)$ is simply replaced by $P_{k_0'}(h)$. Intuitively, to carry out the revision operation, we need more information about k_0 and k_0', and this information is not in the probability distribution functions P_{k_0} and $P_{k_0'}$.

Therefore, even if Jeffrey's rule is used to replace Bayes' rule and structure information is added into the picture, the system still does not have a general way to revise its implicit conditions (i.e., background knowledge behind the probability distribution function). If we want to apply a Bayesian network to a practical domain, one of the following requirements must be satisfied:

1. The implicit condition of the initial probability distribution, that is, the domain knowledge used to determine the distribution initially, can be assumed to be immune from future modifications; or

2. All modifications of the implicit condition can be treated as updating, in the sense that when new knowledge conflicts with old knowledge, the latter is completely abandoned.

From AI's point of view, such domains are exceptions, rather than general situations. In most cases, we cannot guarantee that all initial knowledge is unchangeable, or that later acquired knowledge always completely suppresses earlier acquired knowledge. Usually, revision is a compromise, as in belief change [Voorbraak, 1999] and multiple source information fusion [Dubois et al., 2001].

Once again, this conclusion is only about the version of Bayesianism in AI specified previously. It may not apply to other forms of Bayesianism or other probability-based approaches. Given the possible variations of probability theory, I cannot say that the above problem cannot be solved by some kind of probability theory. However, we have not seen such a solution, that is, a general-purpose reasoning system, which is designed according to probability theory, and works under AIKR.

Furthermore, I haven't seen a convincing argument showing that systems like NARS must be based on probability theory. Existing arguments in this direction, such as Cox's theorem [Cox, 1946], are all derived from assumptions that conflict with AIKR. For example, it is usually assumed that the system's beliefs are consistent, and that the system has enough resources to finish the required calculations.

8.3.2 NARS vs. the Bayesian approach

Since the probability distribution function P is defined on S according to implicit condition k, it provides a summary of available knowledge about propositions in S, but the function says little about k itself. Consequently, the system has no general way to revise and extend k. That is, the Bayesian approach has no general way to represent and handle the uncertainty within the background knowledge and the prior probability function. This is a serious limitation of the Bayesian approach, both in theory and in application.

Though the distinction between explicit and implicit condition is rarely made, the above conclusion, that is, Bayesian approach has limitations in representing and processing uncertainty, is not new at all. From different considerations, many people reached the same conclusion, that is, to use a probability distribution function alone to represent uncertainty is not enough, because it fails to show the *ignorance*, or uncertainty about the function itself. This kind of uncertainty is not the "risk" captured by probability [Ellsberg, 1961].

Several alternative approaches have been proposed to solve this problem. Though these approaches are technically different, they can all (more or less) be seen as attempts of extending the Bayesian approach by using more than one value to represent the uncertainty of a statement, and therefore indicates the ignorance of the system. Some of these approaches will be discussed in the next subsection. In the following I first analyze the response from the Bayesian school against these challenges.

To argue against "more than one number is needed to represent uncertainty," Cheeseman claimed that a point value and a density function will give the same result in decision making [Cheeseman, 1985], which I agree with to a certain extent. However, I believe that he was wrong by saying that standard deviation can be used to capture "the change of expectations" (or revision, as defined previously). If we test a proposition n times, and the results are the same, then the standard deviation of the results is 0, that is, independent of n. But our *confidence* that "the result will remain the same" will obviously increase with n. Actually, what the standard deviation measures is the *variations* among the samples, but what is needed in revision, intuitively speaking, has more to do with the *amount* of the samples.

Pearl said the uncertainty in the assessment of $P_0(e)$ is measured by the (narrowness of the) distribution of $P_0(e|c)$ as c ranges over all combinations of contingencies, and each combination c is weighted by its current belief $P_0(c)$ [Pearl, 1988]. A similar approach is [Spiegelhalter, 1989], where ignorance is treated as sensitivity.

I agree with them that ignorance is the lack of confidence, and confidence can be measured by how much a belief assignment can be modified by possible future evidence (in this sense, it is different from what is measured by the "confidence interval" in statistics). However, in their definitions, they still assume that any relevant future evidence causing a belief change can be represented as an *explicit condition*, and can be processed through conditioning. As a result, their measurement of ignorance (or confidence) cannot capture the ignorance about *implicit conditions*.

When a reasoning system has insufficient knowledge and resources with respect to the task assigned to it, it cannot assume that the initial background knowledge does not need revising, nor that any revising can be treated as complete updating of the probability distribution function. Therefore, the above limitation means that the Bayesian approach is not a normative theory of reasoning wherever AIKR is accepted, such as in NARS.

Now let me list the reasons for NARS to be developed outside probability theory.

- Though a single judgment in NAL by itself can be seen as a probability distribution, different judgments, in this sense, correspond to different probability distributions, because each of them has its own evidence space.

- Because the system has insufficient resources, usually when the truth value of a statement is determined, the system cannot consider all existing evidence, even though it is already in the memory. Similarly, when a piece of new evidence comes, the system cannot afford the resources to re-consider the truth values of all relevant judgments.

- Because of the above two reasons, in each inference step the premises and the conclusion correspond to different probability distributions, and the truth-value functions correspond to "cross-distribution" calculations, which are not defined in classical probability theory. In this sense, the uncertainty calculus of NARS is an extension of probability theory.

Let us see a concrete case. The deduction rule defined in NAL-1 has the following format:

$$\{M \rightarrow P <f_1, c_1>,\ S \rightarrow M <f_2, c_2>\} \vdash S \rightarrow P <f, c>$$

A direct way to apply probability theory would be treating each term as a set, then turning the rule into one that calculates conditional probability $Pr(P|S)$ from $Pr(P|M)$ and $Pr(M|S)$ plus additional assumptions about the probabilistic distribution function $Pr()$. Similarly, the sample size of the conclusion would be estimated, which gives the confidence value.

Such an approach cannot be applied in NARS for several reasons:

- For an inheritance relation, evidence is defined both extensionally and intensionally, so the frequency of "$S \rightarrow P$" cannot be treated as $Pr(P|S)$, since the latter is pure extensional.

- Each statement has its own evidence space, defined by the extension of its subject and the intension of its predicate, so the evidence for a premise cannot be directly used as evidence for the conclusion.

- Since input judgments may come from different sources, they may contain inconsistency.

- When a new judgment comes, usually the system cannot afford the time to update all of the previous beliefs accordingly.

Therefore, though each belief in NARS is similar to a probabilistic judgment, different beliefs correspond to different evidence space, and their truth values are evaluated against different bodies of evidence. As a result, they correspond to different probability distributions. For example, if we treat frequency as probability, the deduction rule should calculate $Pr_3(S \rightarrow P)$ from $Pr_1(M \rightarrow P)$ and $Pr_2(S \rightarrow M)$. In standard probability theory, there are few results that can be applied to this kind of cross-distribution calculation.

NARS is not proposed to replace Bayesian models. If the Bayesian approach can be applied in a situation (i.e., the computational cost and the revision of background knowledge can be ignored there), it may still be better than NARS. It is in situations where the Bayesian approach cannot or should not be applied that approaches like NARS will take over.

8.3.3 "Heuristics and bias" revisited

The study of human judgment under uncertainty reveals systematic discrepancy between actual human behaviors and conclusions of probability theory [Tversky and Kahneman, 1974], that is, between what we should do (according to probability theory) and what we do (according to psychological experiments). Therefore, probability theory is not a good *descriptive* theory for human reasoning under uncertainty, though it is still referred to as a good *normative* theory.

As a result, the research activities in this domain often consist of the following steps [Gigerenzer, 1991, Kahneman and Tversky, 1982]:

1. To identify the problem by carrying out psychological experiments, and comparing the results with the conclusions of probability theory;

2. To explain the result by looking for the *heuristics* that are used by humans and the factors that affect their usage, and to suggest and verify methods to correct the errors.

Heuristics, as methods to assess subjective probability, "are highly economical and usually effective, but they lead to systematic and predictable errors" [Tversky and Kahneman, 1974]. Compared with normative theories, such as probability theory, heuristics are not optimal, not formal, not systematic, and not always correct.

According to this opinion, the fact that probability theory cannot match actual human reasoning is not a problem of the theory. Though the discrepancy is well known, probability theory, especially the Bayesian approach, is becoming more popular as a normative model of reasoning under uncertainty.

However, according to the previous analysis, we see that probability theory in general, and the Bayesian approach in specific, are based on certain fundamental assumptions. When these assumptions are not satisfied, they cannot be applied as normative theory anymore. NARS is based on a different set of assumptions, and is designed to be used when probability theory cannot. Though designed as a normative model, NARS shows some behaviors that are usually explained in term of "heuristics and biases" [Tversky and Kahneman, 1974].

- **Availability**

Availability, "the ease with which instances or occurrences can be brought to mind," is a common heuristic in intuitive judgment of probability. It "is affected by factors other than frequency and probability," therefore "leads to predictable biases" [Tversky and Kahneman, 1974].

The same phenomenon happens in NARS. Because NARS is built under AIKR, the following properties are implied:

1. The system has to base its judgments on the *available knowledge*. Therefore, the estimation of the frequency of a belief is actually about the *experienced* frequency, rather then the *objective* frequency.

2. Judgments must be made with the *available resources*. Therefore, the system often cannot consider all of its beliefs, but only part of it.

3. Which beliefs are consulted is determined by several factors, such as relevance, importance, usefulness, and so on. Therefore, it is not surprising that certain events, like priming and association, influence the availability distribution [Arkes, 1991].

The above properties mean that if the system attempts to estimate the frequency of a type of event, the result may be biased by the events which are readily available to remember and recall, as observed in people. Such a result is different from how frequency, or probability, is defined in probability theory.

Because which belief to use at each step of reasoning is determined by the current context (by priming) and past experience (by association), it is inevitable that some knowledge, necessary for the assessment of uncertainty of a proposition, may be either unknown to the system or cannot be recalled at the time. As a result, the system will have *expectation errors* — i.e., the conflicts between the system's expectations and the system's future actual experience, but this type of error is not caused by mis-designing or malfunction of the system. Under the knowledge and resources constraints, the system has done its best. As long as it can revise its beliefs according to new evidence, there is no error in the system's *operations*, though there may be errors in the *results* of these operations.

- **Representativeness**

Representativeness, or degree of similarity, is often used as probability by human beings. "This approach to the judgment of probability leads to serious errors, because similarity, or representativeness, is not influenced by several factors that should affect judgments of probability" [Tversky and Kahneman, 1974]. The basic difference between probability and similarity is that "the laws of probability derive from extensional considerations," but similarity judgments are based on the sharing of properties, so they are *intensional* [Tversky and Kahneman, 1983].

Here we need to distinguish three different meanings of "probability":

1. As a pure mathematical concept, probability is neither extensional nor intensional.
2. Probability theory is usually interpreted extensionally when applied to a practical domain.
3. In everyday language and intuitive thinking, both extensional and intensional interpretations of probability happen.

Why is only the extensional interpretation referred to as "correct"? There is a historical reason: the normative theories about extension are well developed, but the theories about intension are not. Actually there is no commonly accepted theory about how to define and process the intension of a concept. However, this does not imply that intensional factors should not be taken into consideration when we make predictions about uncertain events.

NARS is an attempt to equally treat extension and intension. When the uncertainty of a judgment is determined, both the extensional factor (shared instances) and intensional factor (shared properties) are considered (as discussed in Section 7.2). By doing this, it does not mean that they are not different, but that their *effects* are the same in the judgment. It is valid to build normative theories to process extension or intension separately, but it is also valid, and maybe more useful, to have theories that process both of them in a unified manner. In the latter case, it is valid to use representativeness and probability indiscriminately for certain purposes.

• **Adjustment and anchoring**

For any system that accepts new knowledge or makes judgments by incrementally considering available knowledge, there must be a rule by which a previous probability judgment is adjusted in light of new evidence or further consideration [Anderson, 1986].

The anchoring phenomenon, or insufficient adjustment from the initial point, is observed in human thinking [Tversky and Kahneman, 1974]. By calling the observed adjustments "insufficient," it is assumed that the correct adjustment rule is Bayes' theorem, or its extension, Jeffrey's rule.

As discussed previously, in NARS, two different cases are distinguished when judgments conflict with each other. If the evidence supporting the two judgments are correlated, the choice rule is applied to do updating, otherwise the revision rule is applied.

In updating, there are also two possibilities: if the confidence of the previous estimation is no lower than the confidence of the new estimation, then nothing is changed, otherwise the former is replaced by the latter. Though the second possibility is the same with Jeffrey's rule, what follows is different: NARS usually cannot afford the resources to update all related judgments, therefore only some of them are updated accordingly, by applying the inference rules and the updating rule of NARS.

In revision, the new frequency is a weighted sum of those of the premises, as discussed previously.

Therefore, in all situations, the adjustment of frequency in NARS is no more than what is required by probability theory. If conditionalization (Bayes' theorem and Jeffrey's rule) is *the* correct way of adjustment, NARS shows the anchoring bias, too. However, as argued above, it is not always valid to use updating as revision, or to assume sufficient resources for global updating.

NARS is not proposed as a descriptive model for actual human thinking, such as Anderson's model [Anderson, 1986]. Its behavior is still different from that of a human being. The approach is not justified by psychological data, but by logical analysis. Therefore there is no psychological experiment conducted to verify the theory. However, psychological observations, as those reported in [Tversky and Kahneman, 1974], do have a strong relation to the study of normative models.

NARS is no less normative than probability theory in the sense that it is developed from some basic principles and assumptions about what a system (human or computer) *should* do under AIKR. It is true that when applied into a practical domain, NARS may produce wrong expectations, but so does probability theory.

A conclusion of the above discussion is that there is no unique normative model for judgment under uncertainty — different models can be established according to different theoretical assumptions. NARS is "less idealized" than the Bayesian approach, because it assumes stronger knowledge–resource constraints. The behavior of NARS is closer to that of people, therefore we have reason to believe that its assumptions are more "realistic" — that is, more similar to the constraints under which the human cognitive mechanism was developed.

Using NARS as an example, we see that it is possible to find a normative interpretation for the "heuristics." They are not necessarily "efficient but biased." Sometimes they indicate the right thing to do, though they do not always succeed.

As for the "biases" and "fallacies" discussed in the psychological literature, the situation is complex. NARS cannot explain all of them, but it does suggest a distinction: some violations of probability theory happen in the situations where probability theory cannot or should not be applied, and they may be explained by other normative theories, therefore they are not necessarily errors. The real errors happen when probability theory should be applied, but the person fails to do so.

Even for the latter case, an explanation is suggested from the study of NARS: because the human mind usually works under some assumptions about knowledge and resources that are quite different from what probability theory assumes, it needs some special effort (which does not always succeed) to suppress the "natural law of thinking," and to learn, to remember, and to follow probability theory.

Now I can say that by analyzing the so called "heuristics and biases," we not only find limitations in human reasoning, but also find limitations in probability theory, especially in the Bayesian approach. Just like no one is born with a digital calculator embedded in their head, brains dont include Bayesian networks for a good cause – in many of our native environments the assumptions made by the Bayesian approach may frequently fail to hold.

8.4 Other probabilistic approaches

"Ignorance cannot be represented in a probability distribution" is an old argument against the Bayesian approach, and a major motivation for alternative approaches. There are several approaches proposed as extensions or variants of probability theory, by taking the uncertainty in probability values into consideration. As stated previously, in a sense NARS is also such an approach. In this section NARS is compared with some of them.

8.4.1 Higher-order probability

Several new measurements are proposed under the assumption that the first-order uncertainty measurement (call it "probability" or "degree of belief") is an *approximation* of a "real" or "objective" probability. Under the frequentist interpretation, the probability of a statement is the limit of frequency, therefore all estimations of it based on finite evidence are not accurate. Even if we take probability as degree of belief, it can still be argued that such a degree should converge to the objective probability if it exists.

If the first-order probability assignment is only an approximation of an unknown value, the need for a higher-order measurement follows naturally — we want to know how good the approximation is, in addition to the approximated value itself.

One natural idea is to apply probability theory once again, which leads to the ideas like "second-order probability," "higher-order probability," and so on [Fung and Chong, 1986, Gaifman, 1986, Paaß, 1991]. In this way, we can assign probability to a probability assignment, to represent how good an approximation it is to the real probability.

However, there are problems in how to interpret the second value, and whether it is really useful [Kyburg, 1989, Pearl, 1988]. For NARS, under the assumption of insufficient knowledge, it makes little sense to talk about the "probability" of "the frequency is an accurate estimate of an objective first-order probability." Since NARS is always open to new evidence, it is simply impossible to decide whether the frequency of a judgment will converge to a point in the infinite future, not to mention where the point will be. If we say that the second-order probability is an approximation itself, then a third-order probability follows for the same reason — we are facing an infinite regression [Savage, 1954].

The confidence c defined in NARS is in [0, 1], can be considered to be a ratio, and is at a "higher level" than frequency f (which is closely related to probability), in the sense that it indicates the stability of f. However, it cannot be interpreted as a second-order probability in the sense that it is the probability of the judgment "f is the (real or objective) probability of the statement." The higher the confidence is, the harder it will be for the frequency to be changed by new evidence, but this does not mean that the judgment is "more accurate," because in an open system like NARS, the concept of a real or objective probability does not exist.

Furthermore, if confidence in NARS is interpreted as second-order probability, then a judgment $S < f, 0 >$ would mean "$P(P(S) = f) = 0$," that is, "The frequency of S is not f," rather than "The frequency of S is unknown," which is the intended interpretation. Consequently, such a measurement would not support the revision operation — we cannot combine a pair of conflicting judgments, given their first-order and second-order probabilities.

With confidence defined in the current way in NARS, there is no "third-order uncertainty" to worry about. While the stability of a frequency value is indicated by the attached confidence value, the stability of the latter can be derived from itself. Because the current confidence is $c_0 = w/(w + k)$, with the coming of new evidence of amount k, the new confidence will be $c_1 = (w + k)/(w + 2k)$, and the change is $c_1 - c_0 = (w+k)/(w+2k) - w/(w+k) = k^2/((w+2k)(w+k))$, which becomes smaller when c becomes larger. Therefore NARS does not need another measurement, and there is no infinite regression.

8.4.2 Probability interval

Another intuitively appealing idea is to use an *interval*, rather than a *point*, to represent uncertainty, and to interpret the interval as the lower bound and upper bound of the real probability [Bonissone, 1987, Grosof, 1986, Kyburg, 1987].

According to these approaches, when the system knows nothing about a statement, the interval is [0, 1], so the real probability can be anywhere; when the real probability is known, the interval degenerates into a point. Therefore, *ignorance* can be represented by the width of the interval.

A related idea in statistics is to calculate the *confidence interval* of a probabilistic estimation [Bernardo and Smith, 1994], which has a high probability (such as 95%) of including within it the real probability. Here, the width of the interval also provides information about the accuracy of the current estimation.

Although the above methods are directly based on probability theory (and thus have a sound foundation), and they are useful for various purposes, they cannot be applied in a system like NARS. The "frequency interval" (defined in Subsection 3.2.3) shares these intuitions, though it does not assume the existence of a "real frequency", or a limit of the experienced frequency.

When the probability interval is interpreted as the interval containing the true probability value, the situation is similar to the case of higher-order probability. For an open system with insufficient knowledge, it cannot be assumed that a frequency always has a limit. Even when such a limit really exists, it is impossible for the system to know how close the current frequency is to it without making assumptions about the distribution of the limit.

If the probability interval is interpreted as an estimation itself, an "interval of the bounds" will follow, so the infinite regression appears again. For the same reason, the "confidence" defined in NARS has different meaning from the "confidence" as in "confidence interval," used in probability theory and statistics, though they do correspond to the same intuition, that is, some frequency estimations are more reliable than others, and their difference can be measured.

The closest probability-based approach to NARS is the "Imprecise Probability" (IP) theory proposed by Peter Walley [Walley, 1991, Walley, 1996]. Walley defines lower and upper probabilities of an event as the minimum and maximum betting rate, respectively, that a rational person is willing to pay for the gamble on the event. Though starting from a quite different place, this theory is related to NARS in an interesting way.

Suppose that an event has a constant (unknown) chance to happen, that the observations of the event are independent to one another, and that the chance has a *near-ignorance* beta distribution as its prior. If the observed relative frequency of the event is m/n, then, according to Walley's theory, the lower and upper probabilities of the event are $m/(n+s_0)$ and $(m+s_0)/(n+s_0)$, respectively. Here s_0 is a parameter of the beta distribution, and it indicates the convergence speed of the lower and upper probabilities. This is exactly the result we get for the lower and upper frequencies previously, though the interpretations of the interval are different. For NARS, the interval $[m/(n+k),\ (m+k)/(n+k)]$ is just where the frequency will be in a *constant near future* (measured by k), and after that it can be anywhere in [0, 1].

Though these two approaches (NARS and IP) define uncertainty measurements differently, they are consistent in the sense that they make the same decisions in situations where both theories are applicable. What makes them different from the other "probability interval" approaches mentioned earlier is: in both NARS and IP, the interval does not bound the limit of the frequency (if such a limit exists).

The major difference between these two approaches comes from the fact that IP is proposed as an extension of probability theory, and therefore the inference is mainly within the same probability distribution. On the other hand, in NARS each piece of knowledge is based on a separate body of evidence, so that the rules introduced previously correspond to inference across different probability distributions. The detailed relationship between these two approaches is an interesting issue for future research.

8.4.3 Dempster-Shafer theory

Evidence theory, also known as Dempster-Shafer (D-S) theory, was developed as an attempt to generalize probability theory by introducing a rule for combining distinct bodies of evidence [Dempster, 1967, Shafer, 1976].

The most influential version of the theory was presented by Shafer in his book *A Mathematical Theory of Evidence* [Shafer, 1976]. In the book, the following postulates are assumed, which form the foundation of D-S theory.

1. *Chance* is the limit of the proportion of "positive" outcomes among all outcomes [Shafer, 1976, pages 9, 202].

2. Chances, if known, should be used as *belief functions* [Shafer, 1976, pages 16, 201].

3. *Evidence combination* refers to the pooling, or accumulating, of distinct bodies of evidence [Shafer, 1976, pages 8, 77].

4. *Dempster's rule* can be used on belief functions for evidence combination [Shafer, 1976, pages 6, 57].

However, I proved in [Wang, 1994a] that under a natural interpretation of the measurement involved, the above four postulates are inconsistent. This proof is briefly summarized in the following.

For a given hypothesis H, assume all pieces of evidence have the same weight w, and the numbers of pieces of positive, negative, and total evidence are t^+, t^-, and t, respectively (so $t = t^+ + t^-$), then after repeatedly applying Dempster rule to combine the evidence, the degree of belief of H, $Bel(\{H\})$, may converge to a limit $Bel_\infty(\{H\})$. However, if that happens, $Bel_\infty(\{H\})$ is usually different from the chance of the proposition, defined as $Pr(H) = \lim_{t\to\infty} t^+/t$. Instead, it is proved that

$$\begin{aligned} Bel_\infty(\{H\}) &= \lim_{w\to\infty} \frac{e^{wt^+}-1}{e^{wt^+}+e^{wt^-}-1} \\ &= \lim_{t\to\infty} \frac{e^{wt^+}-1}{e^{wt^+}+e^{wt^-}-1} \\ &= \begin{cases} 0 & \text{if } Pr(H) < 1 - Pr(H) \\ 1 & \text{if } Pr(H) > 1 - Pr(H) \\ \frac{1}{1+e^\Delta} & \text{if } Pr(H) = 1 - Pr(H) \end{cases} \end{aligned}$$

where $\Delta = \lim_{t\to\infty} w(t^- - t^+)$.

This means that if the chance of the hypothesis exists, then, by repeatedly applying Dempster's rule to combine the coming evidence, if belief function of the hypothesis converge to a value, then that value is usually not the chance of the hypothesis, except in a few special cases.

Therefore, under the above simple interpretation, D-S theory is not a proper extension of probability theory. Because of these problems, as well as several other factors (such as the interpretation of the belief function, the easiness of defining various inference rules on the belief function, the computational cost of the rules, and so on), NARS does not use D-S theory.

To compare NARS with D-S theory, we can see that the *lower frequency* and *upper frequency* in the former are intuitively similar to the *belief function* and *plausibility function* in the latter, respectively, in the sense that

$$\frac{w^+}{w+k} \leq \frac{w^+}{w} \leq \frac{w^+ + k}{w+k}$$

However, in NARS, if the frequency of a statement indeed has a limit, the lower frequency and upper frequency will converge to it. That is

$$\lim_{w \to \infty} \frac{w^+}{w+k} = \lim_{w \to \infty} \frac{w^+}{w} = \lim_{w \to \infty} \frac{w^+ + k}{w+k}$$

Therefore, the previous problem in D-S theory does not exist in NARS.

8.5 Unified representation of uncertainty

Compared with the other approaches, the representation and interpretation of uncertainty in NARS have the following characteristics:

- It satisfies the requirement of NARS, that is, the approach can be applied to an adaptive reasoning system where knowledge and resources are constantly insufficient to deal with the tasks provided by the environment.
- It unifies measurements of different types of uncertainty, such as randomness, fuzziness, ignorance, and so on, into a common framework, and provides them with natural and consistent interpretations.
- It provides a consistent foundation for the uncertainty calculus, which includes several kinds of operations on uncertainty.

In the following, two topics in this unification are discussed in detail.

8.5.1 Interpretations of probability

As mentioned previously, the uncertainty measurements in NARS (*frequency*, *confidence*, as well as their function *expectation*) are closely related to the notion of *probability*.

Probability as a mathematical concept is well defined by the axioms of probability theory [Kolmogorov, 1950]. However, its "interpretation," i.e., how the mathematical concept should be applied to practical problems, has been a controversial topic for a long time. In history, there have been several influential interpretations. Here I will discussion the relation between NARS and each of them.

The Empirical Interpretation treats probability P as the limit of frequency for a given event to occur in a sequence, or for a given hypothesis to be confirmed. This is the traditional interpretation of probability in statistics, and related theoretical discussions can be found in [Reichenbach, 1949, von Mises, 1981]. According to this interpretation, $P(H)$ is an objective property of H, and can only be estimated empirically, given the observed frequency of H.

In NARS, the frequency value of H is the observed confirmation frequency (or proportion) in the past, so it is an empirical result, clearly agreeing with the empirical interpretation. However, in NARS frequency merely records past experience, and says nothing about the limit, neither about its value nor about its existence. For a given judgment in NARS, with the coming of new evidence, its confidence converges to its limit 1, but its frequency may converge to any value, or not converge at all.

The Logical Interpretation treats probability P as a measurement on the relation between a hypothesis H and given evidence E, that is, it should be written as $P(H, E)$, or something like that. Such an opinion can be found in [Keynes, 1921, Carnap, 1950]. According to this interpretation, the probability value for a given pair of H and E is uniquely determined by a logical analysis of the two — see [Carnap, 1950] for a detailed treatment.

In NARS, the truth value (frequency and confidence) of a statement is determined by available evidence, just like the logical interpretation suggests. However, in NARS the evidence supporting a truth value is not explicitly listed as part of the judgment. Furthermore, in NARS the "evidence" comes from the system's experience, while in Carnap's system it is not.

The Subjective Interpretation treats probability P as a measurement on the degree of belief in a system on a hypothesis H. In different systems, $P(H)$ may have different values. The $P(H)$ value in a given system can be revealed by the bet the system will place on the hypothesis in a testing. Such a treatment of probability can be found in [Savage, 1954, Pearl, 1988].

In NARS, truth value is subjective in the sense that it depends on the experience and parameter of the system. However, it is not arbitrary, but fully determined by these factors. When the system must bet on a statement, it will use its expectation value, which is also subjective in the above sense.

In summary, the truth value in NARS partially agrees with each of the major interpretations of probability, but cannot be reduced to any of them. To avoid possible confusions, I do not use the term "probability" in NARS, and mention it only when comparing NARS to other approaches. Nevertheless, theoretically NARS does include an approach of uncertainty management that unifies different interpretations of probability, and its uncertainty calculus partially agrees with probability theory.

8.5.2 Randomness and fuzziness

In practical problem solving, multiple types of uncertainty (such as randomness, fuzziness, and ignorance) usually coexist and merge with each other. In human judgments, such mixtures have been observed between representativeness and probability [Tversky and Kahneman, 1974], as well as between randomness and fuzziness [Smets, 1991]. However, the relationship among them is far from clearly explained, partially due to the lack of a clear interpretation of fuzziness.

Previously in this chapter, I not only propose an interpretation for fuzziness, but also propose a *frequency* interpretation for it (which was claimed as impossible by Zadeh). Therefore, NARS uses a unified representation and interpretation for degree of randomness and grade of membership: both are real numbers in the range [0, 1], and both indicate the ratio of positive evidence among total evidence, that is, w^+/w.

However, this does not mean that fuzziness and randomness cannot be distinguished. For a statement "$S \rightarrow P$" in Narsese, randomness comes from the diversity within the extension of the subject, S, while fuzziness comes from the diversity within the intension of the predicate, P. When S has many instances, and some of them are shared by P while others are not, "$S \rightarrow P$" is a matter of degree, and the uncertainty is randomness; when P has many properties, and some of them are shared by S while others are not, "$S \rightarrow P$" is also a matter of degree, but the uncertainty is fuzziness.

For example, "Birds fly" is uncertain, mainly because some kinds of birds (such as ravens) do fly, but some others (such as penguins) do not. On the other hand, "Penguins are birds" is uncertain, mainly because penguins have some of the properties of (typical) birds (such as having wings), but do not have some others (such as flying).

If these two types of uncertainty are different, why bother to treat them in a uniform way? One reason, as discussed in Section 7.2, is the need to uniformly process extension and intension, and that implies a unified processing of the uncertainty in extension and intension.

Another reason is that in many practical problems these different types of uncertainty are mixed together. Smets stressed the importance of this issue, and provided some examples, in which randomness and fuzziness are encountered in the same sentence [Smets, 1991]. In many situations, what we want to know is not where the various uncertain factors come from, but how they influence the final decision. Since they often appear in mixed form, a unified treatment is necessary.

With a frequency interpretation of these two types of uncertainties, it is not surprising to see that NARS' truth value functions are defined more similarly to probability theory than to fuzzy logic. For instance, the operations used for disjunction and conjunction are *sum/product*, not *max/min*.

Finally, ignorance (and confidence) are brought into the picture by defining as functions of total evidence. Ignorance is different from randomness and fuzziness in that it has little to do with the relative amount of positive and negative evidence, but is mainly about the total amount of evidence. Even so, in NARS it is unified with the other uncertainty measurements, because they are all defined in terms of available evidence.

Chapter 9

Inference Rules

In this chapter, several representative topics are discussed to show the difference between the inference rules of NARS and those in other theories.

9.1 Deduction

Deduction is the type of inference that has been studied most thoroughly. However, there are still problems when the knowledge and resources of the system are insufficient. Here the *reference class* problem is discussed as an example.

9.1.1 Deduction with reference classes

How do we predict whether an individual has a certain property, if direct observation is impossible? A useful method is to look for a "reference class." The class should include the individual as an instance, and we should know something about how often the instances of the class have the desired property, or whether its typical instances have it. Then, the prediction can be done by letting the instance "inherit" the information from the class.

In reasoning under uncertainty, there are (at least) two groups of approaches that use this type of inference: non-monotonic logics [Touretzky, 1986], and probabilistic reasoning systems [Pearl, 1988].

In non-monotonic logics, if the only relevant knowledge is "S is an instance of R" and "Normally, R's instances have the property Q," a defeasible conclusion is "S has the property Q."

In probabilistic reasoning systems, under the subjective interpretation of probability, if the only relevant knowledge is "S is an instance of R" and "The probability for R's instances to have the property Q is p," a plausible conclusion would be "The probability for S to have the property Q is p."

Now a problem appears: if S belongs to two classes R_1 and R_2 at the same time, and the two classes lead to different predictions about whether (or how probable) S has the property Q, what conclusion can we reach? In different contexts, the problem is referred to as "multiple inheritance problem," "multiple extension problem," or "reference class problem." [Grosof, 1990, Kyburg, 1983, Neufeld, 1989, Pearl, 1988, Poole, 1985, Reichenbach, 1949, Touretzky, 1986].

Though the above theories treat the problem differently, they have something in common: None of them suggest a general solution to the problem, though they agree on a special case: if R_2 is a (proper) subset of R_1, R_2 is the correct reference class to be used.

Let us see two examples.

1. "Since Clyde is a royal elephant, and royal elephants are not gray, Clyde is not gray. On the other hand, we could argue that Clyde is a royal elephant, royal elephants are elephants, and elephants are gray, so Clyde is gray. Apparently there is a contradiction here. But intuitively we feel that Clyde is not gray, even though he is an elephant, because he is a special type of elephant: a royal elephant." [Touretzky, 1986].

2. "If you know the survival rate for 40-year old American male to be 0.990, and also that the survival rate for 40-year old American male white-collar workers to be 0.995, then, other things being equal, it is the latter that should constrain your beliefs and enter your utility calculations concerning the particular 40 year old male white-collar worker John Smith." [Kyburg, 1983].

Let us call this principle "specificity priority principle." It looks quite reasonable, and it is not hard to find many examples to show that we do apply such a principle in common sense reasoning. However, the following questions are still open:

1. Why is the principle correct? Can it be justified by more basic postulates?

2. Beside specificity, what are the "other things" that influence the priority of a reference class?

3. When neither reference class is more specific than the other, what should be done?

For the first question, Reichenbach made it a matter of definition by "regarding the individual case as the limit of classes becoming gradually narrower and narrower" [Reichenbach, 1949]; Pearl said it is because "the influence of the remote ancestors is summarized by the direct parents." [Pearl, 1988].

For the second question, Reichenbach said we need to have complete statistical knowledge on the reference class, that is, the probability for R to be Q should be supported by good statistical data [Reichenbach, 1949]. In non-monotonic logics, this corresponds to sufficient evidence which can determine what properties a *normal* instance of the class has.

For the third question, few words are said, except Reichenbach's suggestion to "look for a larger number of cases in the narrowest common class at your disposal." [Reichenbach, 1949]

9.1.2 A thought experiment

Let us reconstruct Kyburg's example in the following way: Imaging that you are working for a life insurance company, and you need to predict whether John Smith can live to 40. You have John's personal information, and for some special reasons (such as you just woke up from a 200-year-long sleep or you are actually an extraterrestrial spy), you have no background knowledge about the survival rates at 40 for various groups of people. Fortunately, you have access to personal files of some Americans, who are alive or died in recent years, and you decide to make the prediction by the "reference class method" defined above.

At first, knowing that John is a male, you begin to build the first reference class R_1 by picking up some files randomly. R_1 consists of two subsets: P_1 includes the positive evidence for John's survival, that is, American males who are more than 40 years old (including those who are already deceased), and N_1 includes the negative evidence, that is, those who died before 40. You should keep in mind that American males who are alive and younger than 40 (including John himself) are neither positive evidence nor negative evidence for the prediction, so they do not belong to R_1.

If you weight everyone equally (and why wouldn't you?), your prediction should be determined by the relative size of P_1 and N_1. Let us say $|P_1| > |N_1|$. Therefore you predict that John Smith can live to 40.

After returning the files, you have a new idea: why not consider the fact that John is, among other things, a white-collar worker? So you build another reference class R_2 similarly. Let us assume, unfortunately, this time you find that $|P_2| < |N_2|$. Here you meet the reference class problem: to see John as a "male" and a "male white-collar worker" will lead to different predictions.

If we apply the *specificity priority principle* here, the result should be dominated by R_2, since "male white-collar worker" is a proper subset of "male." However, it is easy to find a situation to show that sometimes the result is counter-intuitive. If you have looked through 1000 files, and all of them are males and live to 40, and after that you find 1 male white-collar worker who died at 35, will you predict that John will die before 40? It seems very unlikely.

Does this mean that the specificity priority principle is wrong? Of course not. Sample size is obviously one of the "other things" that influence the priority of a reference class. One sample is far from enough to tell us about how a "typical" or "normal" instance looks like, or to support a statistical assertion on the instances. In such a case, the principle is inapplicable, since there is another relevant difference between the two reference classes, beside their specificities.

If you have to make predictions in such an environment, what will you do? Let us consider a simple psychological experiment. Assuming R_1 includes positive evidence only (that is, $R_1 = P_1$; no male is found to have died before 40), but R_2 includes negative evidence only (that is, $R_2 = N_2$, no male white-collar worker is found to be alive at 40).

Even before really carrying out such an experiment on human subjects, I am confident to make the following prediction: If $|P_1|$ is fixed at a big number (say 1000), and $|N_2|$ is increased one by one, starting from 1, the predictions made by subjects will be positive before $|N_2|$ reaching a certain point, and negative after reaching that point. That critical point may vary from person to person, but is always smaller than $|P_1|$.

The "sample size effect" can also be used to answer the following question: If a more specific reference class is always better, why do not we simply use the *most specific reference class*, defined by all available properties of John Smith? The reason is simple: in most situations such a class is *empty* — nobody is similar to John to such an extent. With more and more properties used to define a reference class, the extension of the class becomes narrower and narrower. As a result, fewer and fewer samples can be found to support or discredit our prediction. From this point of view, specificity is not preferred.

Previously, I talked about the reference classes R_1 and R_2, as if they are accurately defined. Obviously this is a simplification. Though we can ignore the boundary cases for "male," the fuzziness in "white-collar worker" cannot be neglected so easily. As argued by fuzzy set theory [Zadeh, 1965] and prototype theory [Rosch, 1973], whether an instance belongs to a concept is usually a matter of degree. This membership function is also related to the current issue: if John can be referred to as a "white-collar worker," but not a typical one, the influence of R_2 will be reduced.

From the above analysis, we can see that the previous solutions from non-monotonic logic and probability theory ignored several important factors when handling deduction with reference classes.

9.1.3 The NARS solution

Now Let us see how NARS treats the reference class problem.

Putting the previous example into Narsese, the premises are:

$$
\begin{aligned}
J_1 &: & \{S\} \rightarrow R_1 \ <f_1, c_1> \\
J_2 &: & \{S\} \rightarrow R_2 \ <f_2, c_2> \\
J_3 &: & (\#x \rightarrow R_1) \Rightarrow (\#x \rightarrow Q) \ <f_3, c_3> \\
J_4 &: & (\#x \rightarrow R_2) \Rightarrow (\neg(\#x \rightarrow Q)) \ <f_4, c_4>
\end{aligned}
$$

Since John shares one property with R_1 ("male") and two properties with R_2 ("male" and "white-collar worker"), we have $w_1 = w_1^+ = 1$ and $w_2 = w_2^+ = 2$. It follows that (assuming $k = 1$) $f_1 = f_2 = 1$, $c_1 = 1/2$, and $c_2 = 2/3$. Under the assumption that R_1 consists of 1000 positive samples, we have $f_3 = 1$ and $c_3 = 1000/1001$. Let us say that R_2 includes negative samples only, but leaves the number of samples, n, as a variable, to see how it affects the final evaluation of $\{S\} \to Q$. Therefore, we have $f_4 = 1$ and $c_4 = n/(n+1)$.

Applying the deduction rule, from J_1, J_3 and J_2, J_4, respectively, we get

$$J_5: \ \{S\} \to Q \ <1,\ c_1c_3>$$
$$J_6: \ \{S\} \to Q \ <0,\ c_2c_4>$$

Since the knowledge that "John is male" is used to evaluate both J_1 and J_2, and they are used in the derivation of J_5 and J_6, respectively, the evidence for J_5 and J_6 is correlated. Therefore, they cannot be merged by the revision rule. Instead, the choice rule is applied to pick the judgment that has a higher confidence as the conclusion. Which reference class will win the competition?

By solving the inequality $c_1c_3 > c_2c_4$ (that is, $(1/2)\times(1000/1001) > (2/3)\times(n/(n+1))$), we can see that

1. When $0 < n < 3$, R_1 is selected. The specificity priority of R_2 is undermined by the fact that the sample size of R_2 is too small.

2. When $n \geq 3$, R_2 is selected. The specificity priority can be established even by a pretty small sample size: with $|R_1| = 1000$ and $|R_2| = 3$, the prediction is still determined by R_2 due to its specificity.

If John is not a typical white-collar worker (i.e., $f_2 < 1$), R_2's confidence is smaller than c_2c_4, so it may need a bigger n for R_2 to be dominant.

Therefore, when NARS is selecting a reference class, several factors are balanced against one another, including specificity, typicality, sample size, and so on. It provides a generalization of the specificity priority principle, by taking more relevant factors into consideration.

NARS' approach is more general than the specificity priority principle in another way. The including of reference classes is only a special case for two judgments to be based on correlated evidence. It follows that the specific priority principle is a special case of NARS' choice rule.

How about competing reference classes that do not involve correlated evidence? Let us say in the previous examples, R_1 is still for "male," but R_2 is changed for "smoker and white-collar worker." If the deduced judgments J_5 and J_6 are not based on correlated evidence in some other ways, the two judgments will be combined by the revision rule of NARS. Other things being equal, R_2 has a higher priority, since it matches better with John's properties. However, in this case a higher priority only means a higher *weight* in determining the frequency of the conclusion. The judgment from the other reference class is not ignored. In this situation, the reference class competing is solved not by *choosing one of them*, but by *combining the two*.

Let us see how NARS treats the "Nixon Diamond" discussed in the study of non-monotonic logics [Touretzky, 1984]. This example assumes we know that Nixon is a Quaker, and Quakers are pacifists. We also know that Nixon is a Republican, and Republicans are not pacifists. From the above knowledge alone, should we predict Nixon to be a pacifist or not? Putting into the previous framework, in this problem we have "Nixon" as S, "Quaker" as R_1, "Republican" as R_2, and "Pacifist" as Q. By deduction, two conflicting judgments J_5 ("Nixon is a pacifist") and J_6 ("Nixon is not a pacifist") can be derived as in the previous example.

Since we can assume the un-correlation of evidence of the judgments (R_1 and R_2 have no known relation), J_5 and J_6 will be combined by the revision rule, and the result depends on the truth value of the premises.

1. If $f_1 = f_2$, $c_1 = c_2$, $f_3 = 1 - f_4$, and $c_3 = c_4$, for the conclusion we will get $f = 0.5$. That is, when the positive evidence and the negative evidence exactly balance with each other, the system is indifferent between a positive prediction and a negative prediction.

2. If $c_1 > c_2$, and the other conditions as in (1), we will get $f > 0.5$. That is, when Nixon shares more property with Quaker, the system will put more weight on the conclusions suggested by the evidence about Quaker.

3. If $f_3 > 1 - f_4$ or $c_3 > c_4$, and the other conditions as in (1), we will get $f > 0.5$. That is, when we have stronger statistical data about Quaker, the system will put more weight on the conclusions suggested by the evidence about Quaker, too.

In any situation, what NARS does is to combine the evidence from both sources. Even if "Quaker" is given a higher priority, the evidence provided by "Republican" still has its effect on the result. On the other hand, this kind of conflict does not always (though sometimes it does) cause complete indifference or ambiguity, as it does in non-monotonic logics [Touretzky, 1986].

In summary, compared with non-monotonic logics and probability theory, the processing of the reference class problem in NARS has the following characteristics:

1. While still following the specificity priority principle, several factors, such as sample size and degree of membership, are taken into account to quantitatively determine the priority of a reference class, and all the factors are projected into a common dimension, that is, the amount of evidence.

2. The specificity priority principle has been generalized into a "confidence priority principle," which will pick a judgment with the highest confidence among the competing ones, supported by correlated evidence. As discussed above, specificity is one way to get a high confidence, while the inclusion relation between reference classes causes evidence correlation.

3. When conflicting judgments come from different sources, the revision rule is applied to combine them by summarizing the evidence. This operation is not directly available in non-monotonic logics and probability theory.

Why cannot similar things be done in non-monotonic logics and probability theory? One of the major reasons is that the *confidence* (or equivalently, *amount of evidence*) measurement cannot be easily introduced there. From the view point of NARS, the confidence of all the default rules (in non-monotonic logics) and probability assignments (in probability theory) is 1, that is, they cannot be revised by accommodating its current evaluation to new evidence.

Therefore, the reference class problem provides another piece of evidence for the previous criticism on non-monotonic logic (Section 8.1) and probabilistic logic (Section 8.3), as general solutions to the reasoning under uncertainty problem.

9.2 Induction

Induction is a major topic in this book, because it has been called "the glory of science and the scandal of philosophy," as well as because the basic ideas of NARS to a large extent were formed during my study on the problem of induction.

9.2.1 The problem of induction

The term "induction" is usually used to denote the inference that derives *general* knowledge from *specific* knowledge. There are some people who call all non-deductive inferences "induction," but in this way the category includes too many heterogeneous instances to be studied fruitfully.

There are three major academic traditions in the study of induction. The *philosophical/logical* study concentrates on the formalization and justification of induction; the *psychological* study concentrates on the description and explanation of induction in the human mind; and the *computational* study concentrates on the implementation of induction in computer systems.

Though Aristotle mentioned induction as the method by which general primary premises can be obtained, he did not develop a theory for this type of inference, as he did for deduction. It was Bacon who for the first time proposed a systematical inductive method, with the hope that it could provide a general methodology for empirical science [Cohen, 1989]. However, such an approach was seriously challenged by Hume, who argued that the inferences that extend past experience to future situations cannot have a logical justification [Hume, 1748]. After Hume, most philosophical and logical work on induction are about the justification of the process. The mainstream approach is to use probability theory, with the hope that though inductive conclusions cannot be absolutely true, they can have certain probabilities [Carnap, 1950].

In recent years, the study of induction has been enriched by AI researchers. With computer systems as tools and platforms, different formalizations and algorithms are proposed and tested. In terms of the formal language used, we can further divide the existing approaches in this domain into three "families."

The first family uses propositional logic and probability theory. Let us say that S is a proposition space and P is a probability distribution function on it. Induction is defined in this situation as the operation of determining $P(H|E)$, where H is a hypothesis and E is available evidence, and both belong to S. The inference — or more precisely, calculation — is carried out according to probability theory in general, and Bayes' theorem in particular. This family is the mainstream of the philosophical and logical tradition of induction study [Keynes, 1921, Carnap, 1950, Good, 1983], and it has been inherited by the Bayesian school in AI [Korb, 1995, Pearl, 1988].

The second family uses first-order predicate logic. Let us say that K is the background knowledge of the system, and E is available evidence (both K and E are sets of proposition). Induction is defined in this situation as the operation of finding a proposition H that implies E and is also consistent with K. Because the inference from H and K to E is deduction, induction thus defined, as the inference from E and K to H, is often referred to as "reverse deduction." This family is very influential in machine learning [Michalski, 1993].

The third family uses term logic. Though Aristotle discussed induction briefly in his work [Aristotle, 1989], it was Peirce who first defined different types of inference in term logic, roughly in the following manner [Peirce, 1931]:

deduction	**induction**	**abduction**
$M \rightarrow P$	$M \rightarrow P$	$P \rightarrow M$
$S \rightarrow M$	$M \rightarrow S$	$S \rightarrow M$
$S \rightarrow P$	$S \rightarrow P$	$S \rightarrow P$

One interesting fact is that though Peirce's distinction of deduction, induction, and abduction is widely accepted, his formalization in term logic is seldom followed. Instead, the above definition is rephrased within the frame of predicate logic [Michalski, 1993]. We will see the subtle difference between these two formalizations later.

Obviously, NARS belongs to the term-logic family. Now let us see how NARS answers the questions about the aspects of induction.

9.2.2 To represent inductive conclusions

As mentioned previously, NARS represents all knowledge, including inductive conclusions, in Narsese. The simplest sentence has the form of "$S \rightarrow P$," with a truth value attached, which is determined according to the past experience of the system.

To decide truth according to the available evidence or according to a set of axioms is fundamentally different. In the former situation, no decision is final in the sense that it cannot be revised by future evidence. Each piece of evidence contributes, to a certain extent, to the evaluation of truth value. Therefore, truth value is always a matter of degree in a system like NARS.

This opinion is against a well-known conclusion proposed by Popper. He claimed that there is an asymmetry between verifiability and falsifiability — "a positive decision can only temporarily support the theory, for subsequent negative decisions may always overthrow it" [Popper, 1959].

The crucial point here is: what is the content of a *general statement*, or, in Popper's words, a *theory*?

According to my opinion, "Ravens are black" is a general statement, for which a black raven is a piece of positive (affirmative) evidence, and a non-black (e.g., white) raven is a piece of negative (rejective) evidence — the former verify an inheritance relation "$raven \rightarrow [black]$" to a certain extent, while the latter falsify it, also to a certain extent. When we say that "All ravens are black," it means that according to our experience, the inheritance relation between the two terms only has positive evidence, but no negative evidence. In this case, the truth value of the statement is still a matter of degree, though the frequency value happens to be at its maximum, 1.

What Popper referred to as theory are *universal statements*. Accordingly, when we say "All ravens are black," we mean that all ravens in the whole universe, known or unknown, are black. Such a statement can only be true or false, and there is no middle ground (if we ignore the fuzziness of the terms). We know the statement is false as soon as we find a non-black raven, but we need to exhaust all ravens in the universe to know it is true.

Such a formalization of inductive conclusions is shared by the Baconian tradition of induction [Cohen, 1989]. According to an approach proposed by Cohen, induction is a sequence of tests with increasing complexity, and the (Baconian) probability of a hypothesis indicates how many tests the hypothesis passed in the process.

If we accept the above definition of scientific theory, all conclusions of Popper and Cohen follow logically. However, why should we accept the definition? As a matter of fact, many empirical scientific theories have counterexamples, and we do not throw them away [Kuhn, 1970]. It is even more obvious when we consider our common-sense knowledge. A general statement like "Ravens are black" works well as our guide of life, even when we know that it has counterexamples. Such a statement can be applied to predict new situations, though its truth value is determined by past experience. We do hope to establish theories that have no known counterexamples, but it does not mean that theories with known counterexamples cannot be used for various practical purposes. Only in mathematics, where truth values are determined according to fixed axioms, do universal statements become available.

The above argument also serves as a criticism to the AI induction projects within the framework of binary logic [Korb, 1995]. To define induction as "finding a pattern to fit *all* data" makes it a luxury that can only be enjoyed in a laboratory. Though such an approach can produce research results, these results are hardly applicable to practical situations. Also, this over-idealization makes the process fundamentally different from the generalizations happening in the human mind. It is not even appropriate to justify this approach as "a preliminary step toward more complex studies," because when giving up the idea that "an inductive conclusion can be falsified once for all," the situation will become so different that the previous results are hardly useful at all.

Because in NARS truth values are determined by available evidence, we need to first precisely define what is counted as evidence and how evidence is quantitatively measured.

Though it is natural to say that a black raven is a piece of positive evidence for "Ravens are black," and a white raven is its negative evidence, Hempel points out that such a treatment leads to counter-intuitive results [Hempel, 1943]. If "Ravens are black" is formulated as $(\forall x)(Raven(x) \rightarrow Black(x))$, a green shirt will also be counted as a piece of positive evidence for the sentence, because it confirms the "logically equivalent" sentence $(\forall x)(\neg Black(x) \rightarrow \neg Raven(x))$ ("Non-black things are non-ravens"). Such a result is highly counterintuitive, and may cause many problems (for example, a green shirt is also a piece of positive evidence for "Ravens are white," for exactly the same reason).

Here I will not discuss the various solutions proposed for this paradox. It is enough to say that almost all of those attempts are still within the framework of predicate logic, whereas in the following we can see that the problem does not appear in term logics like NARS.

As we already know, in Narsese "Ravens are black" can be represented as "$(\#x \rightarrow raven) \Rightarrow (\#x \rightarrow [black])$." For this statement, "black ravens" are positive evidence, "non-black ravens" are negative evidence, and "non-ravens" are not directly relevant (according to the definition of evidence for implication statements in Chapter 5). On the other hand, "Non-black things are non-ravens" can be represented in Narsese as "$(\neg(\#x \rightarrow [black])) \Rightarrow (\neg(\#x \rightarrow raven))$." For it, "non-black non-raven" are positive evidence, "non-black ravens" are negative evidence, and "black things" are not directly relevant.

Comparing the two statements, we see that, in the terminology of NARS, they have the same negative evidence, but different positive evidence (as discussed in Section 5.1.4). In a binary logic, the truth value of a statement only indicates whether there is any negative evidence, so these two statements have the same truth value, or are "equivalent." In a logic where truth value is determined by both positive and negative evidence, they may have different truth values, and are no longer equivalent.

Therefore, what Hempel's paradox reveals is that "equivalent statements" in a binary logic do not necessarily have the same truth value when the system is extended into a multi-valued logic. This problem does not appear in NARS, because here the evidence for "Ravens are black" and "Non-black things are non-ravens" are different (therefore they usually have different truth values). In NARS, the existence of a green shirt is not directly relevant to whether ravens are black, just as our intuition tells us.

9.2.3 To generate inductive conclusions

The induction rule defined in Section 3.3.4 has the following form:

$$\{M \rightarrow P < f_1,\, c_1 >,\; M \rightarrow S < f_2,\, c_2 >\} \vdash S \rightarrow P < f_1,\, \frac{f_2 c_1 c_2}{f_2 c_1 c_2 + k} >$$

This section explains why the rule is defined in this way.

To show how the rule works on a concrete example, let us to back to the example used in Section 3.3.6, and let P be "*swimmer*," and S be "*bird*." To see if the rule makes intuitive sense, let us at first consider the following special situations.

1. When $f_1 = c_1 = f_2 = c_2 = 1$, M is a piece of (idealized) positive evidence for the conclusion. According to the previous definitions, in this case we have $w^+ = w = 1$ for the conclusion — that is, $f = 1,\ c = 1/(1+k)$. For the "Birds are swimmers" example, here M is a swimmer bird, such as a swan.

2. When $f_1 = 0,\ c_1 = f_2 = c_2 = 1$, M is a piece of (idealized) negative evidence for the conclusion. According to the previous definitions, in this case we have $w^- = w = 1$ for the conclusion — that is, $f = 0,\ c = 1/(1+k)$. For the "Birds are swimmers" example, here M is a non-swimmer bird, such as a robin.

3. When $f_2 = 0$, M is not an instance of S. In this case, no matter it is an instance of P or not, it provides no evidence for the conclusion, therefore $w = 0,\ c = 0$, and f is undefined. For the "Birds are swimmers" example, here M is not a bird (but a dolphin, for example).

4. When c_1 or c_2 is 0, one of the premises gets no evidential support, so the conclusion gets no evidential support either. That means

$w = 0,\ c = 0$. For the "Birds are swimmers" example, here either whether M is a bird or whether M is swimmer is completely unknown.

From these boundary conditions of the truth value function for induction, if all the variables take boolean values (either 0 or 1), we get $f = f_1$ and $w = and(f_2, c_2, c_1)$, here *and* is the Boolean conjunction of the arguments.

To generalize the Boolean function into real numbers, *and* is replaced by multiplication, and that gives us $w = f_2 c_2 c_1$. Using the equation $c = w/(w + k)$, finally we get the truth-value function used in the induction rule.

Because in NARS the truth value indicates the relation between a statement and available evidence, induction is "ampliative" in the sense that its conclusions are more general than its premises, but it is also "summative" in the sense that the conclusions claim no more support than they actually get from the premises. Therefore the traditional distinction between these two types of induction does not apply here [Cohen, 1989, Popper, 1959] in its original form.

On the other hand, the distinction between "truth-preserving" and "ampliative" inferences appears in a different form. In NARS, the confidence of deductive conclusions have a upper bound of 1, and we already know that the upper bound for induction is $1/(1 + k)$, which is smaller than 1. If all premises are absolutely certain, so are their deductive conclusions, but this does not hold for their inductive conclusions.

It needs to be stressed again that the truth value of the conclusion indicates the support provided by the evidence, rather than whether the statement corresponds to a fact in the outside world. An adaptive system behaves according to its beliefs, not because they guarantee success (which is impossible, as Hume argued), but because it has to rely on its experience to survive, even though the experience may be biased or outdated — this is what "adaptation" means.

In summary, my solution to Hume's problem is to justify induction (and all other inference rules) according to an experience-grounded semantics and the notion of adaptation.

9.2.4 To conduct inductive inference

Another feature that distinguishes the induction rule of NARS from other induction systems is that the rule is able to generate and evaluate an inductive conclusion at the same time.

Traditionally, the generating and evaluating of inductive conclusions (or hypotheses) are treated as two separated processes. The most well-known arguments on this issue were provided by Carnap and Popper, though they hold opposing opinions on induction in general [Carnap, 1950, Popper, 1959]. The consensus is that from given evidence, there is no effective procedure to generate all the hypotheses supported by the evidence, therefore the discovery of a hypothesis is a *psychological* process, which contains an "irrational element" or "creative intuition." On the contrary, the evaluation of a given hypothesis, according to given evidence, is a *logical* process, following a well-defined algorithm.

The above opinion is in fact implicitly based on the specific language in which the inductive process is formalized. In probability theory, there is no way to get a unique hypothesis H from given evidence E for the purpose of induction, because for every proposition X in the proposition space, $P(X|E)$ can be calculated, at least in principle. In first-order predicate logic, there are usually many hypotheses H that imply the given evidence E, and also are consistent with background knowledge K. In both cases, some heuristics can be used to pick up an inductive conclusion that has some desired properties (simplicity, for instance), but these heuristics are not derived from the definition of the induction rule [Mitchell, 1980, Haussler, 1988].

In term logic, the situation is different. Here premises of an inductive inference must be a pair of judgments that share a common subject, and the premises uniquely determine an inductive conclusion. (Of course, there is also a symmetric inductive conclusion if we exchange the order of the premises.) Therefore, in NARS we do not need an "irrational element" or domain-dependent heuristics, and the discovery of a hypothesis, in the current sense, also follows logic.

In NARS, induction is unified with other types of inferences, in the sense that the premises used by the induction rule may be generated by the deduction (or abduction, and so on) rule, and that the conclusions of the induction rule may be used as premises by the other rules. In particular, the revision rule may merge an inductive conclusion with a deductive (or abductive, and so on) conclusion.

Therefore, though NARS has an induction rule, it is not an "inductive logic," in the sense that it solves problems by induction only. An answer reported by NARS to the user is usually the cooperative result of several rules in a multi-step inference process. Though there are other "multi-strategy" inference models (which combine different types of inference), using first-order predicate logic [Michalski, 1993], attribute-value language [Giraud-Carrier and Martinez, 1995], or hybrid (symbolic-connectionist) representation [Sun, 1995], the term logic model, proposed by Peirce and extended in NARS, puts different types of inference in the same framework in a more natural, elegant, and consistent manner.

From the above discussion, we see that conclusions in NARS are based on different amounts of evidence, and, generally speaking, conclusions based on more evidence are preferred, because of their relative stability. However, since NARS is designed to be an open system, future evidence is always possible, therefore there is no way for the system to get "complete evidence" for an inductive conclusion.

A reasonable retreat is to use all evidence known to the system — the so-called "total evidence" [Carnap, 1950]. Unfortunately, this is also impossible, because NARS has insufficient resource. The system has to answer questions under a time pressure, which makes exhaustive search in knowledge space not affordable.

Moreover, in NARS the time pressure is variable, depending to the request of the user and the existence of other information-processing tasks. In this situation, even a predetermined "satisfying threshold" becomes inapplicable — such a threshold is sometimes too low and sometimes too high.

As described in Chapter 6, the control mechanism used in NARS is similar to an "anytime algorithm" [Dean and Boddy, 1988]. If the system is asked to evaluate the truth value of a statement, it reports the best conclusion (i.e., with the highest confidence) as soon as such a conclusion is found, then continues to look for a better one, until no resources are available for this task. In this way, from the user's point of view, the system may change its mind from time to time, when new evidence is taken into consideration. The system will never say that "This is the final conclusion and I will stop working on the problem."

The above discussion is directly related to the "acceptance" problem in inductive logic [Kyburg, 1994]. As put by Cohen, "what level of support for a proposition, in the light of available evidence, justifies belief in its truth or acceptance of it as being true?" [Cohen, 1989]. In NARS, there is no such a thing as "accepted as being true." Judgments are true to different extents, and the system always follows the best-supported conclusion (compared with its rivals), no matter what its truth value is — the standard is relative and dynamic, not absolute and static. In this way, an inductive conclusion also benefits from the refutation of competing conclusions, which is stressed by the Baconian tradition of induction [Cohen, 1989] — though its truth value may not change in this process, its relative ranking becomes higher.

According to the definition given be Peirce, the difference among deduction, abduction, and induction is the position of the shared term in the two premises. This property of term logic makes it possible for NARS to combine different types of inference in a "knowledge-driven" manner. In each inference step, the system does not decide what rule to use, then look for corresponding knowledge. Instead, it picks up a task and a belief which share a term, and decides what rule to apply according to the position of the shared term (as described in Chapter 6). In general, an inference process in NARS consists of many steps. Each step carries out a certain type of inference, such as deduction, abduction, induction, and so on. These steps are linked together in run-time in a context-dependent manner, so the process does not follow a predetermined algorithm.

Therefore, NARS is not an "inductive machine" which uses an effective algorithm to generate inductive conclusions from given evidence. Carnap's argument against the possibility of this kind of machine [Carnap, 1950] is still valid. However, this argument does not prevent us from building a computer system that can do induction. The system does not have a general purpose induction algorithm, but can solve problems under its knowledge and resource constraints, and in the problem-solving activities there are inductive steps.

9.3 Abduction

Since in NARS abduction and induction are duals of each other (because extension and intension are duals of each other), most of the previous discussions on induction also apply to abduction. In the following, I will not repeat them, but focus on the special issues that distinguish my approach toward abduction from the other approaches.

9.3.1 Two definitions of abduction

Approaches of defining abduction can be classified into two types: syllogistic and inferential. An inferential definition identifies abduction as a type of inference *process* that carries out a certain cognitive function, such as explanation or hypothesis generation, while a syllogistic definition specifies it as a type of inference *step* with a specific pattern [Flach and Kakas, 2000].

As defined in NAL-1 and NAL-5, in NAL the distinction among deduction, abduction, and induction is formally specified at the inference-step level, according to the position of the shared term in the premises. Such a formal definition makes discussions about them clear and concrete.

To use a formal definition to distinguish various inference types does not prevent us from attributing them with different cognitive functions. Given the definition used in NAL, it is valid to say that among the three, only deduction can produce conclusive results, while the other two only produce tentative results. Both abduction and induction can be seen as "reversed deduction," and the former usually corresponds to explanation, and the latter to generalization. These descriptions are similar to the ones proposed as inferential definitions of the three types. However, in NAL these descriptions are *secondary*, derived from the syllogistic definition. This approach has the advantage of avoiding ambiguity and oversimplification in the definition, and at the same time preserve the intuitive meaning of the terms (i.e., deduction, abduction, and induction) associated with different types of inference.

Though abduction defined in NAL usually can be interpreted as "explanation," to define "abduction" as "explanation" at the inference-process level is a quite different decision. This is the case because what we called "explanation" in everyday thinking may include complex cognitive processes where multiple types of inference are involved. Therefore, to abstract such a process into a consistent and non-trivial pattern is not an easy thing to do, if it is possible at all.

For the same reason, to define abduction as "inference toward the best explanation" makes things even harder, because besides the derivation of explanations, this definition further requires evaluation of explanations and comparison of competing candidates. In this process, many other factors should be taken into account, such as simplicity, surprisingness to the system, and relevance to the given context. If we cover all of these issues under "abduction," it becomes such a complex process that few concrete conclusion can be made. Such a definition is not wrong, but not very useful.

9.3.2 Multi-valued vs. binary

In the framework of binary logic, abduction is usually defined formally as "reverse deduction" which starts from a given conclusion and background knowledge to find a premise that is consistent with the background knowledge, and derives the conclusion deductively.

Such a definition is logically sound, and can lead to fruitful results. However, it ignores certain factors that are crucial for a system working with insufficient knowledge and resources.

In empirical science and everyday life, we usually do not throw away theories that have known counterexamples and inexplicable phenomena. If we do that, there is hardly anything left. Since we usually have insufficient knowledge in these domains, we have to live with imperfect knowledge, because they are still far better than random guesses.

When selecting among competing explanations and hypothesis, measurement of (positive and negative) evidence becomes necessary — if no explanation is perfect, then the one with more positive evidence and less negative evidence is preferred, which is what is measured by the *frequency* defined in NAL. Since evidence may come from time to time, incremental revision becomes inevitable, which requires the amount of evidence to be represented in some way, and this is how the *confidence* measurement becomes necessary.

These measurements enrich our understanding of the inference rules. In the truth-value functions, we can see that the fundamental difference between deductive inference and non-deductive (such as abductive or inductive) inference is in the confidence (not the frequency) of the conclusion. In deduction, if both premises are completely true, so is the conclusion. However, in abduction and induction, the confidence of the conclusion is much lower in this situation, meaning that the conclusion is tentative even when the premises are certain, and can be revised by new evidence.

To ignore quantity of evidence means it will be hard for the system to distinguish hypotheses that have a little of negative evidence from those that have a lot. Even for a hypothesis for which only positive evidence has been found, the amount of evidence still matters — a hypothesis confirmed only once is quite different from a hypothesis conformed a million times. For these reasons, to study abduction in binary logic is not wrong, but not very useful. Unfortunately, it is still the most common approach to the topic of "abduction" in the current AI research.

9.4 Implication

This section addresses two topics in higher-order inference, both about the implication relation.

9.4.1 Implication and relevance

A look at the grammar of Narsese reveals the origin of the intuition behind the design: first-order NAL is closely related to set theory,[1] and higher-order NAL is closely related to propositional logic — both contain logical constants for *negation* ("$\neg$"), *conjunction* ("$\wedge$"), *disjunction* ("$\vee$"), *implication* ("$\Rightarrow$"), and *equivalence* ("$\Leftrightarrow$").

Though the intuitive meanings of the above constants are similar in these two logic systems, there is a fundamental difference. In propositional logic, all of the five logic constants are truth-functional operators that form compound propositions, whose truth values are

[1]Their relation will be discussed in the next chapter.

fully determined by those of their components. In NAL, on the contrary, the five constants belong to two different categories. The first three are *term operators* that form compound (higher-order) terms; the last two are (higher-order) *relations* (i.e., *copulas*), which are not purely truth-functional. Consequently, when P and Q are both Narsese statements, so do $(P \Rightarrow Q)$ and $((\neg P) \vee Q)$, but the latter two are no longer equivalent to each other in NAL.

In propositional logic, $P \Rightarrow Q$ is equivalent to $(\neg P) \vee Q$. Though this equivalence is useful for various purposes, it suffers from the well-known "implication paradox," which says that $P \Rightarrow Q$ is true when P is false (Q can be anything) or Q is true (P can be anything). Though logically consistent with propositional logic, this result is highly counter-intuitive, and it gives people a feeling that some important thing is missing in the definition of implication in propositional logic — P and Q should be somehow *relevant* to each other, which is assumed by the "if ... then ..." structure in natural languages [Copi, 1982].

A whole branch of logic, *relevant logic* [Read, 1989], has been developed specially for this issue. I will not review that type of logic here, but to mention a key property of it, that is, as far as I know, all the works in that branch of logic are still within the framework of predicate logic and propositional logic. On the other hand, in NAL, this problem does not appear in the first place.

In NAL, *implication*, in its idealized form, is defined to be a reflexive and transitive binary relation from one statement to another. In its realistic form, it is multi-valued, with its truth value defined according to available evidence. Here evidence is measured by comparing the sufficient and necessary conditions of the two statements.

As a term logic using syllogistic rules, in NAL the two premises of an inference step must share a common component, otherwise no conclusion can be derived. Also, the conclusion shares terms with the premises, respectively. As a result, the three must be related to one another in their meanings, according to EGS — two terms are related in their meanings as soon as they appear in the same belief of the system.

Specially, in the induction rule introduced in NAL-5, $(P \Rightarrow Q)$ can be derived from P and Q only if the two premises are based on the same (implicitly represented) evidence. From an arbitrary pair of statements, nothing will be derived — to know their truth values is not enough.

Even when $(P \Rightarrow Q)$ is derived from P and Q, its confidence is low, because the rule is induction. Only when P and Q have been repeatedly supported by the same evidence for many times (and the evidence is different at each time), can $(P \Rightarrow Q)$ then become more confident (by merging the individual conclusions with the revision rule).

In NAL, $(P \Rightarrow Q)$ and $((\neg P) \vee Q)$ are no longer equivalent, but still related to each other. Especially, they have the same negative evidence (that is, when P is true and Q is false). Here the situation is exactly the same as the one revealed by the previous analysis on the "confirmation paradox," that is, since in binary logic a truth value only indicates the existence of negative evidence, statements with exactly the same scope of negative evidence are treated as equivalent. In multi-valued logic with EGS, however, these statements may have different truth values if they have different positive evidence. In NAL, two statements are equivalent if they have the same scope and amount of both positive evidence and negative evidence.

Now we can see that, in this sense, both "confirmation paradox" and "implication paradox" are problems of *binary predicate logics with MTS*, but not problems of *logic* in general. To properly capture the intuitive meaning of concepts like "confirmation," "implication," and so on, we need a multi-valued term logic with an experience-grounded semantics.

9.4.2 Implication and causation

Causal inference is a very important cognitive function, and it has attracted researchers from AI [Pearl, 2000], psychology [Cheng, 1997], and philosophy [Sosa and Tooley, 1993].

What differs NAL from the other approaches in causal inference is: though NAL also attempts to capture all aspects of causal inference, it does not treat "causal inference," as well as the related concepts like "causal relation" and "causation," as logical constants, in the sense that there are inference rules especially responsible for causal inference.

Unlike the Bayesian interpretation of causal inference [Pearl, 2000] (in which causal relation is formalized by conditional probability), in NAL causal inference is carried out by formal inference rules on a formal language. Even in this framework, how to define "causal relation" is still a controversy. Logically speaking, it has been defined as sufficient condition, necessary condition, sufficient-and-necessary condition, or even something more complicated, by different people [Copi, 1982, Sosa and Tooley, 1993]. In practical applications, there are debates on the "cause" of all kinds of events in every newspapers everyday.

In my opinion, this situation indicates that "causation" is a concept we use to organize our experience, and especially for the prediction of the future [Anderson, 1990]. Since in different domains predictions are made in different ways, the meaning of this concept is context dependent. For this reason, we should not expect a physicist, a biologist, an economist, and a historian to agree on the accurate definition of "cause."

Even when this is the situation, it is still possible to provide a common logical foundation for all the different usage of the concept "causation" (and the related concepts). In NAL, causal inference, or prediction in general, is seen as consisting of two basic aspects, a logical one and a temporal one. The logical factor is represented by the implication relation (and its variant, the equivalence relation), which indicates when a statement can be derived from another one. The temporal factor is represented by the temporal orders introduced in NAL-7. According to the common usage of the term, a "cause" of an event E should be a *pre*condition of it, though it depends on the context whether it is a sufficient one, a necessary one, an equivalent one, or even something more complicated.

In this way, the NAL logical constants provide the "greatest common factor" of all kinds of causal relations, which are treated in NARS as "ordinary relations" (as defined in Section 4.4), with meanings learned from the experience of the system. They can include additional considerations on causation, such as the distinctions between "causation" and "covariation," between "causes" and "enabling conditions," and so on [Cheng, 1997, Sosa and Tooley, 1993], though none of these considerations are included in the logical constants of NAL.

In general, a question with the form of "$? \Rightarrow Q$" is a task looking for an *explanation* of statement Q, and a question with the form of "$P \Rightarrow ?$" is a task looking for a *consequence* of statement P. Causal explanation and causal consequence are just special cases of the above general higher-order inference tasks. As described before, both tasks, as well as yes/no question "$(P \Rightarrow Q)\,?$," can be answered by a judgment "$P \Rightarrow Q < f, c >$." If there are multiple candidate answers, the choice rule is used to pick the best one. If the question is not about the implication relation in general, but about a special causal relation "*cause*," then the questions will be like "$? \circ\!\!\rightarrow (\perp \mathit{cause} \;\diamond\; Q)$," "$? \circ\!\!\rightarrow (\perp \mathit{cause}\ P \diamond)$," and "$((P \times Q) \circ\!\!\rightarrow \mathit{cause})\,?$," respectively, and the answer will depend on the current meaning of "*cause*."

Like the other statements, an implication statement can be derived in more than one way. For example, when a causal judgment "$P \Rightarrow Q < f, c >$" is derived by induction, it may correspond to a causal hypothesis obtained from observed regularity; when it is derived by deduction, it may correspond to a causal hypothesis obtained according to an underlying mechanism; when it is derived by abduction, it may correspond to a causal hypothesis obtained through an explanation. If there is more than one way to support a hypothesis, it will get a higher confidence value after the revision rule merges evidences from different sources, though none of the sources are absolutely necessary for the conclusion to be confident. The same is true for temporal implication/equivalence relations, and the various causal relations obtained in experience. Therefore, in NARS there is no separate rule set for "causal inference" — all inference rules may contribute to the selection and evaluation of various "causes" and "effects."

As an answer to Hume's question on the validity of causal induction [Hume, 1748], in NARS causal inference is a way for the system to organize its experience, rather than a way to find "natural causal laws." The truth value of such a conclusion measures its available evidential support, not its distance to the "objective truth" — even if the system gets such a truth, it cannot confirm the case, given its insufficient knowledge and resources. Under AIKR, all causal beliefs in the system may be revised by future evidence, and none of them fully specifies the causes or consequences of any event. Nevertheless, the causal beliefs still serve a crucial role in the adaptation process of the system.

Chapter 10

NAL as a Logic

As specified previously, NARS is a reasoning system implementing a logic NAL. This chapter discusses the relationship between NAL and other related logics studied in AI and other disciplines.

10.1 NAL as a term logic

In the history of logic, there are the *Term Logic* tradition (TL in the following) and the *Predicate Logic* tradition (PL in the following). The defining differences between TL and PL are in the format of sentences and the format of inference rules. NAL belongs to TL, and it can be seen as an extension of the previous term logics, especially, of Aristotle's Syllogism.

10.1.1 Two traditions in formal logic

In TL, an atomic statement is *categorical*, in the sense that it has the form of "S c P," where S is the *subject term* of the statement, P the *predicate term*, and c is a *copula*. Intuitively, the statement says that "S is a kind of P," or some variation of it, like "S is a P." For example, "Tweety is a bird" can be represented as "$Tweety$ c_1 $bird$," where c_1 means "to be an instance of," and "Birds are animals" can be represented as "$bird$ c_2 $animal$," where c_2 means "to be a kind of."

In PL, an atomic statement (usually called a *proposition*) is *functional*, in the sense that it has the form of "$P(a_1, \ldots, a_n)$," where P is a *predicate*, and $a_1, \ldots, a_n$ ($n \geq 1$) are *arguments*, like in a mathematical function. Intuitively, it represents that "There is a P relation among entities $a_1, \ldots, a_n$ (in that order)." When $n = 1$, "$P(a_1)$" represents that a_1 has property P. Beside constant entities addressed by name, a variable can be used with a (universal or existential) quantifier to represent an entity, too. Compound statements are formed from simpler (atomic or not) statements by truth-value functions, including *not* ($\neg$), *and* ($\wedge$), *or* ($\vee$), *implication* ($\supset$), and *equivalence* ($\equiv$). For example, "Tweety is a bird" can be represented as "$Bird(Tweety)$," and "Birds are animals" can be represented as "$(\forall x)(Bird(x) \supset Animal(x))$." *Propositional logic* is a subsystem of predicate logic, where the internal structures of atomic statements are not explicitly represented.

In summary, the difference between the two traditions in sentence format is whether to stay with a "subject-predicate" pattern, with a "copula" as logical constant to connect the two terms.

The differences in the format of inference rules between these two traditions are stem from their differences in sentence format.

In TL, typical inference rules are *syllogistic*, in the sense that each rule takes two premises sharing one term, and the conclusion is formed between the other two terms. For example, from "$Tweety\ c_1\ bird$" and "$bird\ c_2\ animal$," such a rule can derive "$Tweety\ c_1\ animal$," that is, "Tweety is an animal."

In PL, typical inference rules are *truth-functional*, in the sense that the premise P can derive conclusion C if and only if "$P \supset C$" is true. The truth value of "$P \supset C$" is determined only by the truth values of P and C, and has nothing to do with their content. For example, from "$Bird(Tweety)$" and "$(\forall x)(Bird(x) \supset Animal(x))$," such a rule can derive "$Animal(Tweety)$," again, "Tweety is an animal."

As shown by the above example, in simple situations the functionality of TL and PL are roughly equivalent. However, this is not the case in general.

Because of the features described above, TL is also called "categorical logic" or "syllogistic logic." This tradition was established by

Aristotle's Syllogistic [Aristotle, 1989][1]. Since then, the TL tradition had dominated the history of logic for more than two thousand years, until the coming of "mathematical logic."

In the attempt of putting mathematics on a solid logical foundation, the founders of FOPL (First-Order Predicate Logic) judged categorical sentences and syllogistic rules as to be unnecessarily restrictive, and invented the language of PL, which is more similar to the languages used in mathematics [Frege, 1970, Whitehead and Russell, 1910]. Since then, FOPL has been taken by many people as "the logic," or at least a core that should be included in any logic system. At the same time, TL is usually regarded as only of historical value and as only covering special cases of FOPL. Lukasiewicz analyzed Aristotle's Syllogistic "from the standpoint of modern formal logic," though his goal is to correct certain misunderstandings on Aristotle's work, not to use TL to challenge PL. Actually, he thought that it was a prejudice to say that every proposition has a subject and a predicate [Lukasiewicz, 1951].

In recent years, the only notable voice that has used TL to challenge PL came from the work of Sommers and Englebretsen [Englebretsen, 1981, Englebretsen, 1996, Sommers, 1982, Sommers and Englebretsen, 2000]. The major consideration in their work is the closeness of TL to natural language (given by the subject-predicate structure), and the simplicity of inference in TL (which may be controversial). They extended Aristotle's Syllogism by introducing singular terms, relational terms, unanalyzed statements, compound statements, and so on. The result is a logic system, Term-Functor Logic (TFL, also called Algebraic Term Logic), which is functionally similar to FOPL, while still keeping the defining syntactic properties of TL, that is, categorical sentences and syllogistic rules [Sommers and Englebretsen, 2000]. Though their work is well known, it has not attracted many follow-up or application works.

---

[1] Aristotle's logic does not exactly fit the above description about TL in general, because it did not explicitly use a copula in his representation [Geach, 1968]. Singular term like "Tweety" is not allowed in Aristotle's logic, and each syllogism is an implication statement, not an inference rule [Lukasiewicz, 1951]. However, Aristotle's logic is still the most typical example of TL.

10.1.2 NAL and other term logics

NAL can be seen as an extension and revision of Aristotle's logic.

With respect to sentence format and grammar in general, Narsese has the following new features:

- revising the implicit binary copula in Aristotle's logic into the multi-valued inheritance relation,
- evaluating the inheritance relation according to both extensional evidence and intensional evidence,
- adding variants of inheritance to represent similarity, instance, and property relations,
- adding compound terms to represent sets, intersections, and differences formed from existing terms,
- adding products and images to represent ordinary relations that cannot be treated as copulas,
- adding higher-order relations and statements to represent statements about statements,
- adding temporal attributes to indicate the temporal relations between events,
- adding procedurally interpreted statements to represent operations that can be executed by the system itself,
- adding non-declarative sentences to represent questions and goals.

With respect to inference rules, NAL extends Aristotle's Syllogism in the following aspects:

- extending the inference rules to work on multi-valued statements,
- using an experience-grounded semantics to justify the inference rules,
- adding non-deductive inference rules,

- adding rules to compose/decompose various types of compound terms,
- adding higher-order inference rules,
- using backward inference rules to process non-declarative sentences.

Though Aristotle briefly mentioned induction in his work, it was Peirce who introduced the deduction/induction/abduction trio in the framework of term logic, by defining the other two as "reversed deduction" in different forms [Peirce, 1931]. NAL keeps the usage of the three terms. However, the compound terms and truth-value functions are not from Peirce (though he did try to combine logic with probability theory).

One traditional criticism to TL is on its expressive power, which is usually judged as lower than that of PL. Though this conclusion applies to Aristotle's logic, which can only represent certain relations between simple terms, it does not apply to other logics in the TL tradition, such as TFL [Sommers and Englebretsen, 2000] and NAL. There are some similarity between NAL and TFL in the way they extend the language of term logic (such as introducing singular terms, relational terms, unanalyzed statements, compound statements, and so on). However, TFL is designed to be functionally equivalent to FOPL, so it is still a binary deductive logic, while NAL is not.

10.1.3 NAL and set theory

Though in the previous discussion I presented NAL as a term logic, initially the idea about its language and rules came from set theory, and I did not recognize its relationship with the work of Aristotle and Peirce until several years later.

At the beginning of the project, concepts were treated as a sets, with three basic set-theoretic relations among them: membership ("$\in$"), subset ("$\subset$"), and equal ("$=$") [Wang, 1986].

Though such a representation looked very natural and general, I gradually realized its conflict with AIKR, and moved away from classic set theory. Later, I redefined the relations — *subset* becomes *inheritance*, *membership* becomes *instance* (also called "singular inheritance" previously), and *equal* becomes *similarity* (also called "symmetric inheritance" previously), though the same symbols had been kept for a while (see [Wang, 1995a] for example).

Set theory is closely related to term logic (and that is why I discuss it here). It is possible to map Aristotle's syllogisms into theorems in set theory. In NAL, the intuition behind compound terms in NAL-2 (sets), NAL-3 (intersections and differences) and NAL-4 (products and images) all came from set theory. Also, set theory has been used in the design of NAL, as part of its meta-language.

On the other hand, there are several major differences between set theory and NAL (in its current form). Obviously, classic set theory does not have uncertainty measurement and non-deductive inference in it.

A fundamental and subtle difference between NAL and set theory is that in NAL the *inheritance* relation is considered as more fundamental than the *instance* relation, while in set theory the *subset* relation is defined by the *membership* relation. One consequence of this is that in NAL extension and intension, and therefore evidence, are defined by *inheritance*, not by *instance*.

For a person with set-theoretic intuition, "$S \rightarrow P$" will be seen as a statement between two sets S and P, and interpreted as "S is a subset of P." In this way, the frequency of this statement should be $|(S \cap P)|/|S|$, that is, the proportion of instances of S which are also instances of P. Furthermore, when deriving such a conclusion by induction from premises $\{M \rightarrow P,\ M \rightarrow S\}$, the amount of evidence of the conclusion should be $|(S \cap M)|$, that is, the number of instances of M that are also in S.

Though it looks natural, such an interpretation is not acceptable in the context of NARS, because it is based on a different semantics. According to this semantics, a term is a set, with its meaning fully determined by its instances. On the contrary, in NARS the meaning of a term is determined by its extension and intension, which are sets of terms that are directly linked to it by the inheritance relation. Therefore, in "$\{tomato \rightarrow vegetable,\ tomato \rightarrow plant\}$," the evidence is *tomato* (as a term), not instances of *tomato* (as objects in the world).

Furthermore, what counted as evidence for "*tomato* $\rightarrow$ *vegetable*" and "*tomato* $\rightarrow$ *plant*" are different from what is counted as evidence for "*vegetable* $\rightarrow$ *plant*."

In the terminology of set theory, the extension of a term in NAL is not "the set of its elements," but "the set of its subsets." Furthermore, a meaningful term in NARS only needs to have non-empty extension and intension, and it does not need to be represented as a set of instances. Consequently, uncountable nouns (mass nouns and abstract nouns) are represented and processed more naturally in NAL, because they don't need to be seen as sets anymore. We will return to this topic in more detail a little later.

Such a treatment of extension is necessary for the duality between extension and intension in NAL. This elegant result comes from their definition, as opposite directions of the inheritance relation. The same result would not be obtained if *instance* (or *property*) were taken as the primary relation in NAL.

Since a term in NAL cannot been seen as a set, usually the inference rules of NAL should not be analyzed using Venn diagrams, which is a set-theoretic representation.

It is possible to represent the set-theoretic relations (membership, subset, and so on) in NAL — that is, as ordinary relations introduced in NAL-4. These relations are on a different level from the copulas, which are"built-in" relations. For example, the transitivity of the inheritance relation is implemented by the deduction rule (truth value omitted):

$$\{M \rightarrow P\,,\ S \rightarrow M\} \ \vdash\ S \rightarrow P$$

which can derive judgment "$term_a \rightarrow term_c$" from "$term_b \rightarrow term_c$" and "$term_a \rightarrow term_b$." On the other hand, the transitivity of the subset relation is represented by the following implication statement (truth value omitted):

$$\begin{aligned} & (((\{\#y\} \times \{\#z\}) \rightarrow subset) \wedge ((\{\#x\} \times \{\#y\}) \rightarrow subset)) \\ \Rightarrow\ & ((\{\#x\} \times \{\#z\}) \rightarrow subset) \end{aligned}$$

which can be used, together with "$(\{set_b\} \times \{set_c\}) \rightarrow subset$" and "$(\{set_a\} \times \{set_b\}) \rightarrow subset$," by the (higher-order) deduction rule to derive "$(\{set_a\} \times \{set_c\}) \rightarrow subset$."

Though the above two inferences are intuitively similar, they are carried out quite differently in NARS.

10.2 NAL vs. predicate logic

The mainstream of modern logic is dominated by PL, especially, by FOPL. In the following, let us see how NAL differs from PL in general, and FOPL in particular.

10.2.1 Categorical vs. functional

As mentioned previously, TL uses "categorical" sentences (with the "subject-predicate" format), and PL uses "functional" sentences (with the "predicate-arguments" format).[2]

TL and PL use different formats because they come from different backgrounds. In the age of Aristotle, logic was an attempt to regulate and formalize (to a certain extent) inference carried out in everyday life, usually in a natural language. Since the "subject-predicate" syntax is shared by many natural languages, TL is closer to a "logic of natural language." [Sommers, 1982].

At the age of Frege, the focus of study in logic turned to the foundation of mathematics. Consequently, people preferred a logical language that is more similar to the language used in mathematics. Since $P(a_1, \cdots, a_n)$ is just like the format of a mathematical function, it is more suitable for the job. Also, since "The Greeks defeated the Persians at Plataea" and "The Persians were defeated by the Greeks at Plataea" have the same conceptual content, their syntactic difference (such as which term is the subject) can be ignored in logic [Frege, 1970]. To the mathematical logicians, to force all statements into the "subject-predicate" format seems neither necessary nor possible.

[2]It is important to see that the notion of "predicate" has different (though related) senses in TL and PL, and neither is the same as how the notion is used in linguistics. For example, in the sentence "The morning star is the same as the evening star," in TL "predicate" is the *term* "evening star," in PL it is the *relation* "to be the same as," and in linguistics it is the *phrase* "is the same as the evening star."

The "categorical vs. functional" difference does not only influence the naturalness and expressibility of the language. It also leads to different ontological commitments, or different categorical systems.

The distinction between predicates and arguments in PL is *absolute*, in the sense that nothing can be both a predicate and an argument. On the contrary, the distinction between subject and predicate in TL is *relative*, in the sense that a term can be the subject of one statement and the predicate of another statement. For PL, its domain of application must be perceived as consisting of a set of individuals, with properties (defined on individuals) and relations (defined among individuals). Through an interpretation, the individuals are mapped into arguments, and the properties/relations into predicates, in the logical language. On the contrary, for NAL, its domain of application consists of interrelated concepts, and many of them play different roles in different relations.

To see a domain as consisting of individuals and properties/relations is usually good enough for mathematics, but such an ontological commitment has problems outside mathematics.

It is well known that in FOPL it is hard (though not impossible) to reason on uncountable nouns, such as mass nouns (e.g., "water") and abstract nouns (e.g., "intelligence"), because such a noun should not be treated as a set of individuals [Fox, 2000]. How to represent "water," so that "Water is a kind of liquid" and "Raindrop is a kind of water" derive "Raindrop is a kind of liquid"? Neither a predicate representation nor an argument representation works well in this case. A naive solution would be to represent uncountable nouns as predicates (just like countable nouns), so that the above inference becomes to derive "$(\forall x)(Raindrop(x) \supset Liquid(x))$" from "$(\forall x)(Water(x) \supset Liquid(x))$" and "$(\forall x)(Raindrop(x) \supset Water(x))$" in FOPL. However, in such a solution, it is hard to explain what x stands for semantically, and in practical usage, such a solution ignores the conceptual difference between countable and uncountable nouns.

Though the same problem exists in Aristotle's logic (where a term roughly corresponds to a set), it is easier to be solved in the TL framework. As described previously, the inheritance/instance relations in NAL are intuitively similar to the subset/membership relations in set theory, but in NAL "inheritance" is not defined on "instance" (while in set theory "subset" in defined on "membership"). Consequently, in NAL the above relations are simply inheritance relations among terms "*raindrop*," "*water*," and "*liquid*." Though in NAL a set can be defined, not all terms are eventually reduced to sets, and the domain is not reduced into individuals with their properties and relations.

Another related topic is in the notion of "higher-order" inference. In FOPL, variables can only represent individuals, but not predicates. To represent properties shared by many predicates, a second-order (or higher-order) logic is needed. Here the concept of a "higher-order" directly comes from the absolute distinction between predicate and individual.

In a TL, since the distinction between subject and predicate are relative, the "properties of properties" (as well as "instances of instances") can be represented as other statements, and no "higher-order" structure is needed here. In NAL, the notion "higher-order" inference still exist, but as we have seen, it means "inference about statements on statements," and can naturally be integrated with first-order inference.

For a logic that has to work with insufficient knowledge and to support an evolving categorical structure, it is better to avoid the rigid predicate/argument distinction, and use the more flexible subject/predicate distinction.

10.2.2 Inheritance as the primitive relation

Since a copula in TL can be defined as a predicate in PL, and a non-copula relation can be represented in TL as a "relational" term, there is no difference in expressive power between "categorical" languages and "functional" languages on this aspect. Instead, the difference is in whether to represent the copulas as logical constants.

While in PL all relations are treated as equal, in TL, and especially in NAL, several relations (*inheritance*, with its variants *similarity*, *instance*, *property*, *instance-property*, *implication*, and *equivalence*), are treated as logical constants, or "built-in" relations, of the logic. Their meaning is not experience-grounded, but defined in the meta-language (as described earlier in the book) and completely reflected in their processing by the inference rules.

The most typical inference rules in TL are *syllogistic* in the sense that each takes two premises (in subject-predicate form) that share a common term, and the conclusion (also in subject-predicate form) is about the relation between the other two (unshared in premises) terms. In NAL, such a rule is justified according to the transitivity of the inheritance relation, and each rule has a truth-value function that determines the truth value of the conclusion. In this way, the conclusion and the premises are related both in truth value and in meaning.

In FOPL, the relation among premises and conclusion is purely *truth-functional*, and not semantic, in the sense that in an inference process only truth value matters, and regardless of the contents of the propositions involved. This property leads to counter-intuitive results, as shown by the well-known "implication paradox." As discussed in Section 9.4.1, this problem does not appear in NAL, because of the nature of TL.

A closely related problem is consistency. FOPL cannot tolerate any contradiction in its premise set, because any arbitrary proposition can be derived from a contradiction, and the system will become practically useless. Again, various "paraconsistent logics" are proposed, within the PL tradition [Priest et al., 1989]. Such a problem does not appear in NAL, simply because the inference rules are not purely truth-functional. Not only that in NAL there is a revision rule that deals with contradictory evidence, but also that a contradiction is a "local event" that only influences the terms directly involved in it.

In PL, abduction and induction are often defined as "reversed deduction," in the sense that they are inference processes that produce hypotheses consistent with background knowledge and implying the given conclusion (plus some additional properties) [Flach and Kakas, 2000]. Defined in this way, when the background knowledge and conclusion to be implied are given, the hypotheses may not be fully determined, that is, many hypotheses may satisfy the condition. Again, as discussed in Section 9.2.4, this problem does not appear in NAL, because of the nature of TL.

All the above properties of NAL come from the fact that *inheritance* and its variants are defined as logic constants, and that all inference rules are defined according to the meaning of these constant relations. By treating them in the same way as other relations and defining the inference rules as purely truth-functional, PL does not have the above properties, and many problems appear.

Some people have seen the specialty in what I call "inheritance" in NAL. For example, behind description logic and other class-based representations [Brachman and Schmolze, 1985, Donini et al., 1996], a major motivation is to introduce the inheritance mechanism into the PL tradition [Kurtonina and de Rijke, 1999]. Consequently, such a system has a TL component and a PL component (as shown by the ABox/TBox distinction). By putting everything in the TL framework, NAL provides a more compact and consistent alternative.

Semantically, what makes inheritance and its variants different from the other relations is that they link two terms together, and indicate that one can be used as the other, in certain context. Ulam and Hofstadter have argued that the "as" relation (as in "concept A can be used *as* concept B") plays a central role in cognition and intelligence [Rota, 1989, Hofstadter, 1993a, Hofstadter and FARG, 1995]. The result of NAL provides a piece of concrete evidence for this insight.

10.2.3 Evidence and truth value

Among the factors discussed in this section, the most important one for NAL to be designed in the framework of TL, rather than PL, is the definition of *evidence*.

For a logic that uses insufficient knowledge to answer questions, the number one issue is Hume's Problem: from limited past experience, how to justify the derived conclusions that will be applied to future situations [Hume, 1748].

As discussed previously, the solution to this problem in NAL is partly provided by an "experience-grounded" semantics, in which "truth value" is defined as a function of available evidence, not a "distance" between a statement and an objective "state of affairs." Consequently, a statement is "true" (to a degree) means that it is supported by past experience (to that degree), not that it will be confirmed in the future (even to a degree).

To apply this semantics to a formal language, for any given statement S in the language, we need to define its (positive and negative) evidence, as well as a measurement on the amount of evidence. As discussed in Section 9.2.2, in FOPL such an attempt lead to Hempel's "Confirmation Paradox."

A related problem is Wason's "Selection Task." In this psychological experiment, subjects see four cards showing symbols like E, K, 4, and 7, and know that each card has a letter on one side and a number on the other. The task is to choose the cards that need to be turned over in order to determine whether the following rule is true or false: "If a card has a vowel on one side, then it has an even number on the other side." Most subjects choose E alone, or E and 4, while the correct answer is E and 7, because "Any odd number on the other side of E falsifies the rule in exactly the same way as would any vowel on the other side of 7." [Wason and Johnson-Laird, 1972].

In FOPL, the "rule" to be tested is "$(\forall x)(Vowel(x) \supset Even(x))$." By definition, this proposition is false if there is at least one card that satisfies $Vowel(x)$ but not $Even(x)$, otherwise it is true. Therefore, to determine the truth value of the rule means to check all cards that may *falsify* the rule. What the subjects show is a tendency to find cards that *verify* the rule. In FOPL, cards verifying the rule make no contribution to its truth value. If the 4 card indeed has a vowel on the other side, it does not mean that the rule is true. If the E card and 7 card do not make the rule false, the rule is true, and the 4 card does not need to be turned.

Since this problem has been discussed in detail in [Wang, 2001c], here I only briefly address it. Psychological experiments show that when asked to determine the truth value of a statement, human subjects more frequently look for positive evidence than for negative evidence, though "according to logic," only negative evidence matters. However, according to NAL, what the subjects do is not necessarily an error. If the "logic" under consideration is a binary one like FOPL, then to look for positive evidence is a mistake, but it is not the case if the "logic" is multi-valued, such as NAL.

If what is needed in solving the above problems is a logic where truth value depends on both positive and negative evidence, can it be done in the PL framework? Though there have been many such attempts, such as by combining FOPL and probability theory [Nilsson, 1991], it is not easy for several reasons. Here I only discuss one of them, the scope of evidence.

In FOPL, a general statement is always represented as a universally quantified proposition like "$(\forall x)(Raven(x) \supset Black(x))$," where the variable x can be instantiated by every constant in the domain. Each instantiation, in principle, makes the proposition either true or false, and nothing is irrelevant. This is another problem revealed by the confirmation paradox: in FOPL not only that positive evidence is ignored, but also that it is hard to draw a line between positive evidence and irrelevant things. Though in principle it is possible to use approaches like many-sorted logic to specify different scopes for each variable, it will make the logic very complicated.

As show previously, since NAL is a TL using subject-predicate format for statements, the scope of evidence for a given statement is explicitly given by the extension of the subject term and the intension of the predicate term. Consequently, the distinction between evidence and non-evidence, as well as distinction between positive evidence and negative evidence, can be easily and naturally defined, which provides a foundation for the definition and calculation of truth value in EGS.

Now we can see that, because of the difference in semantics, there is no one-to-one mapping between the sentences in Narsese and the sentences in FOPL, though intuitively approximate mappings can be built. For example, "Ravens are birds" can be represented in NAL as

$$(\#x \rightarrow raven) \Rightarrow (\#x \rightarrow bird)$$

and in FOPL as

$$(\forall x)(Raven(x) \supset Bird(x))$$

Though the two formal representations roughly correspond to each other, they are not defined in the same way. The proposition in FOPL is binary and is about all individuals in the domain; the statement in NAL is multi-valued and is only about terms in the extension of *raven*. Furthermore, the same English sentence can also be represented in Narsese as

$$raven \rightarrow bird$$

which is both extensional and intensional. There is nothing like it in FOPL.

Given this difference (and some similar differences in other structures), it is impossible to accurately say which of FOPL and NAL has more expressive power, since they are simply different. As mentioned previously, FOPL is closer to a mathematical language, and NAL to a natural language, both in syntax and semantics. It is possible to use FOPL as part of the meta-language of NAL (as I have been doing in this book) and to use NAL as part of the meta-language of FOPL (by representing inference rules of FOPL as implication statements in NAL, as introduced in NAL-6), but it does not mean that the two are equivalent at the object-language level.

NAL is not generally better than FOPL, but it is better in situations where the system's knowledge and resources are insufficient. In domains where knowledge and resources can be assumed to be sufficient (with respect to the questions to be answered), FOPL may still be better. Mathematics is such a domain. When NAL needs to do math, it can emulate FOPL (or any other logic) by representing inference rules of that logic as implication relations, and applying them as a special case of deduction, as discussed above in the case of set theory.

10.3 Logic and AI

As Bibel said, "Among the controversies in AI, none is as persisting as the one about logic's role in AI." [Hearst and Hirsh, 2000] Here let us see where NAL is in these controversies.

10.3.1 Different attitudes to logic

In the related fields, FOPL is often taken as the normative model of human inference. Such a belief or assumption can be found in many works in philosophy [Popper, 1959], psychology [Braine and O'Brien, 1998, Wason and Johnson-Laird, 1972], and linguistics [Allwood et al., 1977].

Given this situation, as well as the great success of FOPL in computer science [Halpern et al., 2001], it is quite natural to see FOPL playing a central role in the theoretical foundation of "symbolic/logical AI" [Hayes, 1977, McCarthy, 1988, Nilsson, 1991]. Since the very beginning of the field, AI researchers have used FOPL and its variants for all kinds of works.

In AI, typically a logic is used in a "knowledge-based system," consisting of an inference engine and a knowledge base. Such a system often takes the form of a "production system," which, after be loaded with the proper domain knowledge, becomes an "expert system." Examples of this kind of system include AGI related projects Soar [Newell, 1990] and ACT-R [Anderson and Lebiere, 1998]. In the design of these systems, the inference engine is usually based on FOPL, with a procedural interpretation of certain predicates. The variations among these systems are mostly in the structure of the knowledge base and the control of the inference process. The stress on the content of the knowledge base reaches its extreme form in the AGI system CYC, which is developed under the assumption that the key of intelligence is nothing but huge amounts of common-sense knowledge [Lenat and Feigenbaum, 1991, Lenat, 1995].

It does not mean that everyone is happy with FOPL. On the contrary, there are many well-known problems that cannot be easily handled by FOPL, and there are many attempts of building various types of new logic. Examples of such attempts, both inside and outside AI, include description logic [Donini et al., 1996], non-monotonic logic [Ginsberg, 1987], inductive logic [Carnap, 1952], fuzzy logic [Zadeh, 1983], probabilistic logic [Halpern, 1990], relevance logic [Read, 1989], paraconsistent logic [Priest et al., 1989], and so on. However, these attempts are mostly in the framework of PL, as extensions or revisions of FOPL, and aimed at a single issue. Only a few researchers paid some attention to TL (such as [Dubois and Prade, 1988, Zadeh, 1985]), and even in those works, TL is not proposed as a competitor of PL.

There are people who do not think that the goals of AI can be reached by extending and/or revising FOPL, for various reasons. Such opinions can be found in [Birnbaum, 1991, Dreyfus, 1992, Harnad, 1990, Hofstadter, 1985, McDermott, 1987, Searle, 1980]. Many people believe that the problems are not merely in FOPL, but in the notion of "logic." Consequently, they pursue completely different approaches, where logic plays little role [Brooks, 1991, Holland, 1986, Smolensky, 1988, van Gelder, 1997].

As stated earlier in the book, I still think that a reasoning system with a formal logic provides a proper framework for AI research. However, I believe that FOPL is the wrong type of logic for AI, and all the previous revisions and extensions of FOPL have not gone far enough. What we need for AI is a completely different type of logic.

10.3.2 Different types of logic

As we have seen, concepts in NARS are usually fluid, fuzzy, and elusive; judgments are usually equivocal, conflicting, and fallible; and the system's behaviors are usually unpredictable and irreproducible. The system may be absent-minded, may forget important information, and may change its mind from time to time. How can we still say that NARS works according to a *logic*?

In fact, the concept "logic" has two different senses. Construed very broadly, logic is the set of principles of, and criteria for, valid inference. However, nowadays the concept is used in AI under a narrower construal, which is restricted to FOPL, its variants, model-theoretic semantics, theorem-proving, and so on [Birnbaum, 1991, McDermott, 1987, Nilsson, 1991].

Obviously, NAL does not constitute a logic in this narrower sense. However, it is a logic in the broader sense. Technically, it has a formal language with a semantics theory, and uses a set of formal rules to do inference. Theoretically, it is an attempt to formally capture the principles of valid inference under certain circumstance.

To clarify more sharply the difference between NARS and other reasoning systems, I distinguish three kinds of reasoning systems, based upon their assumptions about the sufficiency of knowledge and resources:

Pure-axiomatic systems. Such systems are designed under the assumption that both knowledge and resources are sufficient (with respect to the questions that will/can be asked), so adaptation is not necessary. A typical example is the notion of "formal system" suggested by Hilbert (and many others), in which all answers are deduced from a set of axioms by a deterministic algorithm, and which is applied to some domain using model-theoretic semantics. Such a system is built on the idea of sufficient knowledge and resources, because all relevant knowledge is assumed to be fully embedded in the axioms, and because questions have no time constraints, as long as they are answered in finite time. If a question requires information beyond the scope of the axioms, it is not the system's fault but the questioner's, so no attempt is made to allow the system to improve its capacities and to adapt to its environment.

Semi-axiomatic systems. Such systems are designed under the assumption that knowledge and resources are insufficient in some, but not all, aspects. Consequently, adaptation is necessary. Most current AI approaches fall into this category. For example, non-monotonic logics draw tentative conclusions (such as "Tweety can fly") from defaults (such as "Birds normally can fly") and facts (such as "Tweety is a bird"), and revise such conclusions when new facts (such as "Tweety is a penguin") arrive. However, in these systems, defaults and facts are usually unchangeable, and time pressure is not taken into account [Reiter, 1987]. Many learning systems attempt to improve their behavior, but still work solely with binary logic where everything is black-and-white, and persist in always seeking *optimal* solutions of problems [Michalski, 1993]. Although some heuristic-search systems look for less-than-optimal solutions when working within time limits, they usually do not attempt to learn from experience, and do not consider possible variations of time pressure.

Non-axiomatic systems. In this kind of system, the insufficiency of knowledge and resources is built in as the ground floor, and the system works accordingly.

According to the working definition of intelligence proposed in Chapter 2, pure-axiomatic systems are not intelligent at all, non-axiomatic systems are intelligent, and semi-axiomatic systems are intelligent in certain respects.

According to the above definition, intelligence is still (as we hope) a matter of degree. Not all systems in the "non-axiomatic" and "semi-axiomatic" categories are equally intelligent. Some systems may be more intelligent than some other systems by having a higher resources efficiency, using knowledge in more ways, communicating with the environment in a richer language, adapting more rapidly and thoroughly, and so on. Though NAL has been based on AIKR from the very beginning at NAL-1, each later layer, from NAL-2 to NAL-8, obviously makes the system "more intelligent" by adding new grammar rules and inference rules.

"Non-axiomatic" does not mean "everything changes." In NARS, nothing is fixed as far as the *content* of knowledge is concerned, but as we have seen in the previous chapters, how the changes happen is fixed, according to the inference rules and control strategy of the system, which remain constant when the system is running. This fact does not make NARS "semi-axiomatic," because the fixed part is not in the "object language" level, but in the "meta-language" level. In a sense, we can say that the "meta-level" of NARS is not non-axiomatic, but pure-axiomatic. For a reasoning system, a fixed inference rule is not the same as an axiom.

Pure-axiomatic systems are very useful in mathematics, where the aim of study is to idealize knowledge and questions to such an extent that the revision of knowledge and the deadlines of questions can be ignored. In such situations, questions can be answered in a manner so accurate and reliable that the procedure can be reproduced by an algorithm. We need intelligence only when no such pure-axiomatic method can be used, due to the insufficiency of knowledge and resources. For similar reasons, the performance of a non-axiomatic system is not necessarily better than that of a semi-axiomatic system, but it can work in environments where the latter cannot be used.

Many arguments proposed as against logicist AI [Birnbaum, 1991, McDermott, 1987, Minsky, 1990], symbolic AI [Dreyfus, 1992], or AI as a whole [Searle, 1980, Penrose, 1994], are actually arguments against a more restricted target: pure-axiomatic systems. These arguments are valid when they reveal many aspects of intelligence that cannot be produced by a pure-axiomatic system (though these authors do not use this term), but some of the arguments seriously mislead by taking the limitations of these systems as restricting all possible AI approaches.

Some critics implicitly assume that because a certain level of a computer system can be captured by FOPL and implemented as a Turing Machine, these axiomatic theories also apply to the entire range of behavior that such a system can exhibit [Dreyfus, 1992, Penrose, 1994]. This is not the case. When a virtual machine M_1 is implemented on a virtual machine M_2, the former does not necessarily inherit all the properties of the latter. For example, it is invalid to conclude that a computer cannot process decimal numbers (because the hardware uses binary numbers), cannot process keyboard characters (because underneath it all, everything is just bits), or cannot use a functional or a logical programming language (because the commands of such languages are eventually translated into procedural machine language).

Obviously, with its fluid concepts, revisable knowledge, and fallible inference rules, NARS violates all the norms of classical logics. However, as a virtual machine, NARS can be built upon another virtual machine that is in fact a pure-axiomatic system, as my implementation demonstrates, but this fact does not in any sense make NARS "axiomatic."

Traditionally, AI has been referred to as a branch of computer science. According to my previous definitions, AI can be implemented with tools provided by computer science, but from a *theoretical* point of view, AI and computer science make opposite assumptions: computer science focuses on pure-axiomatic systems, but AI focuses — or *should* focus — on non-axiomatic systems.

The fundamental assumptions of theoretical computer science can be found in mathematical logic (especially FOPL) and theory of computation (especially computability theory and computational complexity theory). These theories take sufficient knowledge and resources for granted, and therefore adaptation, plausible inference, and tentative solutions to problems are neither necessary nor possible in the typical applications of these theories.

Similar assumptions are often made by AI researchers with roughly the following type of justification: "We know that the human mind works under conditions of insufficient knowledge and resources, but if you want to set up a *formal* model and then implement it as a *computer* system, you must somehow *idealize* the situation."

It is true that every formal model is an idealization, and so is NARS. The key question concerns what to omit and what to preserve in the idealization. In the current implementation of NARS, many factors that should influence reasoning are ignored, but the insufficiency of knowledge and resources is strictly assumed throughout. Why? Because such insufficiency is a *defining* feature of intelligence, so if it were abandoned in the "idealization," the resulting study, whatever its worth might be, would be about something other than intelligence.

10.3.3 Logic and thinking

At the time of Aristotle, the subject matter of logic was human thinking in general. The goal of logic is to find the abstract patterns in valid inference that apply to all domains of human thinking. It remained to be the case until the time of Frege, Russell, and Whitehead, whose major interest was to set up a solid logic foundation for mathematics. For this reason, they were not satisfied by Aristotle's term logic, and developed a new logic in the PL framework, which is focused on valid inference in mathematics, typically the binary deduction processes that derives theorems from axioms and postulates.

Frege (and other mathematical logicians) argued strongly against the so-called "psychologism," and stressed the difference between logic and psychology, as the difference between a *normative* theory of valid inference and a *descriptive* theory of actual inference. Though this "mathematical logic" has achieved great successes in mathematics and computer science, it moves away from actual human inference and knowledge represented in natural languages [Englebretsen, 1981].

In AI, when "reasoning" is mentioned, in most of the cases it is about reasoning in domains other than mathematics. In this field, what is actually happening is a (reversed) change in the subject matter of logic, this time from mathematics back to thinking in general. Even so, most people in AI fail to realize that there are fundamental differences between "valid inference on mathematical knowledge" and "valid inference in empirical knowledge," and continuously use logics built for the former as models for the latter.

This "moving back" in subject matter will not lead logic to psychologism, because in this discussion we still can distinguish *logic* from *psychology of reasoning*. NAL is proposed as a normative theory about how valid inference should be, not a descriptive theory about how actual human inference works. However, it is a normative theory that is different from the traditional "axiomatic logics," because it is based on a different set of assumptions and principles. It is not "better" or "more powerful" than FOPL, but is fundamentally different from it. It competes with FOPL and its variants in the sense that for certain practical problems outside mathematics, its solutions are better justified, as described previously.

Though NAL is not a descriptive theory of human reasoning, it is closer to one than FOPL is, because the human mind is evolved to work with insufficient knowledge and resources [Medin and Ross, 1992]. Therefore, the *basic assumptions* of NAL are selected to be a descriptive model of the *basic principles* of human inference, though at the level of concrete *behaviors*, NARS is still different from a human being, due to many other factors.

The confusion between "mathematical logic" and "empirical logic" is not an error only made in AI. The current descriptive theories of human reasoning are strongly influenced by mathematical logic, especially FOPL. It is well-known that FOPL is a poor descriptive model for human reasoning, but many psychologists still use it as the normative model to be compared with.

For example, in Mental Logic theory [Braine and O'Brien, 1998], some "psychologically realistic" inference rules are proposed, which are described in a form close to FOPL. In the competing Mental Model theory [Johnson-Laird, 1983], the focus is semantics, and the normative theory guiding the research is model theory. According to our previous discussion, when studying "non-mathematical" reasoning, FOPL is not even the proper normative theory to use. Consequently, the related psychological theories need to be re-evaluated accordingly.

A directly related topic is the study on "human heuristics, bias, errors, and fallacy." Typically, researchers compare human behavior to a "normative theory," and whenever the two are different, the human behavior is judged as wrong. In Section 8.3.3, I discussed the case where the "normative theory" in question is probability theory. Here we have a similar situation, where the "normative theory" is FOPL.

As mentioned earlier (Section 10.2.3), in Wason's Selection Task, looking for positive evidence is not necessarily a human error. Actually, NARS will make the same choice as most subjects, if it faces the same problem. Here the key is to realize that there are different types of logics. For one type, truth value only depends on the existence of negative evidence, but for another type, truth value depends on both positive and negative evidence.

A similar situation happens in non-deductive inference. If the uncertainty in the conclusions is omitted, many NAL inference rules produce nothing but various types of "logical fallacy" listed in logic textbooks.

For example, from "Swans are birds" and "Swans fly" to derive "Birds fly" is called "Hasty Generalization" [Copi, 1982], but it is exactly what the NAL induction rule does. Here the key is once again in truth value. In a binary logic, the conclusion will be taken as "true," which is of course wrong. In a logic like NAL, a conclusion can be "true to a degree," indicating the amount of available evidence. Such a conclusion is valid, as far as it is attached with the right truth value.

One problem in psychological research is that many people implicitly assume that there is only one normative theory for reasoning, which is FOPL (and its extensions and variations), with model theory providing its semantics. After another normative theory is established, a lot of conclusions will need to be revised.

10.3.4 NARS as a network

By working on a reasoning system, one does not necessarily commit oneself to the assumptions of traditional logic-based AI theories. Designed as a reasoning system, but not a "logicist" one [McCarthy, 1988, Nilsson, 1991], NARS shares many theoretical assumptions with the *subsymbolic* movement [Hofstadter, 1985, Rumelhart and McClelland, 1986, Smolensky, 1988], despite the fact that I chose to formalize and implement these assumptions in a framework that on the surface looks closer to the traditional symbolic-AI tradition.

In fact, NARS can be naturally described as an *inheritance network*. We can see each concept as a *node*, each belief (and task) as a (directed) *link* between two nodes, and the corresponding truth value as the *strength* of the link. In this way, the memory of NARS is a network.

This inheritance network is different from a semantic network and other symbolic networks, because all links represent the inheritance relation or its variants, whose meanings and functions are precisely defined. There may be two or more links between a given pair of nodes, and their truth values might conflict with each other.

The network has both *active links* (tasks) and *passive links* (beliefs). Priorities are defined among nodes, active links, and passive links. In each atomic step of processing, an active link interacts with an adjacent (i.e., with a common node) passive link to generate new links, and different types of inferences correspond to different combinations of the two component links, as suggested in [Minsky, 1985].

In such a network, to answer a question means to determine the strength of a link, given its type and its beginning and ending nodes ("yes/no" questions), or else to locate that node that has the strongest link the system can find to a specified node ("what" questions); to achieve a goal means to obtain a link between given nodes by executing operations. Since during the processing, not only the topological structure of the network but also the strengths of its links and its priority distribution are all constantly changing, what the system does is much more than search a static network for the desired link or node.

As a network, NARS shares properties with various subsymbolic approaches [Hofstadter and FARG, 1995, Holland, 1986, Smolensky, 1988], such as parallel processing, nondeterminism, self-organization, distributed representations, and so on. Let us consider the last property in more detail here.

At first glance, the internal representations of NARS seem to be local, since the knowledge "Swans are birds" is stored explicitly in the link that runs between the node "*swan*" and the node "*bird*." However, when the system is told "Swans are birds," the judgment is treated both as a passive link (to be set up between the given nodes) and as an active link (to interact with adjacent links); therefore, it will have effects on other nodes and links. On the other hand, when the system is asked "Are swans birds?," it will not only seek the direct link between nodes "*swan*" and "*bird*," but will also try to get an answer from other related nodes and links. Therefore, in both cases the operations are not localized to nodes "*swan*" and "*bird*."

As a result, NARS' internal representations become *distributed* in the following senses: (1) an input task may have non-local effects, (2) an output result may have non-local sources, and (3) local losses of information (such as those caused by forgetting) may be partially recovered from information kept in other parts of the system.

When watching NARS running, if we focus our attention to the priority distribution among concepts, we can observe an *activation spreading* phenomenon, as in a neural network. In a term logic like NAL, if the concept under processing is C_M, and the new task just generated is between the concepts C_S and C_P, it must be the case that the former was already linked to the latter two before the current time. Also, priority of the former will be decreased after the step, while the other two increased (because of the adding of the new task). It looks *as if* activation (measured by priority) flows from the former to the latter two, though this is only a partial and approximate description about what is actually happening.

On the other hand, there are some important differences between NARS (when interpreted as a network) and existing artificial neural networks. The NARS network has the following features.

- The network structure changes as the system is running — new nodes and links are created, and old ones are removed.
- The nodes and links have semantics — meaning and truth are explicitly defined with respect to the system's experience.
- There is no global updating — in each inference step, the system only updates a small number of priority values (similar to activation of nodes) and truth values (similar to the weight of links).
- The inference process does not necessarily converge to an attractor, but can stop after any number of inference steps, when no resources are left for it.

Since NARS can be naturally described as a network, why is a "logical" description still preferred? Indeed, the terminology of networks, using concepts such as "node," "link," "strength," "activation," and so on, can be used for many different purposes without stirring up any controversy, whereas philosophical concepts like "meaning," "truth," and "induction" tend to stir up hornet's nests of argumentation.

So why didn't I present NARS as a network? By constantly using the terminology of logic, what I want to show is: to model intelligence faithfully, what really matters is the set of *underlying theoretical postulates*, not the terminology or the technology. Once a new working definition of intelligence is accepted, a reasoning system can still have many interesting and unexpected properties, despite having the trappings of a formal language and truth-preserving inference rules. The basic troubles with traditional logic-based symbolic AI stem from its fundamental assumptions about sufficiency of knowledge and resources, and its pursuit of completeness, consistency, and decidability, rather than from detailed decisions about how AI systems work.

It seems to me preferable to consider NARS as a reasoning system governed by a logic, due to its preciseness, during the design process and also in explaining the basis for its design, whereas the "network" image seems preferable, due to its vividness, for occasions when the system must be presented in a very short time to other people. Also, the network terminology may work better in the future when a detailed inference control theory is established to describe the dynamics of NARS.

Chapter 11

Categorization and Learning

Though NARS is usually presented as a reasoning system, it can also be seen as a computational model of categorization and learning.

11.1 Concept and categorization

We start by seeing the model of categorization provided by NARS.

11.1.1 Concept in NARS

In NARS a concept C_T consists of a *name* and a *body*. The former is a term T, and the latter, roughly speaking, is a collection of Narsese sentences, including all beliefs and tasks with T as subject or predicate (see Section 6.2.2).

According to EGS, the meaning of a term is its experienced relations with other terms. Similarly, the meaning of a concept is its relations with other concepts, as assembled within its body. Since all these "relations" are the inheritance relation and its variants, we can say that the meaning of a concept consists of its *extensional* relations and *intensional* relations to other concepts.

For example, "$raven \rightarrow bird$" states that the term "*raven*" is in the extension of term "*bird*," and the term "*bird*" is in the intension of term "*raven*." For the corresponding concepts, we can say that the concept C_{raven} has an intensional relation to the concept C_{bird}, and the concept C_{bird} has an extensional relation to the concept C_{raven}.

Intuitively, the extensional relations of a concept link it to its specializations, instances, or exemplifiers; the intensional relations of a concept link it to its generalizations, properties, or attributes. For a higher-order term (a statement), its meaning also includes statements that imply it and statements it implies (i.e., its sufficient conditions and necessary conditions), as defined in Section 5.1.2.

Since in NARS every belief has a truth value attached to show the evidential support it got from the experience of the system, each of the above extensional/intensional relations is "true to a degree." Therefore, for a given concept, there are "typical" instances/properties (with larger frequency and confidence values), and "periphery" ones.

Limited by available resources, each time a concept is used (during the system's processing a task), only part of its meaning is involved, selected according to the priority distribution maintained within its body (as described in Chapter 6). For this reason, we should distinguish the "general meaning" of a concept and its "current meaning" (with respect to a given task at a given time). Usually, the latter is a very small part of the former, and in different times, the same concept may be used with quite different "current meanings."

New beliefs and tasks in NARS come from two sources: either input from the user or another system, or derived from existing beliefs and tasks. The former is the system's direct experience, and the latter can be seen as indirect experience (since it is derived from the former). In both cases, a belief contributes (more or less) to the intension of the subject and the extension of the predicate.

The tasks in a concept corresponds to the system's motivations related to the concept, and it indicates the direction in which the concept may evolve. When the system is asked to evaluate the truth value of a given statement "$S \rightarrow P$," the task can be carried out in several ways. For example, the system may happen to recall a belief with the form "$S \rightarrow P < f, c >$" in concept C_S or C_P, or it may derive such a conclusion by inference (it may be deduction, induction, abduction, analogy, or something else). More likely, it will get candidate conclusions from several paths, then use the revision rule or the choice rule to decide the answer. The same is also true for questions with variables to be instantiated (the "what" questions).

In general, NARS has multiple strategies when a categorical relationship needs to be judged, and the actual result depends on the current meaning of the involved concepts, consisting of some extensional and intensional relations of them. In the following, this model is compared with other models of categorization.

11.1.2 Classical theory

The "classical theory" of categorization can be traced back to Aristotle, and it is roughly identified with the following opinions:

1. Every concept (or category) has a meaning, which remains the same at different times and places.

2. The meaning of a concept is specified by a "definition," which gives the sufficient and necessary condition for an entity to be an instance of the concept. A definition can be given as a logical structure of other concepts, or as an exhaustive list of instances or properties of the concept.

3. Whether an entity is an instance of the concept or not is either true or false, with nothing in between.

Such a theory works well in mathematics, and people used to believe that it is also how categorization works in other domains. In everyday life, we are often asked to follow (or give) definitions of concepts.

However, though "to use concepts according to their definitions" is an effective way for practical purposes, there are many problems in this theory. Even if we ignore the practical difficulties (how to avoid circular definition, how to choose among competing definitions, ...), there are theoretical issues caused by the above opinions. In psychological and linguistic research, many counter evidences of such a theory have been found, which show that the concepts in the human mind do not fit this picture [Laurence and Margolis, 1999]. Currently few people still associate themselves with such a theory in psychology and philosophy.

Nevertheless, such a theory is still very popular in AI. In knowledge representation, new concepts are often introduced by definitions; in machine learning, new concepts are often learned by identifying their sufficient and necessary conditions. In both fields, the membership relation between a possible instance and a concept is still mostly binary (true or false). For examples, see [Kurtonina and de Rijke, 1999, Gärdenfors and Williams, 2001]. Furthermore, the influence of the classical theory can be seen in notions like "class," "frame," and "script," which represents an instance of a concept by specifying its values on a set of predetermined attributes.[1]

If we focus on AI, and do not care much about what is going on in the human mind, should we just keep the classical theory of categorization?

NARS gives a negative answer to this question. Here the classical theory is rejected, not because NARS tries to simulate human behaviors, but because this theory conflicts with the fundamental principle of the project, that is, the system has to adapt to its environment with insufficient knowledge and resources.

In NARS, since future entities and events may be different from the ones the system encountered in the past, but the system has no other guidance beside its experience, it must somehow *treat different things as the same*. Since the system cannot afford the time and space to represent and process its experienced events and entities in all details, it must classify them into groups and ignore some subtle differences. Therefore, categorization becomes necessary in NARS. Because this process is based on experience, there is no way for the system to get a set of well-defined concepts in advance, and expect them to fit all possible experiences nicely. Instead, the system has to try different ways to categorize its experience, which eventually leads to the model of categorization described above.

Therefore, NARS does not accept the classical theory of categorization, mainly because such concepts are not available in an adaptive system working with insufficient knowledge and resources.

However, this does not means that there is no value in the classical theory. Actually, NARS favors "well-defined" concepts. If a concept has a small and stable "core meaning," which consists of a few relations with other concepts, implies most other relations of the concept, and clearly distinguishes instances from non-instances, then usually this

[1]There are some recent attempts to extend object-oriented languages to handle uncertainty, such as [Cao et al., 2002].

concept is more useful than a concept whose meaning is fuzzy, messy, and unpredictable.

According to EGS, the meaning of a concept is determined by its (experienced) relations with other concepts. However, the relations do not contribute equally. Usually, some relations are more "essential," in the sense that many others can be derived from them. If for a given concept such essential relations can be identified, they can be called the "definition" of the concept. However, such a definition, even when it can be obtained, is not perfect in the sense that all other relations can derived from it. Also, it changes as new experience is obtained.

In summary, the classical theory does not fit with the usual categorization process in NARS, though it does provide a desired limit situation for categorization, as well as a good approximation for certain situations.

11.1.3 Similarity-based models

Unhappy with the binary concept of the classical theory, several theories are proposed, based on the opinion that whether something belongs to a concept or not is generally a matter of degree. Furthermore, they define this "degree" as some kind of *similarity*.

According to "prototype theory," a concept is characterized by the *properties* possessed by most of its members [Rosch, 1978]. Therefore, the degree for an entity to belong to a concept is reduced to the degree of similarity between that entity and the prototype of the concept.

According to "exemplar theory," a concept is determined by a set of *exemplars* [Nosofsky, 1991]. Therefore, the degree for an entity to belong to a concept is reduced to the degree of similarity between that entity and the given exemplars of the concept.

Both above theories correspond to (different) special situations in NARS.

When the truth value of a statement is determined in NARS, both extensional evidence and intensional evidence are taken into account. When NARS is asked to decide whether an entity e belongs to a concept C, the system may compare e with known instances of C, check whether e has the properties associated with C, or do both — it depends on the system's beliefs, and which of them are recalled at the time.

For a given concept C_T, if in the experience of the system it is mainly learned through its *intensional relations*, then its core meaning consists of statements "$T \rightarrow P_1$," "$T \rightarrow P_2$," and so on (truth values omitted), where P_1 and P_2 are the properties satisfied by most of the instances of the concept. When the system needs to evaluate "$e \rightarrow T$," the abduction rule is used, which checks the properties one by one (by deriving "$e \rightarrow T$" from "$T \rightarrow P_i$" and "$e \rightarrow P_i$"), then merges the conclusions together according to the revision rule. This is roughly equivalent to comparing e with a prototype that has all the properties.

For a given concept C_T, if in the experience of the system it is mainly learned through its *extensional relations*, then its core meaning consists of statements "$e_1 \rightarrow T$," "$e_2 \rightarrow T$," and so on (truth values omitted), where e_1 and e_2 are the exemplars of the concept. When the system needs to evaluate "$e \rightarrow T$," the analogy rule is used, which compares e with the known instances (by deriving "$e \rightarrow T$" from "$e_i \rightarrow T$" and "$e \leftrightarrow e_i$"). Since the analogy rule usually produces more confident conclusions than the abduction rule, the system may only compare e with one exemplar, if their similarity is sufficient for a confident conclusion.

In both of the above situations, the inheritance relation ("$\rightarrow$") between e and T can be replaced by the instance relation ("$\circ\!\!\rightarrow$"), and the conclusion still holds.

Using the terminology of NARS, we can see that exemplar theory defines the meaning of a concept as its extension, and prototype theory (as well as "feature list" theory) defines the meaning of a concept as its intension. Consequently, the comparison of the two is like the "chicken-and-egg" question — do we get instances first, then generalize properties from them, or get properties first, then determine instances according to them? The answer provided by NARS is: both.

Whenever the system gets a piece of new knowledge, since its form is "$S \rightarrow P$" (with a certain truth value), it always adds something new to the extension of P, and to the intension of S. Therefore, to the system as a whole, extension and intension are symmetric, and are developed together — they are just two opposite directions of a link.

On the other hand, in a concrete concept, it is quite possible that its meaning is mainly determined by its extension, while in another concept, by its intension. It depends on how the concept is learned and used in the past. Consequently, some concepts fit better with the description of prototype theory, while some others fit better with the description of exemplar theory. In general, however, both extension and intension contribute to the meaning of a concept, though not necessarily to the same extent.

The coordination of extension and intension of a concept is a very important aspect of the NARS categorical dynamics. The system does not simply decide which of the two is more essential, but rather keeps a dynamic balance between them. As a result, a change in extension usually causes changes in intension, and vice versa. Though a perfect coherence is almost impossible, the system does try to reduce the internal inconsistency as much as possible.

11.1.4 Relation-based models

To define the meaning of a concept as its relation with other concepts is not really a new idea. For example, this is also the idea behind semantic networks [Quillian, 1968], concept role semantics [Harman, 1982], and the "theory theory" of categorization [Murphy and Medin, 1985]. A typical presentation of this opinion is: "In order to characterize knowledge about and use of a concept, we must include all of the relations involving that concept and the other concepts that depend on it." [Murphy and Medin, 1985] The categorization model of NARS agrees with these approaches on a general level.

Even so, there are still several major differences between NARS and these relation-based theories on categorization.

Since NARS uses a term logic, all statements belong to the inheritance relation and its variants, therefore are *categorical* statements — "$S \rightarrow P$" says that S is a sub-concept of P, and P is a super-concept of S. As a result, the meaning of a concept in NARS can be simply defined by the extension and intension of the concept. In this way, NARS provides a detailed model for how a concept is related to other concepts, and how these relations change by the inference rules when the system runs.

Also, NARS stresses the role of experience in determining the meaning of a concept. When I say that in NARS the meaning of a concept is determined by its *relations* with other concepts, I do not mean "possible relations," "potential relations," or "factual relations," but "relations that have been obtained/derived from the experience of the system." As experience stretches in time, meanings of concepts change accordingly.

Furthermore, context selectively forms a short-term "current meaning" from the long-term general meaning of a concept, so that a concept can be used more or less differently in different situations. In these situations, the coherence of the concept is not absolute, but a matter of degree.

Therefore, the categorization model of NARS is not infinite or holistic. Though it is in principle possible to relate a concept to all other concepts within the system, or to put a concept into infinitely many relations with another concept [Coulson, 2001, Murphy and Medin, 1985], in practice this will never occur under any realistic set of resource constraints.

The "theory theory" of categorization stresses the role played by causal knowledge in the meaning of concepts [Rehder, 1999]. In NARS, this may also be the case for some concepts. In Narsese, causal knowledge is a special case of the higher-level relations (*implication* and *equivalence*), with additional information (such as temporal order, predication power, and so on). For certain concepts, it is quite possible that their core meaning consists of mainly causal knowledge, or knowledge in an intuitive theory about the environment. However, this is not necessarily the case for all concepts in NARS.

Though in the current version of NARS each concept corresponds to a term in Narsese, the categorization model of NARS is not limited to abstract concepts with linguistic labels. In the future, when the system is extended to have sensorimotor capacity, some concepts may mainly correspond to mental images produced by perception, or operation sequences that can be executed within or outside the system. Even with these additional components, the principle for concept representation remains the same, and the meaning of a concept is still determined by its relations with other concepts, including the images and operations. Of course, at that time the system will have difficulty in explaining (in Narsese) the meaning of such a concept to another system, partially because procedural knowledge cannot be fully expressed declaratively.

During the development of NARS, I spent several years at Indiana University as a member of FARG (Fluid Analogies Research Group) lead by Douglas Hofstadter. Therefore, NARS bears certain "family resemblance" with other FARG projects, such as Copycat and Tabletop [Mitchell, 1993, French, 1995, Hofstadter and FARG, 1995]. One of the important themes shared by these projects is their focus on *fluid concepts*. The concepts in NARS are "fluid," in the sense that they are "concepts with flexible boundaries, concepts whose behavior adapts to unanticipated circumstances, concepts that will bend and stretch — but not without a limit" [Hofstadter and FARG, 1995]. These systems solve problems by "analogy" — in a broad sense, it means to use one concept as another concept, which is close to what "inference" means in a term logic. Of course, the conclusions obtained in this way are not always correct, but nevertheless this ability is crucial for cognition and intelligence.

11.1.5 Compositionality

As described previously, in NARS there are two types of terms: atomic and compound. The former is an identifier without internal structure, while the latter is formed by a logical operator from other terms.

NARS has the following types of compound terms:

Sets: Compound terms are introduced by enumerating its instances or properties.

Intersection and difference: Compound terms are introduced by taking the intersection (or difference) of the extension (or intension) of existing terms.

Product and image: Compound terms are introduced according to their (non-copula) relations with other terms.

Statement: A statement, consists of two terms linked together by a copula, is taken as a term.

Compound statement: The negations, conjunctions, and disjunctions of existing statements are introduced as compound statements (therefore compound terms).

NARS allows new concepts to be named by compound terms, so as to get productivity and systematicity from compositionality [Fodor, 1998, Rips, 1995]. However, as discussed previously, the meaning of a concept usually is not fully determined by such a definition, even if it exists.

The meaning of a concept corresponding to a compound term is determined just like the other concepts, that is, by its relations with other concepts. The only specialty here is that among these relations, there are special ones that link the compound term to its components, with the logical operator attached. As a result, its "literal definition," as given above, is part of its meaning, though usually not the whole meaning.

Therefore, the compound terms in NARS are *partially compositional.* On one hand, the meaning of a compound term does heavily depend on the meaning of its components in a way determined by the operator. If there is no other relation available, its meaning can even be reduced into that of its components. On the other hand, as soon as there are other relations, the meaning of the compound term is no longer completely determined by its literal definition, nor is it reducible to its components, though still more or less related to them.

For example, the compound term $([black] \cap board)$ ("blackboard") is related to terms "*black*" and "*board*" by definition, though its relation with the concept of "writing surface" cannot be derived from the above "defining" relations. Instead, it mainly comes from the system's experience, in which "blackboard" often appears as a whole.

The same is true for the truth value of a compound statement. the truth value of $(P \wedge Q)$ can be calculated from the truth values of P and Q. However, $(P \wedge Q)$ as a whole can be evaluated in other ways, so that in the long run, when evidence is collected in multiple ways, it is possible for it to get a truth value that cannot be reduced to that of its components. This is especially the case when the truth values of P and Q are not independent of each other (i.e., they are based on overlapping evidence).

11.1.6 Categorical dynamics

Though some authors have argued for the fluid and instable nature of concepts [Barsalou, 1987, Hofstadter and FARG, 1995], most current models of categorization do not provide a clear picture of the dynamics. A major feature that distinguishes the NARS categorization model from many other models is to treat the meaning of a concept as changing over time, and to accurately specify how such changes happen.

As described previously, the meaning of a concept is nothing but the collection of beliefs and tasks associated with it. As a result, when a new input sentence is inserted into the bodies of the relevant concepts, it also changes their meaning, though usually not too much. Derived beliefs and tasks have the same effect.

Of course, meaning change does not always appear as adding new relations. When a new (input or derived) belief conflicts with an old belief, the revision rule is applied to merge them. Consequently, though no new statement is added into the concept body, the truth value of an existing statement has been changed, which also causes the meaning of the relevant concepts to be changed. Since NARS has insufficient resources, each concept has a certain capacity, and when its body is full, some old beliefs/tasks with low priority will be deleted, which is another form of meaning change.

Since the above insertions, revisions, and deletions happen all the time when NARS is running, the related concepts in the system change their meanings from time to time. In NARS, the distinction between "concept enrichment" and "concept change" is highly relative — it is just that some changes are more radical than the others. Some changes are so little that they can be ignored practically. Large or small, such changes are not arbitrary or random. On the contrary, they are caused by the system's external communication activity and internal inference activity.

Another aspect of the conceptual dynamics in NARS is the adjustment of priority values of beliefs, tasks, and concepts. As described in Chapter 6, in each inference step, the involved items (concept, task, and belief) are selected probabilistically according to their priority values, and these values are adjusted after each step according to the current result. Therefore, though the meaning of a concept is defined as the collection of beliefs and tasks in the body of the concept, it does not mean that each of them contributes equally. Instead, items with higher priority are used more frequently, and therefore are included more often in the "current meaning" of the concept. Especially, some concepts may form a stable "core meaning" over time, which is almost always involved whenever these concepts are used. On the contrary, some other concepts may lack such a stable core, and their meaning tend to change more radically from situation to situation.

Priority adjustments occur not only *within* concepts, but also *among* concepts. The priority value of a concept is increased when a new task is inserted into it, which may later derive another task to be inserted into a related concept. Therefore, in NARS, among concepts there is also an "activation spreading" process (though the technical details are different from those found in neural networks). Within a single concept, this process also increases the priority of the relations those "targets" (the linked concepts) have high priority. Consequently, among the beliefs and tasks within a concept, the ones more relevant to the current context (defined by the current active concepts) get more chance to be used.

A concept with unbalanced extension/intension is usually not very useful — a concept with a big extension and a small intension is like a collection of objects with few common properties, and a concept with a big intension and a small extension is like a collection of features which few objects can satisfy. A concept with "balanced" extension/intension tends to be close to the "basic level" of categorization [Rosch, 1978], which, according to typical human experience, include many instances with many common properties, therefore are usually more useful for various situations.

If for a concept the "current meaning" triggered in a certain context happens to be quite different from its "core meaning," then we may observe a "metaphorical use" of the concept [Lakoff, 1987]. When such things happen often enough, the previous "periphery meaning" may become part of core meaning, and we say that the concept has evolved in meaning.

Such a conceptual evolution process has special importance to concepts associated with compound terms. Typically, such a concept initially is used according to its literal definition, that is, its relation with components, as specified in its name. However, if later the system gets more direct knowledge about the concept as a whole, and such knowledge cannot be derived from its definition (and may even conflict with it), we get a concept which cannot be understood literally anymore.

For example, when the compound term "$([black] \cap board)$" ("blackboard") appears for the first time in the experience of NARS, it will be understood "literally," that is, as a "board" that is "black" (under the assumption that both concepts already exist). Later, when most blackboards the system encountered are writing surfaces, this relation becomes the core meaning of the concept, while the relations with "black" and "board" become periphery, though still there.

Since the experience of NARS includes its communication with other systems, its usage of a term is influenced by how the same term is used by other systems. If several NARS-based systems form a multi-agent community, the agents tend to develop a consensus on the meaning of the concepts, though the same concept may never have exactly the same meaning in each agent. In this sense, we say that a concept has an "objective" meaning, which is not arbitrary or idiosyncratic. If an individual in the community departs from this consensus, it runs into the risk of misunderstanding, though such a departure may also correspond to a creative usage of a concept, which may later spread in the community as the new meaning of the concept.

11.1.7 Factors in categorization

In summary, the meaning of a concept in NARS at a given time depends on the following factors:

- the collection of beliefs and tasks in the body of the concept,
- the truth value of each belief,
- the priority distribution among the items in the concept body.

The truth value and priority value of a belief are related, but not the same thing. Generally speaking, more confident judgments are more useful than guesses supported by little evidence, and affirmative judgments are more useful than negative judgments, but priority depends on other factors beside truth value, as mentioned previously. Especially, the priority distribution is highly context-sensitive, while a truth value is mostly determined by the long-term experience of the system.

In general, the change of meaning in a concept is caused by two major factors: history (past experience) and context (current experience).

The former causes *permanent* changes in meaning and makes the system *adaptive*, while the latter causes *temporary* changes in meaning and makes the system *situated*. Of course, the distinction between these two factors is relative, not absolute.

In the long-term, the process corresponds to concept learning and evolving. Instead of assuming the learning process converges to a "correct representation" or "true meaning" of the concept, in NARS the process is a never-ending adaptation, and the result is determined by the experience of the system.

In the short-term, the system shows context-sensitivity. Under the assumption of insufficient resources, NARS almost never tries to use all relevant beliefs to process a task. Instead, only "partial meaning" will be used, and which part to select is determined by the priority distribution. Roughly speaking, the selected ones tend to be "useful" and "relevant," judged by the system according to past experience and current context.

As a result, it is quite normal in NARS for a concept to be used with different meanings, or for the same input to be categorized ("perceived") differently.

Since in NARS all categorization related processing is carried out by the inference rules, *categorization is reasoning*. On the other hand, since NARS is a term logic, and in each inference step the premises and conclusions are all categorical statements, *reasoning is categorization*. The two notions just focus on different aspects of the same process.

On the contrary, in predicate logics, the inference rules are not based on categorical relations, but on truth values. Consequently, knowledge-based systems with such logics usually use separate mechanisms for categorization (using "terminological knowledge") and reasoning (using "factual knowledge") [Brachman and Schmolze, 1985]. As a result, in these systems categorization and reasoning are not unified as in NARS.

11.2 Learning in NARS

Generally speaking, "learning" means "changing according to experience to improve performance." Obviously, learning plays a central role in NARS.

11.2.1 Multiple types of learning

As described in the previous chapters, there are several types of learning going on in NARS:

- The inference rules generate new beliefs, which are added to the knowledge base of the system.
- The revision rule will merge beliefs with the same content but come from different experience, therefore changing the belief of the system according to new evidence.
- As new beliefs are generated and old beliefs forgotten, the system learns the meaning of the terms according to its experience with them.
- The compound-term composition rules generate new terms from time to time. At the beginning, their meaning is determined by the meaning of their components. However, as the system gets more direct experience on them, they gradually become independent, and are treated for their own sake.
- By adjusting the priority distributions among terms, tasks, and beliefs, as well as by deleting useless ones, the system also learns what is important and relevant and so should be considered first when processing time is insufficient to consider everything. This kind of "structural knowledge" is not declaratively expressed in Narsese, but embedded in the memory structure of the system.

In this sense, all activities in NARS can be referred to as learning, and NARS is nothing but a multi-strategy learning system. Similar to the case of categorization, in NARS "reasoning" and "learning" are merely two different ways to describe the same underlying process.

In the current research of AI and cognitive psychology, there is a strong tendency of treating intelligence and cognition as a "toolbox," i.e., a collection of "functions" like reasoning, categorization, learning, perception, and so on. The justification of such a treatment is that by studying each function in isolation, the task will become easier, and in the future we can integrate the results into a hybrid system consisting of many modules, each of which is responsible for a certain function. What I suggest here is another possibility: all these functions may actually be different aspects of the same underlying process, which can be reproduced and implemented as one piece.

NARS is an attempt to provide a unified (not "hybrid") normative theory of intelligence for both humans and computers. As a reasoning system, it unifies various types of inference. Here we see that it also unifies learning and reasoning.

11.2.2 NARS and machine learning

Though NARS is a learning system, it is quite different from the mainstream "machine learning" research, with respect to the concrete problems they are working on.

A machine learning system is often described as a "learning algorithm," which takes raw data and background knowledge as input, and produces some output, usually a representation of a concept that was learned from the given data.

An "algorithm" is a computational process that, for the same input, always follows the same path, takes the same amount of computational resources, and produces the same output at the end.

NARS does not fit the above description, because in it "learning" is a *life-long process* [Thrun and Mitchell, 1995] that takes different forms and is influenced by many factors. Consequently, even for the same input, the process at different times may follow different paths, cost different resources, and get different results.

To build a learning system in this way is very different from building an algorithm which takes certain input and produces the desired output:

- By working in real time, it allows different response-time requirements to be attached to a task. One task may need a quick solution, while another task may prefer a more carefully considered solution.

- New knowledge can be added from time to time, as well as by the request of the system. The system revises its beliefs incrementally, rather than restart whenever new knowledge arrives.

- When it is impossible to consider all relevant knowledge, the system can make a rational selection according to experience and context.
- The selection of inference rules is data-driven, so neither the designer nor the user needs to specify how to process a concrete task in advance.
- The learning process is integrated with reasoning and categorization — actually they are different names of the same process.
- It is more similar to the learning process of human beings — we seldom learn new ideas by following a predetermined algorithm.

Of course, this does not mean that NARS is always better than the learning algorithms. Actually, whenever a learning algorithm is available and affordable, it usually gives a more reliable and efficient solution than NARS. On the other hand, something like NARS should be used when such an algorithm is not available (due to insufficient knowledge) or not affordable (due to insufficient computational resources).

In the current machine learning study, the closest work to NARS is the Inferential Theory of Learning (ITL) [Michalski, 1993]. Both NARS and ITL are inferential system that carry out multi-strategy learning, and they also share many theoretical and technical assumptions about machine learning, such as to understand learning as "a goal-guided process of modifying the learner's knowledge by exploring the learner's experience." [Michalski, 1993].

These two approaches have similar major components, but the technical decisions on each of them are quite different:

Knowledge representation: ITL uses first-order predicate logic for knowledge representation, while NARS uses a kind of term logic.

Semantics: In ITL, "truth" is defined according to model, while in NARS it is according to experience.

Inference rules: The two systems have different rule sets. Though both include deduction, induction, abduction, and analogy, the exact definitions are not the same.

Knowledge organization: In NARS, priority distributions are maintained among tasks and beliefs, so that tasks are processed at different rates and beliefs have different probabilities of being used. By adjusting the priority distributions, the system learns control and context information. There is no such mechanism in ITL.

Control mechanism: ITL characterizes a learning process as a goal-guided search through a knowledge space. NARS processes its tasks by interacting them with beliefs at different rates to find matching answers and to derive new knowledge and tasks. This process does not follow a predetermined algorithm.

11.2.3 Learning and intelligence

Though few people deny the importance of learning in AI research, there are many "AI systems" that have little learning capability. Many people just treat learning as an *additional* function that can be later added into the system to improve its performance. To them, "working" and "learning" are two separate phrases in the life cycle of an AI system.

Even in the machine learning community, it is a common practice to formalize learning as a normal computation process that maps given inputs into corresponding outputs, according to a predetermined algorithm, function, or sample set.

As we have seen, NARS treats learning differently, in several aspects:

- According to the working definition of intelligence of NARS, a system without learning capability is not intelligent at all. In a sense, "machine learning" is not merely part of AI, but AI itself.

- For a system, learning is not simply a procedure call, but is everywhere in the system's life-long history.

- Learning takes different forms, and changes various components and aspects of the system.

In NARS, the learning process shows the two aspects in Piaget's theory [Piaget, 1963]: *assimilation*, by which new experience is perceived and interpreted according to the internal status of the system, and *accommodation*, by which internal status of the system changes according to the new experience.

NARS is a *general-purpose* system in the sense that its learning capability is not designed to suit any specific practical problem. Like a baby, it has the *potential* to become experts in many different domains. However, also like a human being, its experience will cause bias in its belief structure and skill set, so it won't become an expert in all the domains at the same time. In discussions on AI, arguments against a unified theory are based on the (correct) observation that different techniques work better for different problems [Minsky et al., 2004]. However, it is often ignored that at a lower lever these techniques may be based on a common technique, which is not problem-specific. Without such a common foundation, attempts to integrate multiple techniques have been difficult [Roland and Shiman, 2002].

As described previously, on the object-language (Narsese) level, everything can be learned (i.e., changed according to experience), including beliefs, tasks, and concepts, with their attributes (truth value, priority, meaning, and so on).

This does not mean that everything changes in NARS. Instead, in the meta-language (the design language of NARS used in this book), there are many things which remain unchanged in NARS, such as the grammar of Narsese, the semantic principles of Narsese, the meaning of the logic constances (the built-in relations and operators), the inference rules (with their truth-value functions), the working cycle, and the control strategy. These are the "innate" components of the system.

When an implementation of NARS is "born," only the above innate parts are required, and the memory of the system may be empty. After the system starts to communicate with the environment, it begins to learn beliefs, tasks, and concepts (with their attributes). Then, the system's behaviors will be determined both by its "nature" (the innate components) and "nurture" (experience), but not by either of the two alone.

For practical purposes, it is possible to start a NARS with a "preloaded" memory. To do this, we can simply start a NARS with an empty memory, educate it, then at a later time save its memory into a file or database, which can be copied into another NARS to be used as the memory-at-birth. This method will improve education efficiency of the system. We may even allow direct "memory-editing" to get a desired memory. All these techniques do not change the principle that everything on the object-language level is learnable, given proper experience.

NARS is *creative*, but not in the sense that all the results of such behaviors are of benefit to the system, or excellent according to some outside standards. Nor does it mean that these behaviors come from nowhere, or from a "free will" of some sort. In contrary, it means that the behaviors are novel to the system, and cannot be attributed either to the designer (who determines the system's initial state and skills) or to a tutor (who determines part of the system's experience) alone. Designers and tutors only make the creative behaviors possible. What turns the possibility into reality is the system's experience, and for a system that lives in a complex environment, its experience is not completely determined by any other systems (human or computer). For this reason, these behaviors, with their results, are better to be attributed to the system *itself*, than to anyone else [Hofstadter, 1979].

In the discussions on future AI, there is the opinion that a real AI should be able to achieve "complete self-modifying," include changing its own source code. Technically, it is possible to let NARS change something on the meta-level, however, that approach is not adopted at the current stage, for the following reasons:

- "Complete self-modifying" is an illusion. If we allow NARS to modify its meta-level knowledge, we need to give it meta-meta-level knowledge to specify how the modification happens. As flexible as the human mind is, it cannot modify its own "law of thought." As Hofstadter put it, "below every tangled hierarchy lies an inviolate level" [Hofstadter, 1979].

- Though meta-level self-modifying will give the system more flexibility, it does not necessarily make the system more intelligent. It is often dangerous, and should be used only when the same effect cannot be produced in the object-level. To assume "the more radical the changes can be, the more intelligent the system will be" is unfounded. It is easy to allow a system to modify its own source code, but hard to do it right.

- In the future, I will explore the possibility of meta-level learning in NARS, but will not attempt to do so until the object-level learning is mature. To try everything at the same time is just not a good engineering practice.

In summary, at the current stage, the learnability of NARS is determined accurately by the level of language: everything on the object-language level is learnable, while everything on the meta-language level is not.

Chapter 12

Control and Computation

This chapter discusses the control mechanism of NARS, which is based on AIKR. Consequently, NARS behaves quite differently from traditional computing systems.

12.1 NARS and theoretical computer science

12.1.1 Turing Machine and computation

The word "computation" has two different senses. In a broad sense, it refers to whatever a computer does; in a narrow sense, it is a concept defined in theoretical computer science.

A *Turing Machine* M has a finite number of states, and among them there is one initial state and at least one final state. At each state q_i, M moves into another state q_j, according to the given input data. A *computation* is a finite sequence of moves by which M transforms from its initial state q_0 to one of its final states q_f, in response to the input data d_i. In the final state, M provides the output data d_o as the result of the computation. We can equally well say that M is a *function* that maps d_i to d_o, or that M is an *algorithm* with d_i and d_o as input and output, respectively [Hopcroft and Ullman, 1979].

It is very important to see that Turing Machine is not defined merely as a *system*, but as a *process* in that system. According to the definition of computation, the process has the following features:

1. There is a unique *initial state* in which the system can accept input tasks, and tasks are processed in a one-by-one manner. If a task arrives when the system is still busy with another task, the new task has to wait.[1]

2. The system always yields the same result for a given task, no matter when the task is processed.

3. The amount of resources spent on a task is a function of the task, depending on the complexity of the algorithm, but independent of when the task is processed.

4. There is a predetermined set of *final states* in which the system will stop working on a task and provide a result, no matter whether there are other tasks waiting to be processed.

Usually, an algorithm is defined on a problem *class*, which has more than one concrete problem *instances*. In each time the algorithm is used, it is applied on a problem instance. A problem (class) is *computable* if there is an algorithm that generates a correct result in finite number of steps for each instance of the class. In that case, the algorithm (or the corresponding Turing Machine) is referred to as the solution of the problem.

The amount of time or space used by an algorithm is called the (time or space) *complexity* of the algorithm, and is represented as a function of the "size" of the problem instance. According to the category the function belongs, we call the complexity to be *constant*, *logarithmical*, *polynomial*, *exponential*, and so on. In particular, a problem is *feasible*, or *tractable*, if it has a polynomial solution or better. The problems without polynomial algorithms are *intractable*, because the cost of a solution will become astronomical figures for large instances of the problem [Rawlins, 1992].

According to computability theory and computational complexity theory, using a computer to solve a problem usually follows this procedure:

[1]Even if interrupt mechanisms are taken into consideration, the picture is fundamentally the same.

1. to define the problem by accurately specifying the valid inputs, and for each of them, specifying the required output;

2. to design an algorithm that correctly generates output for each valid input;

3. to analyze the complexity of the algorithm, and to select the most efficient one if there are multiple algorithms for the problem;

4. to code the algorithm in a programming language, and to load the executable code into a computer system.

If the algorithm is correct and the program is efficient enough, the problem is considered as solved — for each instance of the problem, the program will produce a correct solution. The output and its (time-space) cost are determined only by the algorithm and the input.

Conceptually, a computer system can be seen as a collection of algorithms, and it works by repeating the following cycle:

> to wait for the user to input a new problem instance;
> to call the corresponding algorithm when an input comes;
> to execute the algorithm on the given input;
> to report the output;
> to reset the working environment.

Though modern computer systems allow multiple problem instances to be processed in parallel by time-sharing, the above conceptual picture remains unchanged. When an algorithm is working, whether there are other algorithms running should make no difference in the result, unless it has communication with other algorithms (which should be taken as part of the input). If the same problem instance is asked again later, the result should be exactly the same.

12.1.2 Algorithm and intelligence

Since AI is usually taken as a branch of computer science [Newell, 1990], it has also, to a large extent, inherited the theoretical heritage associated to the concept of computation. Most people in AI conduct their research according to the previous procedure, by defining problem, designing algorithm, analyzing computational complexity, and so on. A well known example is Marr's classification of AI research into the levels of computation, algorithm, and implementation [Marr, 1982].

For a given problem, this approach has three possible results:

No algorithm is found. If we do not have the knowledge to design an algorithm to solve a given problem, then of course there is no solution, at least at current moment. For certain problems, it can even be proved that they are not computable.

No tractable algorithm is found. If we know an algorithm, but it is too resource demanding, then it cannot be used except on very simple instances of the problem. When a solution does not "scale up" to complicated cases, in AI it is usually not counted as a solution. For example, in principle many problems (such as playing chess) can be solved by exhaustive search, but we cannot afford the time it requires — exponential algorithms easily produce "combinatorial explosion," so they make no practical sense. That is why some authors treat tractability as a central issue in AI [Bylander, 1991, Levesque, 1989].

A tractable algorithm is found. By definition, in this situation the problem has a known solution that can be practically used. For computer scientists, this is the end of the story (unless there is the need to improve the algorithm). However, for the purpose of AI, some people are unhappy — "Where is the intelligence? This is just programming!" The concept of "intelligence" is intuitively related to creativity and flexibility, so to many people, solving a problem by accurately following a predetermined algorithm cannot be it.

So AI is in a weird situation — it either fails to solve a problem, or solves a problem "without intelligence." Therefore, it seems that AI never works! This is what Hofstadter calls *Tesler's Theorem* — "AI is whatever hasn't been done yet." [Hofstadter, 1979].

Different people have different attitudes toward this situation. Some people take it as an argument for the impossibility of real AI; some people blame the von Neumann computer, and believe that AI needs a fundamentally different hardware which is not a Turing Machine; some people do not take this as an issue — they proudly see AI as the expanding frontier of computer science.

For several reasons, this is an issue. We do not need to run psychological experiments to know that the problem-solving processes in the human mind rarely follow a predetermined algorithm. On the other hand, one important motivation behind AI research is to introduce flexibility, autonomy, and creativity into computer systems. If AI still follows the common practice of computer program development, then "intelligence" is simply a fancy label on old stuff, and the systems developed will continue to be "brittle," in the sense that it cannot handle any event that is not fully anticipated when the system is designed [Holland, 1986].

Some people suggest that the problem is caused by the narrowness of the concepts in theoretical computer science [Hofstadter, 1985, Kugel, 1986, Sloman, 2002]. However, the majority of the AI researchers continue to apply the theory of computation to AI, and analyze problems in terms of computability and computational complexity, as exemplified by [Bylander, 1991, Edmonds, 2000, Hutter, 2001, Levesque, 1989, Littman et al., 1998, Valiant, 1984].

12.1.3 NARS at different scales

The relationship between NARS and the notion of "computation" (in the narrow sense of the word) is subtle, since the system can be analyzed on different scales of description.

Let us first identify the "problems" NARS attempts to solve. From the previous description, it is obvious that each "task" in NARS can be seen as a "problem instance." If a task is taken as input, what NARS does to it is not computation, because almost all components in the previous definition of "computation" are missing here:

- Though NARS does solve problems, it processes each problem instance (i.e., a task) in a case-by-case manner, without a general algorithm for the "problem class" as a whole. As a result, it may give a pretty good solution to a task, but may fail on a similar one.

- In NARS, there is no unique "initial state" in which the system waits for and accepts new tasks. At any moment when the system is running, tasks can be accepted, in many different internal states.

- Similarly, there is no "final state" for a task. For instance, if a task's priority is low (relative to other tasks), it is even possible for it to be completely ignored. If a tentative answer to a question is reported, usually neither the system nor its human designer can predict whether a better answer will be reported later, since that will depend on events still to take place in the future, such as whether the system acquires new knowledge related to the task, or whether more time winds up being spent on it.

- For a given problem, whether a result is a "solution" become a matter of degree. Under AIKR, NARS cannot give a "perfect solution" to a problem.

- When processing a task, the inference steps are chained together into an inference process in run-time. There is no predetermined algorithm to follow for a given task.

By slightly changing the meaning of the term, one might say that NARS has an initial state — namely, when its memory is completely empty (the system can be "born" without any innate domain knowledge). Its state changes as soon as it interacts with its environment and begins processing tasks. The system never will return to its initial state, until and unless a user terminates the processing and erases all of its memory. In such a case, the system can of course be "reborn" with the same "genetic code" — its sets of inference rules, control mechanisms, personal parameters, and so on. However, unless the experience of the system perfectly repeats its experience in its "previous life," the system's behaviors will be different.

In summary, the system's behaviors are determined by its initial state and its experience, but not by either one of the two alone.

Now we can see that NARS can be observed on (at least) three scales, in term of what is referred to as its *input* and *output*.

- In the scale of each *inference step* (i.e., the "execution cycle" defined in Section 6.3.1), the system's activity is computation, where the input is the memory before that step, and the output is the memory after that step. In NARS, there is an explicitly coded algorithm for this process.

- In the scale of each *task-processing activity*, where the input is a task, and the output is the result of the processing of the task, what the system does cannot be captured by concepts like computation, function, or algorithm, as discussed above.

- In the scale of each *whole-life cycle*, with its "birth" state (which can be any state, since the system can be born with a preloaded memory) as an "initial state," and a "death" state (which can be any following state of the birth state) as a "final state," what NARS does can be seen as computation, with its "life-long" experience as the input, and its all behaviors as the output.

In summary, the behavior of NARS can be described on different "scales" or "levels." NARS is computing on some, but not all, of them. This state of affairs has been articulated by Hofstadter in the following way: "something can be computational at one level, but not at another level," [Hofstadter, 1985] and by Kugel as "cognitive processes that, although they involve more than computing, can still be modeled on the machines we call 'computers'." [Kugel, 1986].

In contrast to this, conventional computer systems, while also describable at these levels, are computing in all of them. Let us use an ordinary sorting program as an example: you can take either a single sorting problem, or a *sequence* of such problems, as the input, and the processes in both cases are computation — the program's response to a given sorting task is fully determined (by the algorithm and the input data) and does not depend on its experience and context (i.e., the processing of other sorting tasks). Even if the exactly same problem is given the system twice, the system will repeat the same procedure, rather than directly report the remembered solution in the second time.

Among the three levels, the most important one is the one in the middle, where a "problem" or "task" is the input, and its solution is the output. It is this level on which the computability and computational complexity are defined and studied. Since what NARS does on this level is not computation, the traditional theoretical computer science cannot be directly used in the system on this level.

12.1.4 Beyond computation

Though still following algorithms at a certain level, NARS is creative and autonomous in the sense that its behavior is determined not only by its initial design, but also by its "personal" experience. It can generate results never anticipated by its designer, and can produce them by its own choice. A "tutor" can "educate" it by manipulating its experience, but cannot completely control its behavior due to the complexity of the system. From a pragmatic point of view, this is neither necessarily a good thing, nor necessarily a bad thing. It is simply the case that an adaptive system with insufficient knowledge and resources has to behave in this way.

Compared to the computing processes studied by theoretical computer science, the inference control mechanism of NARS has the following properties:

- It does not define a "problem" as a set and use the same method to solve all of its instances. Instead, it treats each "problem instance" as a problem on its own, and solves it in a case-by-case manner. For example, when evaluating a proposed inheritance statement, NARS sometimes checks the extension of a term in the statement, and sometimes its intension.

- For a problem, it does not draw a sharp line between solutions and non-solutions, and treat all solutions as equal. Instead, each solution is good to a degree, and the system compares candidate solutions to decide which one is the best (that the system has found so far).

- It does not insist on the "one problem, one solution" format. Instead, for a problem, the system may generate zero, one, or a sequence of solutions, each of which is better than a previous one.

- It does not depend on a predetermined algorithm to solve a problem. Instead, it cuts a problem-solving process into steps. Each step may still follow an algorithm which takes constant time to finish, but the who process is linked together at run time.

- It processes tasks (and subtasks) in parallel, but at different speeds, according to their priority values.
- It does not attempt to use all relevant beliefs to solve a problem. Instead, in each step it only considers a single belief, selected according to their priority values.
- In each step, it lets the selected task and belief decide how the task is processed.
- It does not throw away the intermediate results at the end of a problem-solving process. Instead, it keeps them for future tasks, and lets all tasks to interact with the same knowledge base.
- When memory is full, it removes items with the lowest priority.
- It adjusts the priority distributions according to the experience of the system and the current context, so as to give important and relevant items more resources.

Built in this way, NARS shows many novel features:

- Knowledge is accepted by the system as sentences in a formal language. The user can assign the system any question or goal that can be phrased in the formal language, and the system will not be paralyzed by tasks beyond its current capability. Neither the designer nor the user needs to provide the system with task-oriented algorithms.
- The user can assign initial priority and durability value to a task to influence (though not to determine) the system's resource allocation to that task.
- The system may provide a quick answer to a question, then refine the answer incrementally. In this sense, NARS can "change its mind" when new beliefs are taken into consideration.

- The system usually concentrates on the most important and promising tasks, but it also pays some attention to other "peripheral" tasks.

- The response to a question depends not only on what the system has been told, but also on what the system has been asked. For example, the system may spend a long time finding an answer, but if the same question (or a similar one) appears again later, the answer usually comes sooner.

These properties distinguish NARS from other reasoning systems. It reproduces many properties displayed by human thinking processes, and it shows the potential of working in situations where no other approaches can be applied, because of their assumptions.

This approach is not just a picturesque new way to see things, but has important methodological implications for AI research. When a system like NARS is designed, the designer should not try to decide what answer the system should produce in response to a given question — that should be decided by the system itself at run time; the designer simply cannot exhaustively consider all possible situations in advance (the designer, hopefully, is also an intelligent system, thus limited by insufficient knowledge and resources). For similar reasons, the designer cannot decide in advance how much resources to spend on a certain task, for this is totally context-dependent. Thus, the designer is no longer working on either domain-specific algorithms or general-purpose algorithms (like GPS), but rather on *meta-algorithms* or *micro-algorithms*, which carry out inferences, manage resources (like a small operating system), and so on. In this way, the problems solved by the designer and the problems solved by the system itself are clearly distinguishable from one another — all problems expressed in Narsese are solved by the system, while the problems expressed in the meta-language of NARS are solved by the designer.

Although general-purpose algorithms and meta-algorithms are both independent of specific domains, there is still a fundamental difference between the two types of algorithm, with respect to how a domain problem (i.e., a question asked by the user) is solved. In the former cases, the problem-solving process is still computation. As described previously, the system accepts the problem at its initial state, processes it according to predetermined procedure, then stops at the final state and reports the solution. We already know that NARS does not work in this way. The fact that NARS still consists of a set of algorithms does not mean that the system's problem-solving (or task-processing) activities follow any algorithm. Of course, the algorithms in NARS do facilitate the problem-solving activities, but in a different way.

For example, Section 6.3.1 actually describes the algorithm that controls an execution cycle, which invoke other algorithms, like the put-in and take-out procedures of various bags. However, none of these algorithms takes user-provided problems (i.e., input tasks) as its input. In fact, armed with these algorithms, NARS deals with input tasks *without* (task-oriented) algorithms. We call them meta-algorithms, because they are not ready-made methods for user problems, but (ready-made) methods by which the "object-level" methods can be formed dynamically in run time.

These ideas allow us to explain why *Tesler's Theorem* ("AI is whatever hasn't been done yet" [Hofstadter, 1979]) applies to many AI projects: in those projects, the designers usually use their own intelligence to solve domain problems, and then implement the solutions in computer systems in the form of problem-oriented algorithms. The computer systems then executes the algorithms on specific instances of the problems, an activity that can hardly be referred to as "solving problems intelligently."

For example, many "expert systems" have no learning ability. Such systems are designed by "knowledge engineers," who extract domain knowledge from experts in a particular field, and then implant this knowledge into a computer system, so as to reproduce the experts' problem-solving ability. According to my working definition of intelligence, both the domain experts and the knowledge engineers are intelligent — they work with insufficient knowledge and resources, and they learn from their experience — and yet the expert system itself is not intelligent, because, ironically, when it faces a problem, it faithfully follows the predetermined algorithms that were extracted from the experts' intelligent behaviors, thanks to the intelligence of the knowledge engineers.

This new way of describing AI also changes what we usually referred to as a "solution." Let us take the "combinatorial explosion" problem as an example. If, for a particular problem, there is an algorithm that takes an amount of time that grows exponentially with some parameter in the problem, usually such an algorithm is useless in actual practice — the time expense will rapidly increase to astronomical figures, and the system will simply be paralyzed. The traditional way to deal with this problem is to look for a faster algorithm, even if that implies sacrificing the quality of the solution. Since in NARS problem-oriented algorithms are not used, the very concept of "computational complexity" disappears, and the situation becomes very different.

In this case, a new interpretation and solution to the "scaling up" problem becomes available. It is well-known that many AI projects work fine at experiment stage with small knowledge sets, but fail to work in real-world situations. This is often caused by the "sufficient resource" assumption accepted by the approaches. Such an assumption is usually made implicitly in the operations carried out by the approach, such as to exhaust possibilities for a specific purpose. These operations are affordable on small amount of knowledge, but become inapplicable when the knowledge base is huge. For the systems based on the assumption of insufficient resources, such as NARS, the situation is different. These systems do not take advantage of the small size of the knowledge base by exhausting possibilities, and also do not attempt to do so when the knowledge base is huge. Consequently, the resource management mechanisms used by these systems do scaling up. The system's performance still becomes not as good when the problem is hard and the knowledge base is huge, but the degradation happens in a *graceful* way, just like what happens to the human mind in similar situations. Especially, the system will not be trapped by hard problems, because it knows when to give up and turn to other things, while still let the problem wait for new possibilities.

Finally, there are some disclaimers: I am not arguing that the traditional theoretical computer science is wrong, but that it does not apply to many situations in AI. Similar conclusions can be found in a recent collection [Scheutz, 2002], though my argument is different in many aspects. Also, NARS is not the first system that goes beyond the narrow sense of "computation" — many approaches to be discussed in the following subsection are in that territory, too, though this issue has rarely been raised in theoretical discussions in the field.

12.2 Various assumptions about resources

Any problem-solving or information-processing process will cost computational resources, especially processor time and memory space. A theory about these processes must address the issue of resources. However, different theories, for different purposes, make different assumptions about resources. In this section, the assumptions about resources behind NARS is compared with the assumptions of other theories.

12.2.1 Computability and complexity

In the study of computability, the attitude to resource can be summarized as: the time spent in the problem-solving activity can be ignored, given that it is finite. People can wait a period with an arbitrary-but-finite length for a solution. Also, what counts as a "solution" for a given problem is resource-independent. In other words, whether a state is final, or whether a solution has been found, is defined by some criterion, in which the resource cost of the solution is not taken into account.

Though the above abstraction is necessary for certain purposes, people are often unsatisfied by it. Obviously, for almost all practical purposes, time is valuable, and people hope to solve problems as soon as possible.

Under the assumption that the system's hardware (processors and memory) remains constant, the time spent on a problem depends on two factors: the problem itself and the system's method for the problem, and the latter often takes the form of an algorithm.

When the computational complexity of an algorithm is analyzed, resource expense is taken into consideration in a quantitative manner. We are no longer satisfied by the knowledge that the cost will be a finite number — we want that number to be as small as possible. However, the solution of a given problem is still defined in a resource-independent way — a bad result cannot be referred to as a "solution" just because it is easy to get.

If the task is processed by a traditional algorithm, the processing stops at predetermined final states. Therefore the time cost is determined completely by the computational complexity of the algorithm and the size of the task. It has nothing to do with the user or the context. Though in a time-sharing system, the response time depends on the load of the system, the processor-time that is really spent on the task remains roughly the same.

The difference between NARS and theory of computation on the assumption about resources has been discussed previously.

12.2.2 Time pressure in problems

For many problems, we still do not have polynomial algorithms, and for many others, even polynomial algorithms are too slow. It is often the case that a *time requirement* is an intrinsic component of a problem, and a result must meet the requirement to be accepted as a solution of the problem. Consequently, the system works under a *time pressure*. In such a situation, the time needed for a perfect solution usually exceeds the time requirements of the problems — otherwise the time requirement can be ignored. This can happen even if the algorithm used by the system only takes constant time. For example, if an algorithm solves a problem by searching a file thoroughly, which takes two seconds in the given hardware, then such an algorithm cannot be used when a solution must be provided within one second.

For a system to work in this situation, the criterion for a result to be a "solution" of the given problem have to be relaxed. In fact, for many problems, whether a result can be counted as a solution is a matter of degree. Instead of saying whether a result is a solution or not, here people need to compare, or even measure, the qualities of different solutions. Usually, the quality of a solution depends on its resource cost, which includes its processor-time expense. To work under a time pressure means to give up best solutions, and to find a trade-off between solution quality and time cost [Good, 1983].

To let a system make trade-off between solution quality and time cost, this is not a new idea. Approximation algorithms and heuristic algorithms are all motivated by this consideration [Rawlins, 1992]. Similarly, we can first decide the time request, usually in the form of a *deadline*, then look for an algorithm which can meet the deadline, and can also provide a solution as good as possible. This approach leads to the concept of *real-time algorithm* [Laffey et al., 1988,

Strosnider and Paul, 1994]. However, in these algorithms, the trade-off is determined when an algorithm is designed. As a result, the problem-solving process is still computation, only on a relaxed version of the problem. It is still the problem-oriented algorithm that decides which step to take at each instant, and where to stop at the end.

Though these approaches take time pressure into consideration, they are still inappropriate for NARS, for the following reasons:

1. The system cannot depend on the environment to assign such deadlines, because the resulting time requirements may exceed the system's capability.

2. In general, the system cannot anticipate how much time it ought to spend on a task when it is accepted (from the environment) or generated (by the system itself), because that depends on future events — for example, on whether an answer will be found soon, and on how many new tasks will show up in the near future.

3. The concept of "deadline" implicitly assumes a step function of the utility of the answer by requesting an answer at a certain time, t — that is, an answer provided before t does not get extra credit, and an answer found after t is completely useless. Such a rigid, black-and-white attitude is not suitable for many situations.

NARS works in real time, and the bag structure (see Section 6.2.1) shares ideas with techniques used in real-time systems, such as pruning, ordering, approximation, and scoping [Strosnider and Paul, 1994]. However, NARS does not stop at deadlines, because it often results in a waste of resources if the system idles afterwards. For an adaptive system, the tasks that appeared in the past may happen again in the future, with some variants. Therefore, even when the user no longer needs a result after a certain amount of time, the system still has reason to work on it, if there are resources available.

In many situations, it is better to treat time pressure as a variable and context-dependent factor, because the time requests of problems, the desired quality of solutions, and the system's time supply for a problem (in a multi-task environment) may change from context to context. It is inefficient, if not impossible, to equip the system with a family of algorithms for each possible context. For this situation, we hope to take the time pressure into consideration in the *run time* of the system.

One instance of this approach is to use an interruptible algorithm. In the simplest case, a "trial and error" procedure can be used to "solve" a uncomputable problem [Kugel, 1986]. Suppose we want to check whether a Turing Machine halts, we can use such a procedure. It reports "NO" at the very beginning, then simulate the given Turing Machine. When the Turing Machine halts, the trial-and-error procedure reports "YES" and halts. Such a procedure is not an algorithm (according to the strict definition of the concept) because it may not stop, but it can be implemented in ordinary computers, and its *last* report is always a correct one, though the user may not have the time to get it, or cannot confirm that it is really the last one when it is "NO."

A more general concept along this path is "anytime algorithm," [Dean and Boddy, 1988] which is a program that provides approximate answers to a problem in such a way that: (1) an answer is available at any point in the execution of the program; and (2) the quality of the answer improves with an increase in execution time.

Such an "algorithm" no longer corresponds to a Turing Machine. Because there is no predetermined final states, the program is stopped by an external force, rather than by itself. Consequently, the amount of time spent on a problem is completely determined by the user (or a monitor program) at run time, and no result is "final" in the sense that it could not be revised if the program had spent more time on the problem.

In this way, the time pressure on a problem-solving activity is no longer a constant. The user can either attach a time request, such as a deadline, to a problem at the beginning, or let the program run, then interrupt it at the end. Under different time pressure, the same program may provide different solutions for the same problem.

In NARS, there are three factors that altogether determine the time spent on a given task: the time request of the user, the design of the system, and the current context. As described in Section 6.1.2, the user can assign a priority value and a durability value to a task, otherwise default values are used. Because both of the values are given in relative forms, the actual time allocated to the task is decided by the allocation mechanism and the current situation of resource competition. In this way, neither the designer nor the user have complete control, and the system *itself* participates in the decision, by taking its past experience and current context into account.

This approach is more flexible than traditional algorithms, because the user can influence the resource allocation at run time. It often happens that the same type of problem may have different time requests, which is hard to be satisfied by a predetermined standard for final states.

Such an approach is also more similar to the resource management mechanism of the human mind. Obviously, the human mind is a real-time system that responds to different time requests, but it seldom stops thinking about a problem at a deadline, even if such a deadline exists, such as in an examination.

12.2.3 Resource allocation

A more complex situation happens when the idea of anytime algorithm is used at the sub-problem level.

If a task can be divided into many subtasks, and the system does not have the time to process all of them thoroughly, it is often possible to carry out each of them by an anytime algorithm, and to manage the processing time as a resource. According to Good's "Type II rationality," [Good, 1983] in this situation an optimum solution should be based on decision theory, by taking the cost of deliberation and the expected performance of the involved algorithms into account.

To do this, the system needs a *meta-level* algorithm, which explicitly allocate processing time to object-level procedures, according to the expected effect of those allocations on the system's performance. This idea is developed in AI under the names of "deliberation scheduling" [Boddy and Dean, 1994], "metareasoning" [Russell and Wefald, 1991b], and "flexible computation" [Horvitz, 1989]. These approaches focus on planning for resource allocation, therefore the quality of the management depends upon the quality of the expectations, though run-time monitoring is also possible [Zilberstein, 1995]. However, if the information about object-level procedures mainly comes at run time, the meta-level planner may have little to do before the procedures actually run — its expectations will be very different from the reality revealed later. To be efficient, the resource allocation has to be adjusted dynamically when the system is solving object-level problems, and the advanced planning become less important (though still necessary). This is particularly true for adaptive systems.

Though NARS shares certain intuitions with the above approaches, it is different from them in technical details, by the use of *asynchronous parallelism* — it does not give each task a fixed budget at the beginning of its processing, but processes them concurrently, and lets them compete for resources.

Though the initial idea comes from the concept of time-sharing, the controlled concurrency of NARS is still very different from ordinary time-sharing. In NARS, the parallel processed tasks (1) consult a shared knowledge base, (2) access beliefs according to the current priority distribution in the memory, (3) change the priority distribution after each step, and (4) can be stopped after any number of steps. As a result, the mutual influence among the tasks become very strong. The coexistent tasks not only influence the processing speed of a task (this is also true for ordinary time-sharing systems), but also strongly influence its processing depth (i.e., when the processing terminates) and path (what beliefs are consulted, and in what order).

In spirit, the "controlled concurrency" in NARS is very similar to Hofstadter's "parallel terraced scan" [Hofstadter and FARG, 1995] (though their implementation details are quite different). When exploring an unfamiliar territory to achieve a goal under a time pressure, it is usually impossible to try every path to its end. Without sufficient knowledge, it is also impossible to get a satisfactory plan before the adventure. However, if somehow the system can investigate many paths in parallel, and the intermediate results collected at different levels of depth can provide clues for the promise of the paths, the system may be able to get a relatively good result in a short time.

This kind of terraced scan moves by stages: first many possibilities are explored in parallel, but only superficially. Then the system reallocates its time resources according to the preliminary results, and let the promising ones to be explored more deeply. Stage by stage, the system focuses its attention to less and less good paths, which hopefully lead the system to a final solution.

Putting it differently, we can think of the system as exploring all the possible paths at the same time, but at different *speeds*. It goes faster in the more promising paths, and the speeds are adjusted all the time according to the immediate feedback on different paths. The system usually does not treat all paths as equal, because that means to ignore available information about them; the system usually also does not devote all its resources to a path that is the most promising one at a certain time, because in that way the potential information about other paths cannot be collected. In between these two extreme decisions, the system distributes its time resource *unevenly* among its tasks, and dynamically adjusts its bias according to new results.

A similar idea can be find in evolutionary computing, where multiple solutions to a problem are explored in parallel, with the more promising ones having higher chance to be preserved for the future [Holland, 1992].

12.2.4 NARS vs. production systems

In some aspects NARS is similar to Soar [Laird et al., 1987, Newell, 1990] and ACT-R [Anderson and Lebiere, 1998], both of which are AGI systems built using the "production system" technique. For instance, an inference step of NARS roughly corresponds to the firing of a production rule, and the inference control process of NARS roughly corresponds to "conflict resolution" in a production system, where a production rule (or an operator) is selected to be applied, among all candidates whose firing conditions are satisfied. All three systems try to make the selections according to experience and context, so as to achieve the maximum efficiency in problem-solving processes.

However, the design of the control mechanism of NARS are quite different from the other two systems [Johnson, 1997]. Beside the fact that NARS is a reasoning system, not a production system, the differences are mainly caused by the assumptions made by the systems about their resources. Though all three systems are designed with resource limitation in mind, the concrete restrictions applied to the design are quite different.

- NARS is designed to be a real-time system, while the other two are not. NARS allows the user to specify initial priority and durability to input tasks, and allows the system to stop working on a task before its logical end. In the other two systems, the user does not attach time requirement to tasks (goals), and each of them is processed until a satisfying solution is found.

- As an open system, NARS lets new tasks be accepted when old tasks are still being processed, and lets multiple tasks share the same memory (knowledge base) and compete for resources. The other two systems work on one task at a time (though it may derive multiple subtasks). Consequently, problem-solving in NARS is more context-sensitive.

- In NARS, since in each step only a single belief is taken into account, and the system works in real time, usually a solution is only derived from partial relevant knowledge. In the other two systems, in each step all relevant production rules are checked to see if it should be fired — "The system takes actions to attain its goals, using all the knowledge that it has."[Newell, 1990] In NARS, though the same is desired (usually the more knowledge that is considered, the better the conclusion is), it is rarely affordable in practice.

Of course, NARS is also very different from Soar and ACT-R in its logical part, as described and discussed in the previous chapters.

12.3 Dynamic natures of NARS

Finally, let's see the implications of the control mechanism of NARS in the behaviors of the system.

12.3.1 Context-sensitive processes

It is well known that reasoning processes in the human mind are context dependent. For the same task, different results are obtained in different situations. Now the problem is: how to represent a "context" or "situation"?

The previous solutions proposed to this problem often treat a context as a *static* entity that can be identified by a name. Then, to represent context-sensitive reasoning process, what is needed is to explicitly include a reference to the context. Such ideas induce "contextual reasoning," [McCarthy, 1993] "situation theory," [Barwise and Perry, 1983] and "micro-theory." [Guha and Lenat, 1990] For example, McCarthy used $holds(p, c)$ to represent "Proposition p holds in context c."

On the contrary, in NARS the "context" of a task refers to the *internal circumstance* in which a task is processed. It is not named, but represented implicitly. When I say that a process in NARS is context-dependent, what I mean is that it may (though not necessarily) have a different result when carried out by the system at different times.

With the control mechanisms described in Chapter 6, it is easy to see that the processing triggered by a task given to NARS is context-sensitive. Even for the same task, with the same priority and durability values, the result may be different. How a task is treated depends on what knowledge the system has, how the knowledge is organized, and how much of the system's resources the task gets — put simply, it is determined by the system's experience, which includes not only events that take place before the task showed up, but also events that happened after that moment. The contents, the order, and the timing of events all matter. Furthermore, a question may result in no answer, one answer, or more than one answer.

This context-sensitivity is another feature that differs NARS from other similar ideas, like anytime algorithm [Frisch and Haddawy, 1994] or parallel terraced scan [Hofstadter and FARG, 1995]. In NARS the processing of a task becomes unpredictable and un-repeatable (from the initial design of the system and the task itself), because the context plays a central role. It should be understood that the system is nondeterministic in above sense, rather than because it takes out items from bags according to a probabilistic distribution — that is simply a way to allocate resources unevenly, and can be implemented deterministically [Hofstadter, 1993b].

From the user's point of view, the most distinguished nature of NARS' control mechanism is non-determinism. Even if the user provides the same task to the system, with the same priority and durability values, the task may be processed differently: when the system is busy (that is, there are many other tasks with higher priority), the task is only briefly processed, and some "shallow" implications or answers are found; when the system is idle (that is, there are few other tasks), the task is processed more thoroughly, and deep results can be obtained. Generally speaking, a task can be processed for any number of steps, as in anytime algorithms. The actual number of steps to be carried out is determined both by the initial assignment of priority and durability, and by the resources competition in the system. Furthermore, the processing procedure and result depend on the other tasks existing currently and recently. Every task changes the knowledge structure while being processed, and therefore influences how other tasks will be processed. For example, if the system just processed a task T_1, and then begins to work on a related task T_2, the beliefs that contribute to T_1's processing will get a higher chance of being used again.

12.3.2 Flexible resource consuming

NARS has the ability to learn from its experience and to adapt to its environments, even though the system provides no guarantee regarding the absolute answer quality and response time for a certain task. What can be said about it is: if the system spends more time on a task, the quality of the answer, mainly measured by its *confidence* value, usually will improve.

However, how much resources will be actually spent on a task and the quality of the answer are determined only at the *end* of the processing, not at the *beginning* of it (as in other "flexible computation" approaches).

In NARS, the time spent on a given task is determined by the time request of the user, the design of the system, and the current context. These factors are combined in the control functions that calculate the priority values and durability values of the concepts, tasks, and beliefs in the system.

What makes these functions different from the heuristic functions used in expert systems is their domain-independent nature. In the design of these functions, no assumption was made about the *content* of the task or the knowledge. However, this does not mean that NARS uses a general-purpose, context-independent algorithm, which always uses the same predetermined method to solve problems, and is insensitive to available domain knowledge. Indeed, there is a pervasive influence of domain knowledge on the control of inference at all times when the system is running. As mentioned previously, the system's choice at each step strongly depends on the available and activated domain knowledge, which is certainly not predetermined by the designer. On the other hand, the *design* of the system, including all the functions discussed above, is independent of any particular domain.

With all these control-related quantities adjusted dynamically, some tasks are processed faster, and some beliefs are more accessible, while others slowly slide into dormancy. And because storage space is limited, concepts, tasks and beliefs with sufficiently low priorities may wind up being permanently forgotten. This, though sometimes disadvantageous from a practical point of view, is the price any system has to pay when its knowledge and resources are insufficient.

By dynamically allocating resources among tasks at run time, what the system is optimizing is not the quality of solution for any particular task, but the overall performance of the system on all existing tasks, under the restriction of available resources. Since the resource budget for each task is not predetermined, this control mechanism allows more flexibility in the resource consuming of a task.

12.3.3 Task and motivation

NARS is *goal-directed* in the sense that all internal activities are driven by the tasks (new beliefs, questions, and goals). Since there are many of them under processing at the same time, the overall behaviors of NARS are determined by the "resultant of forces" of its internal tasks.

Initially, the system is driven only by input tasks. The system then derives subtasks recursively by applying inference rules to the tasks and available knowledge.

Tasks compete for resources. Given constant total amount of resources, to give one task more attention means to give the others less.

Goals may directly or indirectly conflict with each other. When the system tries to achieve multiple goals at the same time, operations may also get conflicting evaluations from different goals.

In NARS, it is not guaranteed that the achievement of the derived tasks will turn out to be really helpful or even related to the original tasks, because the beliefs, on which the derivation is based, is revisable. On the other hand, it is impossible for the system to always determine correctly which tasks are more closely related to the original tasks. As a result, the system's behavior will to a certain extent depend on its *own* tasks, which are actually more or less independent of the original tasks, even though historically derived from them. This is the *functional autonomy* phenomena [Allport, 1937, Minsky, 1985]. In the extreme form, the derived tasks may become so strong that they even prevent the input tasks from being fulfilled. In this way, the derived tasks are *alienated.*

The alienation and unpredictability sometimes result in the system to be "out of control," but at the same time, they lead to *creative and original* behaviors, because the system is pursuing goals that are not directly assigned by its environment or its innateness, with methods that are not directly deduced from given knowledge.

NARS has a "life-time of its own." [Elgot-Drapkin et al., 1991] When the system is experienced enough, there will be lots of tasks for the system to process. On the other hand, new input can come at any time. The system usually works on its "own" tasks, but at the same time, it is always ready to respond to new tasks provided by the environment. Each input task usually attracts the system's attention for a while, and also causes some long-term effects. The system never reaches a "final state" and stops there, though it can be reset by a human user to its initial state. In this way, each task-processing activity is part of the system's life-time experience, and is influenced by the other activities. In comparison with NARS, traditional computer systems take each problem-solving activity as a separate life cycle with a predetermined end.

12.3.4 Reasonable results

Like all other approaches that take the limitation of resources into consideration, NARS gives up the requirement for optimum results, and turns to look for the best results the system can get under the constraints of available resources — what Good calls "type II rationality." [Good, 1983].

Though there are many approaches attempting to deal with the problem of "insufficient resources," their concrete specifications of the problem are very different. By interpreting "insufficient knowledge and resources" as being finite and open, and working in real time, the constraints assumed by NARS are stronger than the assumptions accepted by other approaches. For example, as discussed previously, only a few systems are designed to deal with variable time pressure or unexpected (both in content and timing) questions.

Many features of NARS directly follow from this assumption. For example, the results are usually only derived from part of the system's knowledge, and which part of the knowledge base is used depends on the context at run time. Consequently, NARS is no longer "logical omniscient" [Fagin and Halpern, 1988] — it cannot recall every piece of knowledge in its knowledge base, not to mention being aware of all their implications.

To select the data items (concept, task, and belief) to be processed, what happens in NARS is similar to the *priming* and *association* happened in the human mind. The system uses what is being processed as the *center of attention*, and sends activation to the neighborhood to gradually bring related concepts, tasks, and beliefs into consideration.

NARS uses a *forgetting* mechanism to manage its memory. Though many systems release memory when running, usually it is done when the data there is no longer useful. The challenge to the forgetting mechanism in NARS is to decide what to ignore or delete, even when it may be useful in the future.

This reminds us of human memory. On one hand, we all suffer from forgetting information that became needed later; but on the other hand, it is not hard to image what a headache it would be if every piece of knowledge was equally accessible — that is, equally inaccessible. Like it or not, properties such as forgetting are *inevitable consequences* of the insufficiency of resources.

This theme appears from time to time in the discussion about NARS — though many of its properties are usually judged as unwelcome in AI systems, they become inevitable as soon as we want a system to work under AIKR. Furthermore, they are often produced by the same mechanism that generates the desired results, so that we cannot get the latter without the former.

In this sense, we say that many mistakes made by the system are *reasonable*, given its working environment. The only way to inhibit these mistakes is to limit the problems that the systems are exposed to, like in most computer systems. However, in this way computer systems lose the potential to conquer real hard problems, because the problems we call "hard" in everyday life are precisely the problems for which our knowledge and resources are insufficient.

In summary, "correct results" (resource-independent) and "reasonable results" (resource-dependent) are often different, and are produced by different mechanisms. Though for practical purposes the former is often preferred, intelligence is needed to provide the latter when the former cannot be obtained.

Part IV

Conclusions

Chapter 13

Current Results

Though NARS is still an on-going research project, there are already many interesting results, as described in the previous chapters. In this chapter, I will summarize them, on each of the three levels (theory, model, and implementation).

13.1 Theoretical foundation

Compared with the other AI projects, the most important theoretical features of NARS are the working definition of intelligence and the framework of a reasoning system.

13.1.1 The working definition of intelligence

As stated previously, NARS is based on the working definition of intelligence as *adaptation with insufficient knowledge and resources.* After all the descriptions and discussions, now we can evaluate this definition according to the requirements set up in Section 1.1.3:

Faithfulness. Obviously, certain natural information-processing systems (i.e., humans and some animals) are adaptive, and they have to work with insufficient knowledge and resources. With higher adaption capability, human beings are much more intelligent than other animals. By contrast, though traditional computing systems also have limited knowledge and resources, they are usually only used on a carefully limited class of problems, chosen so that their knowledge and resources will in fact be *sufficient* for those problems. Therefore, the definition draws a line between intelligent and non-intelligent systems that is faithful to the common usage of the word "intelligence."

Sharpness. The definition is sharp, because whether a system is adaptive can be determined by testing whether its behaviors depend on its experience. For a computer system, whether it is designed under the assumption of insufficient knowledge and resources can be determined by checking for the three properties: *finiteness* (Can the system manage its own memory?), operation in *real time* (Can the system work under a range of different time constraints?), and *openness* (Does the system restrict what it can be told or asked?).

Fruitfulness. As the foregoing chapters have demonstrated, the definition has yielded fruit by inspiring the major components of NARS, which are fundamentally different from existing approaches, and the system has exhibited many desired properties. Indeed, rooted in this definition of intelligence, NARS addresses many facets of AI in a consistent manner, and provides consistent solutions to more AI problems than many other AI theories do, as described in the previous chapters.

Simplicity. The definition is quite simple, making it easy to discuss and to apply to research. Its direct outcome, NARS, is also relatively simple in its structure (compared with other AGI systems), though the system's behavior can be very complex, due to its interaction with its environment.

Because of these considerations, I believe that the working definition of intelligence introduced in this project is preferable to many others accepted by AI researchers. Compared to this approach, the other AI schools (as introduced and categorized in Chapter 1) miss the essence of intelligence in various ways, though still guiding scientific research that is valuable for other purposes.

- The *structure* approach contributes to neuroscience by building brain models, but fails to reveal the regularities of the *mind*, independent of the details of the *brain*.

- The *behavior* approach contributes to psychology by providing explanations of human behavior, but fails to distinguish the factors that are mainly due to intelligence in general from those that mainly due to the "implementation details" of these factors in the human brain.

- The *capability* approach contributes to application domains by solving practical problems, but fails to distinguish AI from conventional computer application, and as a result, still mainly depends on the intelligence of the designer (rather than that of the computer system) to solve the problems.

- The *function* approach contributes to computer science by producing new software and hardware for various computing tasks, but fails to base all these functionalities on a consistent foundation, and as a result, can hardly integrate them into an AGI system.

Compared to the other similar definitions of intelligence in which adaptation and knowledge/resource restriction play central roles, NARS is different in its concrete specification of the restriction (that is, AIKR), as well as in its insistence on following this definition completely and thoroughly in the theory, model, and implementation of intelligence.

For example, Simon's "bounded rationality" allows systems to make choices without consulting all relevant knowledge, and therefore can be finished within limited time [Simon, 1983]. However, such a system does not need to work in real time, or be open to all possible knowledge.

13.1.2 The reasoning system framework

As described in Section 2.2.1, there are three major schools of formalization: dynamic system, reasoning system, and computing system. Being developed within the framework of a reasoning system, NARS has the following desired properties:

- The major components of the system (language, semantics, inference rules, memory structure, and control strategy) can all be designed according to the working definition of intelligence.

- The design of the system is domain independent, though the concrete behaviors of the system is determined by its domain-specific experience.

- When properly extended, the concept of "reasoning" can capture various types of cognitive functionalities (such as learning, categorization, planning, problem solving, decision making, and so on), and therefore NARS is indeed a model for general intelligence, rather than merely for "logical reasoning" in the traditional sense.

- The separation of the logic part and the control part allows the system to have both *rigid* steps (each corresponds to an inference rule) and *flexible* processes (formed by chaining the steps in a context-sensitive manner in run).

- The interface language (Narsese) is rich enough to allow complicated input and output.

Compared to the reasoning system framework, the dynamic system framework lacks a rich representation. In principle, it is possible to represent the state of a NARS-like system as a point in a multi-dimensional space, and the activities of the system as trajectories within the space. However, with such a representation, it would be very difficult to specify the dimensions and to justify the state changes, because this representation is so "low level" that the important regularities would be drowned by the details in the system.

On the other hand, though the computing system framework allows more complicated representation and processing, the logic part and the control part of such a system are usually mixed together, and it is not easy to stop the working process at any point without destroying the integrity of the knowledge. As a result, it is more difficult (though not impossible) to implement a NARS-like system, with justifiable steps and flexible processes.

Logic-based AI has been widely criticized for its rigidness, and its failure is a major driving force behind the competing approaches, such as neural network, evolutionary computing, reactive robotics, and dynamic system. The practice of NARS shows, however, many of the problems can be solved within the framework of a reasoning system by using the *ideas*, but not the *techniques*, of the other approach, and the result is a system that is more flexible and more powerful than the competing approaches can produce.

13.2 Formal model

NARS, as a general-purpose reasoning system, has been developed according to the above theory about intelligence. Because of its theoretical assumption, all major components of the system are fundamentally different from those of the conventional reasoning systems, and altogether, they provide a unified model of intelligence.

Beside the capability of this model, hopefully some reader can also recognize and appreciate its unique elegance (i.e., conceptual simplicity) in many aspects (such as the dual between extension and intension, the related measurements of uncertainty, the unified form of various kinds of inferences, and the isomorphism between first-order and higher-order inference).

13.2.1 Experience-grounded semantics

According to the working definition of intelligence, the semantics of NARS defines the truth values of statements and the meanings of terms in Narsese as functions of the system's experience.

The fundamental difference between this semantics and the popular model-theoretic semantics is that the former can be applied to adaptive system with insufficient knowledge and resources, while the latter cannot.

To avoid circular definition, first a binary language, Narsese-0, is introduced, and is given an experience-grounded semantics. Then, using this language to specify the idealized experience of the system, the semantics of Narsese is specified. Finally, each piece of the actual experience is treated as equivalent to certain idealized experience. Overall, the semantics supports the interpretation of the input and output of the system, as well as the justification of the inference rules.

Concretely, the truth value of a statement consists of a frequency value and a confidence value, where the former measures the proportion of positive evidence among all available evidence, while the latter measures the proportion of available evidence among all evidence to be available in the near future. The meaning of a term consists of its extension and intension, including all of its experienced relations with other terms.

This semantics provides a unified representation of various types of uncertainty in a statement, such as randomness, fuzziness, and ignorance, as well as a elegant duality between extension and intension of a term. It also provides a unified justification for various defeasible inference, such as induction, abduction, analogy, and so on.

The semantic theory can be applied into situations where the sentences in the language have procedural interpretation. Therefore, it can also be used after the system is extended into a system with sensorimotor capability.

13.2.2 Extended term logic

NAL belongs to the term logic tradition, not the currently dominating predicate logic tradition.

A main reason for this choice is that the notion of "evidence" can be naturally introduced in term logic, but it is hard to do in predicate logic. Another reason is the close relationship between term logic and categorical hierarchy. As a result, Narsese is more similar to a natural language in both grammar and semantics than other formal languages are, and reasoning in NARS is the same process as categorization.

The traditional criticism to term logic is focused on its poor expressive power. In NAL, this problem is solved by the introduction of compound terms. Step by step, the following compounds are added into Narsese as terms: sets, intersections and differences, products and images, as well as statements and compound statements. Finally, by adding questions, goals, events, and operations, as well as the related operators and relations, the language is no longer purely declarative anymore.

All inference in NARS is about inheritance relations among terms. NARS' inference rules are constructed by taking both extensions and intensions of the involved terms into account, and by considering all possible types of combinations of premises. As a result, NARS has a set of inference rules for choice, revision, deduction, induction, abduction, exemplification, comparison, analogy, compound terms composition and decomposition, higher-order inference, and backward inference. These different types of inference are carried out in a uniform format, are justified by the same semantics, and are used in similar ways altogether.

Defined in a syllogistic format and justified by an experience-grounded semantics, all inference rules in NAL guarantee the semantic relevance among the premises and the conclusions in each inference step.

Though predicate logic may still be better for pure-axiomatic reasoning systems working in domains where knowledge and resources can be assumed to be sufficient, a (properly designed) term logic is better for an adaptive reasoning system working with insufficient knowledge and resources.

13.2.3 Dynamic resources allocation

NARS' memory is organized into a two-level structure: the concept level and the task/belief level, where the concepts are named by terms, and the tasks and beliefs are clustered into concepts according to the terms appearing in them. On each level, items are stored in "bags," which are fixed-sized probabilistic priority queues. A concept is a large-scale unit of resource allocation and inference activity. Concepts cooperate by sending messages (tasks) to one another.

The system works by repeatedly executing atomic inference steps. In each step, a concept is selected, and within it a task reacts with a belief to generate new tasks. Each step takes only a very short time to finish. New tasks can be accepted at any time, and their processing will depend on the current state of the system's beliefs and resources.

To work in real time and with insufficient resources, NARS processes many tasks in parallel. The processor time is distributed among the tasks unevenly, and the distribution is dynamically adjusted when the situation changes. More important and promising tasks are given more processing time, so that in effect, they are processed faster, and their implications (in the case of beliefs) or their potential solutions (in the case of questions and goals) are explored further than the other tasks.

The system produces the best solution it can get for each task, and continues to improve them whenever there are still resources available for the task. Consequently, one question may get zero, one, or more than one answer, depending on the current situation within the system. The task-processing processes do not follow predetermined algorithms, and cannot be analyzed in terms of computability and computational complexity.

The derived tasks are treated in the same way as the input tasks. Given insufficient knowledge and resources, the solving of a "child" task will not necessarily contribute to the solving of its "parent" task. As a result, the overall behavior of NARS is determined by the "resultant of forces" of all its existing tasks. In this way, the system becomes not only autonomous and creative, but also unpredictable (from its initial design alone).

The memory structure and inference control mechanism provide a unified explanation of cognitive phenomena including memory, attention, association, forgetting, motivation, and so on.

13.3 Computer implementation

The formal model of NARS described previously can be fully implemented on a computer system, using currently available hardware and software technology. The resulting system is not complex technically, but does produce complex behaviors, as predicted by the theory.

13.3.1 Historical versions of NARS

I started working on NARS in 1983 at Peking University, and some ideas were mentioned in my Bachelor Thesis on Intelligent Database (in Chinese), though there was no actual program written.

The first prototype of the system, NARS 1.0, was written in Prolog in 1985-86, and the system was described in my Master Thesis

[Wang, 1986]. That version included many of the rules of first-order NAL (NAL-1 to NAL-4 in this book), though the language was not as complete as the Narsese presented in this book, and some truth-value functions are different from the current form, due to the fact that the idea of an experience-grounded semantics was in its premature stage at that time.

The next groups of prototypes, NARS 2.0, 2.1, and 2.2 were written in Scheme, and were developed partially as a course project at Indiana University in 1992-93. The inference rules were roughly the ones in NAL-1 and NAL-2, and the semantics was more clear. However, the control mechanism was highly simplified. This work was described in [Wang, 1993b, Wang, 1993c].

NARS 3.0 was developed in C++ in 1994-95, and was described in my Ph.D. Dissertation [Wang, 1995a]. The inference rules were still the ones in NAL-1 and NAL-2, but the memory structure and control mechanism were more complicated, and for the first time, the system had a graphic user interface for interactive communication.

Shortly after NARS 3.0, a version 3.1 was finished, which contained some minor revisions. This version was not described in any publication.

Since 1998, a customized version of NARS was developed and integrated into a commercial software, Webmind, as its inference engine. Though the company was dissolved in 2001, before the software was completed, the inference engine had shown its capability in solving reasoning problems, as well as in supporting other cognitive functionality of the software, such as natural language processing.

13.3.2 Recent versions of NARS

Since 1997, NARS development has been mainly in the programming language Java. NARS 4.0 was finished in 1998, and since then the most recent version of NARS was put at the website of the project[1] as an applet that provides on-line demonstration.

NARS 4.1, finished in 1999, provided a better demonstration of the project. It came with a brief User's Guide, as well as examples showing

---

[1]Currently at `http://www.cogsci.indiana.edu/farg/peiwang/NARS/`.

simple inference processes. It was presented among the Intelligent Systems Demos of AAAI-2000 [Wang, 2000b]. NARS 4.1 implemented a complete First-Order Non-Axiomatic Logic, though there are some minor difference between what was implemented in the program and what are defined as NAL-1 to NAL-4 in this book. The Applet is in a JAR file of about 50K. It runs in a browser, and opens several windows for input/output, process tracing, and real-time control. The user can either communicate with the system interactively, or run a pre-edited script consisting of input tasks, separated by numbers indicating the timing between adjacent input, in terms of inference steps.

NARS 4.2 was finished in 2004, and has been on-line since then. This extension included rules of NAL-5 and NAL-6, that is, higher-order inference with variables. Consequently, it can be referred as a complete Non-Axiomatic Logic on non-temporal knowledge. The expressive and inferential capability of this version is comparable with Predicate Calculus, though the two are neither equivalent, nor included by each other.

Finished together with NARS 4.2 was a stand-alone NAL inference engine in Prolog, NAL 1.0. In this program, all NAL rules in NAL-1 to NAL-6 are included (with the associated truth-value functions), and each call to it carries out a single inference step, that is, to generate conclusions from given premises. It does not include the NARS control mechanism, which is needed for multi-step inference processes. In the future, the Prolog versions of the NAL inference engine will be logically identical to the inference engine in the Java versions of NARS, but much simpler in code. It will be used for testing and demonstrating, and also has the potential of being embedded by other systems for inference service.

NARS 4.3 is currently under development. It contains the rules of NAL-7 and NAL-8, that is, temporal and procedural inference. When finished, everything described previously in the book should have been implemented.

This book makes no attempt to describe the implementation details of NARS. Instead, that task is left to the project website, as well as to future publications. Especially, since working examples of problem-solving in NARS will inevitably require such details, they will be handled better by the website.

Chapter 14

NARS in the Future

As a research project, NARS has achieved many important results, as summarized in the previous chapter. However, its story is still far from complete. In this last chapter, I will briefly discuss the future plan of the project, as well as its implications.

14.1 Next steps of the project

Once again, the topics to be addressed in the project can be divided into the three levels: theory, model, and implementation (from high to low, in terms of abstraction). Usually, results on a higher level guides the work on a lower level, and results on a lower level expose problems to be solved on a higher level. Therefore, normally works on a higher level progress farther than those on a lower level, though none of the three can be finished alone.

14.1.1 NARS, by itself

As mentioned before, an implemented NARS consists of a logic part and a control part. The former includes a language and a set of inference rules, and the latter is mostly about resources allocation.

The development of the logic part of NARS has been following an incremental approach, and in each new version, only part of the formal model is added into the system, based on the part already established in the current version. Consequently, each version is more intelligent than the previous one, according to the working definition of NARS — with a richer language and more inference rules, the system is more adaptive, and more efficient in using available knowledge and resources.

After NAL-8 is implemented, the logic part of the system, the "inference engine," will be mostly finished. After that, it will be the time for massive testing, with the following tasks:

- To check the expressive power of Narsese. After NAL-8, Narsese should roughly have the expressive power of a natural language. For testing purpose, various sample texts in different domains will be manually encoded into Narsese. This task may lead to revisions in the grammar of Narsese.

- To check the inferential power of NAL. Similar to the previous task, sample human inference cases will be analyzed, and compared to what NAL will produce for the same case. NAL does not have to accurately duplicate human reasoning behavior, but the differences should be documented and explained. This task may lead to revisions in the inference rules of NAL.

Compared to the logic part, the design of the control part of the model is less mature — in the previous chapters, I draw the big picture without the details. Overall, the design of the control part is deliberately postponed until the design of the logic part is finished. This is because there is an one-way dependency between the two parts of the model. When designing the logic part, the control part can be ignored, except the general principles (such as that no conclusion can be based on all beliefs of the system, and that the same evidence should not be repeatedly used to support a conclusion, and so on). On the contrary, each time a new set of inference rules are introduced, many details of the control part have to be changed accordingly. Therefore, it does not make much sense to fine tune the control part before the logic part becomes stable.

When the logic is largely in place, it will be the time to pay more attention to control. Unlike the logic, which is mainly designed by theoretical analysis, the control part has to be designed by both theoretical analysis and empirical experimenting.

Based on the control principles introduced in the previous chapters, I will attempt to provide a more detailed model for all the resources allocation strategies and functions. It will borrow intuitions from psychology and economy, as well as from previous research in AI and cognitive sciences (such as the work on credit assignment [Holland, 1986], rational analysis [Anderson, 1990], evolutionary economics [Baum, 1998], and so on) and computer science (such as resources allocation in operating system).

The above theoretical analysis will inevitably leave certain parameters (that depend on the hardware/software host systems) to be determined by actual experimenting in the implemented system. The plan for this part is to first select a set of benchmark testing cases, with certain evaluation criteria. Then, parameters will be manually tuned to find the settings that lead to the best performance. It is very likely that for each parameter there is no "optimum value" but a "normal range," and different values in the range give systems different "personalities," and none of them is always the best in all situations. Unlike the logic part, in the control part the design will never be "done," and there will always be space for refinements. Especially, the system's capability on procedure inference may be used within the system itself, so as to NARS can achieve self-monitoring and self-control to certain extent.

14.1.2 Additional capabilities

Beyond the design of NARS *per se*, there are the following directions that future NARS-based research may explore (though I may not pursue all of them myself):

Education theory. The design of NARS, no matter how complete, only determine the initial state of the system. Though it is possible to implant some "innate beliefs" into the system, its behaviors will still inevitably be determined by its experience. Therefore, NARS as designed is like a baby that has great potential, but little built-in skill. To really make the system to accomplish any practical task, extensive "education" (or call it "training") is needed, which is nothing but external control of the system's (initial) experience. Unlike the training of current AI systems (like most connectionist models), NARS cannot be trained to "converge" to certain determined behaviors. Instead, the situation will be more like the education of human beings — the tutors will have influence, but not complete control, of the system's behaviors. Therefore, I expect the education theory for NARS to be similar (though not identical) to that for human beings.

Sensorimotor subsystem. As mentioned before, NAL-8 provides a general interface for adding sensorimotor capability into NARS. Under procedural interpretation, certain goals in the system will invoke operations defined outside NARS. These operations can be procedure calls to other software, or commands to other hardware. As results, new input tasks are presented to the system, and there are side effects within the system or in the environment. The ability to use "tools" is not required in my working definition of intelligence, but if a system has this capability, its interface language will *ipso facto* be greatly extended, and it will therefore be more intelligent than a system that has only a language interface. In this way, NARS can be customized into either an "intelligent operating system" (which can flexibly integrate all kinds of software tools for various tasks), or a "mind" of a robot with particular sensorimotor mechanism (which is not necessarily similar to that of a human being).

Special hardware. I never believed that the past failure of AI was due to von Neumann computer, and I have been building NARS on the conventional hardware/software platform. Even so, specially designed hardware will surely improve the efficiency of the system. Given the fact that all inference activities happen within individual concepts, and each inference cycle consists of several fixed steps, it is quite possible to design a special computer with multiple processors, specialized to carry out inference in parallel. However, it is important to realize that even in such a hardware, AIKR will still be true, and the system will still needs to allocate limited resources among a larger number of items and activities — the system will never have so many processors that each task will get one. Therefore, though special hardware will improve efficiency, it will not change the principles on which the system is based.

Natural-language processing. Though to be able to use a natural language is not a function of NARS by design, it is often desired for various purposes. NARS has the potential of using its general-purpose learning mechanism to learn different languages. According to my current idea, the major difference between the language processing in NARS and the current approaches is that the boundary between syntax and semantics will be broken, and linguistic concepts will be handled just like other concepts. However, given the fundamental difference between human experience and the experience of an implemented NARS, I do not expect it to use a natural language as a native speaker of the language.

Multi-system community. It will be interesting to implement multiple copies of NARS, and let them communicate (in Narsese or other languages) with each other. Due to differences in system parameters, in hardware/software, and in experience, they will develop different beliefs. On the other hand, since communication creates shared experience, we can also expect some consensus to be developed among the systems. In such a setting, we can study topics like cooperation, negotiation, speech action, game playing, self/other distinction, and so on.

System evolution. The learning capability of NARS lets the system change its belief structure according to its experience, and the self-control capability will allow the system to change its internal processes to a certain extent. Still, there are some changes that will break a system's integrity. To achieve higher intelligence, we can let a NARS community evolve, using ideas like genetic algorithm. To me, *intelligence* and *evolution* are two forms of *adaptation*. The former is experience driven and within an individual system, and the latter is experience independent and across generations. The two can be combined to achieve more complicated adaptation.

Though the details of the above tasks remain to be worked out, I have reason to believe that it makes more sense to tackle these challenges from a non-axiomatic point of view (rather than from a pure-axiomatic or semi-axiomatic point of view), since they are all closely related to adaptation under insufficient knowledge and resources. Therefore, I feel that the current NARS model constitutes a necessary step toward these goals. On the other hand, I do not include them as parts of NARS, because of the belief that they are additional features, rather than essential natures, of intelligent systems. Unlike some other AGI projects, I have always taken a "minimalist" approach when designing NARS, by only equipping the system with what I believe to be absolutely necessary for an AI system.

14.1.3 Theoretical speculations

Though in this book I have already discussed many issues in AI and cognitive sciences, there are still some important topics that may be fully addressed by the progress of NARS in the future (after the previously mentioned extensions get implemented). Though now is too early to give any conclusion to these issues, I would like to briefly speculate on how to handle them in NARS.

After NAL is fully implemented and the control mechanism of NARS becomes more complicated, the system will display certain phenomena, which are not designed into it as a separate process or function, but are produced as emergent epiphenomena produced by the mass low-level events [Hofstadter, 1979]. *Emotion* may be such an example, which corresponds to different internal modes. A certain emotion may be produced by the evaluation of the general situation of the system according to the desire values of the related statements, and lead to adjustment in resources distribution. Of course, emotions in the system will not be based on biological mechanisms, but it will serve a similar function in the system as in the human mind. It is reasonable to assume that emotion plays an important role in intelligence, but it is wrong to conclude that it cannot appear in an AI system.

As mentioned previously, NARS runs continuously, without resetting itself after a task is finished. To prevent the system from being trapped in dead-ends or ignoring unusual possibilities, it may be a good idea to periodically put the system into "rest" or even "sleep," by blocking its input channels, and reducing the activation level of the concepts and tasks. In this way, when it is "woken up," the system may try very different approaches for the old problems. Furthermore, we may even allow the system to "dream," that is, to carry out some internal activity during sleeping. Since the resources competition is much weaker in this state, the system may follow paths that are judged as too unlikely to be explored in the "sober" state. Again, though their biological functions are gone and they often cause undesired results, notions like "sleep" and "dream" (like "forget") may be found to serve important information-processing functions.

A related topic is *imagery*. After the system gets sensorimotor capability, the internal representation of sensory patterns ("mental image") will be linked to concepts as parts of their meaning, according to EGS. These links will be used in both directions. If the system has some kind of visual device, then, on one hand, a cat in the visual field will provide the sensory material for the production of a mental image within the system, which is related to the concept "cat" as the result of perception. On the other hand, the processing of the "cat" concept will activate the mental image (as part of its meaning). In this way, "reasoning by image" may become just a special case of the general "reasoning by concept" process.

The "task alienation" phenomena in NARS may produce many interesting results. For example, we may find the system carries out certain activities that are not directly related to any input task, and seem have little practical utility. The system seems to be doing it "just for fun," like when humans play games. Similarly, the system may prefer certain sensation/perception patterns and processes, while the information perceived has little practical utility, like when human beings enjoy various types of art. If we analyze these processes step by step from the very beginning, I believe that originally they do serve other purposes. However, with insufficient knowledge and resources, derived tasks gradually become independent of the original ones. This tendency will be reinforced by the socialization process in multi-system society, where a system may learn to pursue a goal without knowing why, except that it is pursued by the other systems.

After NAL-8 is fully implemented, the system can be given certain sensorimotor capability, through it the system interacts with its environment directly (without using Narsese). Here the "environment" can be the inside of the system itself, which means the system will have certain self-monitoring and self-control capability. At this point, we may begin to touch the notion of *consciousness*. In principle, internally-oriented sensorimotor will be just like externally-oriented, except with very different "sensors" and "operators". The system will be able to answer questions like "What are you thinking?" according to what is actually happening in the system (rather than according to how people usually answer this question, like many "chatbots" do). Given limited knowledge and resources, the self perception of the system cannot be complete — there are lots of "subconscious" processes going on, beyond its vision. It will get ideas that seem pop up from nowhere ("inspirations"), and beliefs that cannot be traced back to their supporting evidence ("intuitions"). The system may also have its *mind-body problem*, because its internally-oriented and externally-oriented sensorimotor mechanisms will correspond to different "vocabulary" and therefore lead to different categorization. When the system attempts to describe the cause or effect of an internal process according to an external observation, it will need to cross the gap between the two, which is not always easy.

Finally, related to the above topics are the moral and ethical issues of AI research. Since the tasks (or call them motivations, drives) of NARS are determined both by the initial design (there may be built-in tasks) and by the system's experience, it will not necessarily generate "evil" ones, such as "to dominate the world." On the other hand, it will also not necessarily be friendly to human beings. Like other technology, AI is "morally neutral," and can either lead to good results or bad ones. In this sense, it is not more dangerous than other technologies. Given its flexibility, there is no way to put some foolproof safety device into NARS to prevent bad results. Our attitude toward AI should be the same as toward all scientific and technical research goals, that is, to explore them carefully, and to predict their practical consequences (though we can never be absolutely certain). Since I currently see much more reason to continue my research than reason to stop it, I will keep going on.

14.2 What NARS is not

Even after all the extensions and refinements mentioned above, there are still things that NARS cannot do, simply because it is not designed for them.

14.2.1 NARS and the other AI schools

Obviously, NARS will not reach the aims set up by the other AI schools.

NARS is not designed to simulate the human brain. I still believe that what we call "intelligence" (and the related notions like "thinking," "mind," "cognition," and so on) can be abstracted from their realization in the human brain. Of course, function and structure are not completely independent of each other, so if certain aspects of NARS turn out to be similar to what is observed in the human brain, it is not a coincidence. For example, it is not hard to recognize the similarity among the induction rule of NARS, Hebbian Learning, and Pavlovian Conditioning.

NARS is not designed to be an accurate model of human reasoning. The system should follow the same *principles* as does the human mind. However, it is not necessary to have the same internal *structure* and *mechanisms* as in the human mind, since computer hardware is fundamentally different from human wetware. Moreover, since NARS' experience will always be different from that of a human being, it is not necessary (though it is still possible to a certain extent) to have the same external behaviors as the human mind, such as exactly reproducing some psychological data or passing the Turing Test.

Though NARS does have great potential for various applications where no existing technique can be used (due to insufficient knowledge and resources), it is not designed to solve particular domain problems. When NARS is used to solve practical problems, it cannot guarantee that its results will be *correct* or *optimal* (in the eye of an omniscient observer); judgments in NARS are always subject to being revised by the system or refuted by future experience. For any given problem, it is always possible to design a special-purpose system that works better than NARS. It is like the relation between hands and tools: for any given job, it is always possible to design a tool that works better than our bare hands. However, I will not trade my hands for any tool, because of their generality and flexibility. Of course, a hand/tool combination is better than either of the two alone. For the same reason, for a given problem, it is better to let NARS use a special "tool" (i.e., special-purpose software/hardware) if it is available, rather than directly handle the problem by the system. However, I will not build these tools into NARS, just like I will not implant a hammer into my hands.

As discussed in the previous chapters, NARS has many cognitive functions, but they are usually specified quite differently from those in the other AI projects. We have discussed such cases as "learning," "induction," "planning," and so on. In NARS, they are interwoven processes without clear-cut ending points, rather than algorithms that generate certain output from certain input. Therefore, accurately speaking, what makes NARS different from many other AI projects is not the *solutions* it provides, but the *problems* it aims at. NARS is not aimed at any of those traditional AI problems, because I don't think those problems are actually related to the essence of intelligence. To me, in terms of the notion introduced in this book, they are "alienated subtasks" of AI that are derived from the original AI problem through biased and distorted beliefs.

Since NARS works with insufficient knowledge and resources, it is impossible for it to have properties that only a pure-axiomatic system can have, such as *consistency*, *completeness*, *decidability*, and so on (though the system does its best to move toward these aims when it is running). At the same time, the system is not bounded by the restrictions of pure-axiomatic systems, neither. For example, Gödel's Theorem is not directly applicable to NARS. Some people claim that Gödel's Theorem has set the limitation of AI [Lucas, 1961, Penrose, 1994], by implicitly assuming that an AI system has to be consistent. To me, the situation is just the opposite (that is, an intelligent system must have internal contradictions and conflicts), and therefore their conclusions are wrong.

The above "limitations" of NARS are easy to deal with — we can just ignore them. This is not to say that the attempt to overcome them is not a valuable goal for research, but simply that such a goal is fundamentally different from (though still related to) my current goal — exploring the essence of intelligence. These limitations of the NARS project mean that if someone is looking for a computer model with these properties, then NARS should not be a candidate, having been designed with other goals in mind.

14.2.2 How to criticize or reject NARS

As a scientific research project, the theory, model, and system of NARS can all be criticized or rejected, given valid evidence and argument.

Unfortunately, most of the criticisms I received so far are invalid, because they try to evaluate NARS in a school that it does not belong to. According to the previous discussion, it should be clear that NARS cannot be criticized for "being biologically unrealistic," "cannot pass the Turing Test," "hasn't solved any practical problem," "not working on the problems as formally specified by the AI community," "often making mistakes," "cannot find an optimum solution," and so on.

For such a system, what are valid ways to criticize it? The following is a list for a critical reader, that goes along the logical path of this book:

- You can challenge the four criteria (borrowed from Carnap) on good working definition.
- You can suggest a better working definition of intelligence, according to the above criteria.
- You can argue that a reasoning system is not the best way to formalize my working definition of intelligence.
- You can propose better selections for the components of the formal model (its formal language, semantics, inference rules, memory structure, control strategy, and so on) to implement my working definition of intelligence.
- You can design a computer system that implements the formal model in a better way.
- You can identify inconsistency among my descriptions and discussions about NARS.

Even if a valid criticism of NARS is accepted, it does not mean that the theory/model/system has been "falsified," [Popper, 1959] and should be completely rejected. Instead, it usually leads to a revision of NARS.

When will I give up this project, and move to a completely different approach? It will happen only when that approach works better in general in explaining human cognition and producing a thinking machine.

Until such a moment, the NARS project will be continued.

14.3 General implications

Finally, I will briefly mention some general implications of this research.

14.3.1 For AI

Wolfram said in an interview that "I'm convinced that after it's understood, it really won't be difficult to make artificial intelligence. It's just that people have been studying absolutely the wrong things in trying to get it." [Stork, 1997a] I quite agree with him on this comment (though we have very different opinions on what is the "right thing" to do).

From what was described in this book, it can be observed that there is nothing complicated or fancy in the technical details of NARS. However, the philosophy and methodology of this research is quite different from, and sometimes even exactly opposite to, what is accepted by the mainstream AI. It is mainly on "what to do," rather than on "how to do it" that NARS differs from the other AI projects, and this partially explains why it is hard for this research to get accepted by the AI community.

Though the problem of AI is still not completely solved, I believe that I have given it a better clarification, and that my work finished so far has shown the potential of this research direction. Especially, it is shown to be possible to treat many problems according to the same theoretical foundation, and in the same formal model. In this way, we can give AI its identity as a scientific discipline.

As a science, this theory can explain many phenomena in human thinking. Especially, it will provide a unified picture about how the mind works and why it does not work in another way.

As a technology, NARS does not really compete with existing computer techniques, but is aiming at a domain where no current computer system can be used.

14.3.2 For cognitive sciences

Though NARS is mainly a research project in AI, its results nevertheless have contributions to other disciplines in cognitive sciences.

The most directly related discipline is *logic*. NAL can be seen as an attempt to move the subject matter of logic from the foundation of mathematics back to the regularity in thinking. In this sense, it is just the reverse of what Frege did when founding mathematical logic.

In this book, I have addressed several issues in *psychology*, such as in the previous study of reasoning, learning, categorization, human errors, and so on. Designed as a normative model, NARS does not compete with the descriptive models of aspects of human cognition. Instead, it reveals some misunderstanding and confusion in psychology when normative models are related to psychological observations.

At the current stage, the contribution of NARS to *linguistics* is mainly in the field of semantics. I believe that EGS can be used to explain many linguistic phenomena. In the future, the research may produce more results on syntax, pragmatics, and other fields.

In this book, I have discussed many topics in *philosophy*, especially, in philosophy of mind (how the mind works), philosophy of logic (what logic is about), and philosophy of science (what is a good scientific theory). Since every philosophical theory is based on certain opinions about intelligence/mind/thinking, we can expect a revolution in philosophy when the AI problem is finally solved. AI may have a larger impact in philosophy than the ones caused by Newton, Darwin, and Einstein.

Part of this project is the attempt to establish a scientific theory of intelligence. It is my opinion that a "scientific theory" is nothing but a system of shared beliefs and tasks in a community, and it is formed and changed according to the same principle of intelligence as those of an individual system, as discussed in this book. It follows that a good theory should have the following properties:

objective. A good theory should be consistent with the experience of the individuals in the community (so it cannot merely be idiosyncratic opinions),

structured. A good theory should summarize the beliefs into a simple knowledge structure (so it cannot just be a collection of unrelated judgments),

instructive. A good theory should provide concrete predictions for the future (so it cannot be vague, ambiguous, vacuous, or tautological).

The theory about intelligence presented in this book is developed to meet these criteria.

Eventually, I believe that this research will lead us to a general theory on intelligence, which covers the fields of human intelligence, computer intelligence, animal intelligence, group intelligence, extraterrestrial intelligence, and so on. We will find that they are special cases of the same underlying principles.

Bibliography

[Allen, 1984] Allen, J. F. (1984). Towards a general theory of action and time. *Artificial Intelligence*, 23(2):123–154.

[Allport, 1937] Allport, G. (1937). The functional autonomy of motives. *American Journal of Psychology*, 50:141–156.

[Allwood et al., 1977] Allwood, J., Andersson, L.-G., and Dahl, O., editors (1977). *Logic in linguistics.* Cambridge University Press, Cambridge.

[Anderson, 1990] Anderson, J. (1990). *The Adaptive Character of Thought.* Lawrence Erlbaum Associates, Hillsdale, New Jersey.

[Anderson and Lebiere, 1998] Anderson, J. and Lebiere, C. (1998). *The Atomic Components of Thought.* Lawrence Erlbaum Associates, Mahwah, New Jersey.

[Anderson, 1986] Anderson, N. (1986). A cognitive theory of judgment and decision. In Brehmer, B., Jungermann, H., Lourens, P., and Sevón, G., editors, *New Directions in Research on Decision Making*, pages 63–108. Elsevier Science Publishers, Amsterdam.

[Aristotle, 1989] Aristotle (1989). *Prior Analytics.* Hackett Publishing Company, Indianapolis, Indiana. Translated by R. Smith.

[Arkes, 1991] Arkes, H. (1991). Costa and benefits of judgment errors: implications for debiasing. *Psychological Bulletin*, 110:486–498.

[Barsalou, 1987] Barsalou, L. (1987). The instability of graded structure: implications for the nature of concepts. In Neisser, U., editor,

Concepts and Conceptual Development: Ecological and intellectual factors in categorization, chapter 5, pages 101–140. Cambridge University Press, Cambridge.

[Barsalou, 1999] Barsalou, L. (1999). Perceptual symbol systems. *Behavioral and Brain Sciences*, 22:577–609.

[Barwise and Perry, 1983] Barwise, J. and Perry, J. (1983). *Situations and Attitudes*. MIT Press, Cambridge, Massachusetts.

[Baum, 1998] Baum, E. (1998). Manifesto for an evolutionary economics of intelligence. In Bishop, C., editor, *Neural Networks and Machine Learning*, pages 285–344. Springer-Verlag, New York.

[Bellman and Giertz, 1973] Bellman, R. and Giertz, M. (1973). On the analytic formalism of the theory of fuzzy sets. *Information Sciences*, 5:149–157.

[Bernardo and Smith, 1994] Bernardo, J. and Smith, A. (1994). *Bayesian Theory*. John Wiley & Sons, Chichester, England.

[Bhatnagar and Kanal, 1986] Bhatnagar, R. and Kanal, L. (1986). Handling uncertain information. In Kanal, L. and Lemmer, J., editors, *Uncertainty in Artificial Intelligence*, pages 3–26. North-Holland, Amsterdam.

[Birnbaum, 1991] Birnbaum, L. (1991). Rigor mortis: a response to Nilsson's "Logic and artificial intelligence". *Artificial Intelligence*, 47:57–77.

[Bocheński, 1970] Bocheński, I. (1970). *A History of Formal Logic*. Chelsea Publishing Company, New York. Translated and edited by I. Thomas.

[Boddy and Dean, 1994] Boddy, M. and Dean, T. (1994). Deliberation scheduling for problem solving in time-constrained environments. *Artificial Intelligence*, 67:245–285.

[Bonissone, 1987] Bonissone, P. (1987). Summarizing and propagating uncertain information with triangular norms. *International Journal of Approximate Reasoning*, 1:71–101.

[Bonissone and Decker, 1986] Bonissone, P. and Decker, K. (1986). Selecting uncertain calculi and granularity. In Kanal, L. and Lemmer, J., editors, *Uncertainty in Artificial Intelligence*, pages 217–247. North-Holland, Amsterdam.

[Brachman, 1983] Brachman, R. (1983). What is-a is and isn't: an analysis of taxonomic links in semantic networks. *IEEE Computer*, 16:30–36.

[Brachman and Schmolze, 1985] Brachman, R. and Schmolze, J. (1985). An overview of the KL-ONE knowledge representation system. *Cognitive Science*, 9:171–216.

[Braine and O'Brien, 1998] Braine, M. and O'Brien, D., editors (1998). *Mental Logic*. Lawrence Erlbaum Associates, Mahwah, New Jersey.

[Brooks, 1991] Brooks, R. (1991). Intelligence without representation. *Artificial Intelligence*, 47:139–159.

[Bylander, 1991] Bylander, T. (1991). Tractability and artificial intelligence. *Journal of Experimental & Theoretical Artificial Intelligence*, 3:171–178.

[Campbell, 1997] Campbell, M. (1997). "An enjoyable game": How HAL plays chess. In Stork, D., editor, *HAL's Legacy: 2001's Computer as Dream and Reality*, pages 75–98. MIT Press, Cambridge, Massachusetts.

[Campbell et al., 2002] Campbell, M., Hoane, A. J. J., and Hsu, F.-H. (2002). Deep Blue. *Artificial Intelligence*, 134:57–83.

[Cao et al., 2002] Cao, T. H., Rossiter, J. M., Martin, T. P., and Baldwin, J. F. (2002). On the implementation of Fril++ for object-oriented logic programming with uncertainty and fuzziness. In Bouchon-Meunier, B., Gutierrez-Rios, J., Magdalena, L., and Yager, R. R., editors, *Technologies for Constructing Intelligent Systems 2: Tools*, pages 393–406. Physica-Verlag, Heidelberg.

[Carnap, 1950] Carnap, R. (1950). *Logical Foundations of Probability*. The University of Chicago Press, Chicago.

[Carnap, 1952] Carnap, R. (1952). *The Continuum of Inductive Methods.* The University of Chicago Press, Chicago.

[Chalmers et al., 1992] Chalmers, D., French, R., and Hofstadter, D. (1992). High-level perception, representation, and analogy: a critique of artificial intelligence methodology. *Journal of Experimental & Theoretical Artificial Intelligence*, 4:185–211.

[Chater and Oaksford, 1999] Chater, N. and Oaksford, M. (1999). Ten years of the rational analysis of cognition. *Trends in Cognitive Science*, 3:57–65.

[Cheeseman, 1985] Cheeseman, P. (1985). In defense of probability. In *Proceedings of the Eighth International Joint Conference on Artificial Intelligence*, pages 1002–1009.

[Cheeseman, 1986] Cheeseman, P. (1986). Probabilistic versus fuzzy reasoning. In Kanal, L. and Lemmer, J., editors, *Uncertainty in Artificial Intelligence*, pages 85–102. North-Holland, Amsterdam.

[Cheng, 1997] Cheng, P. (1997). From covariation to causation: A causal power theory. *Psychological Review*, 104(2):367–405.

[Cherniak, 1986] Cherniak, C. (1986). *Minimal Rationality.* MIT Press, Cambridge, Massachusetts.

[Cohen, 1989] Cohen, L. (1989). *The Philosophy of Induction and Probability.* Clarendon Press, Oxford.

[Coletti et al., 1993] Coletti, G., Gilio, A., and Scozzafava, R. (1993). Comparative probability for conditional events: a new look through coherence. *Theory and Decision*, 35:237–258.

[Copi, 1982] Copi, I. (1982). *Introduction to Logic.* Macmillan Publishing Co., Inc., New York, 6th edition.

[Coulson, 2001] Coulson, S. (2001). *Semantic Leaps: Frame-shifting and Conceptual Blending in Meaning Construction.* Cambridge University Press, Cambridge.

[Cox, 1946] Cox, R. (1946). Probability, frequency, and reasonable expectation. *American Journal of Physics*, 14:1–13.

[Cruse, 1986] Cruse, D. (1986). *Lexical Semantics*. Cambridge University Press, Cambridge.

[Dean and Boddy, 1988] Dean, T. and Boddy, M. (1988). An analysis of time-dependent planning. In *Proceedings of AAAI-88*, pages 49–54.

[Dempster, 1967] Dempster, A. (1967). Upper and lower probabilities induced by a multivalued mapping. *Annals of Mathematical Statistics*, 38:325–339.

[Deutsch-McLeish, 1991] Deutsch-McLeish, M. (1991). A study of probabilities and belief functions under conflicting evidence: Comparisons and new methods. In Bouchon-Meunier, B., Yager, R. R., and Zadeh, L. A., editors, *Uncertainty in Knowledge Bases: Proc. of the 3rd International Conference on Information Processing and Management of Uncertainty in Knowledge-Based Systems, IPMU'90*, pages 41–49. Springer, Berlin, Heidelberg.

[Diaconis and Zabell, 1983] Diaconis, P. and Zabell, S. (1983). Some alternatives to Bayes's rule. In *Information Pooling and Group Decision Making: Proceedings of the Second University of California, Irvine, Conference on Political Economy*, pages 25–38.

[Donini et al., 1996] Donini, F. M., Lenzerini, M., Nardi, D., and Schaerf, A. (1996). Reasoning in description logics. In Brewka, G., editor, *Principles of Knowledge Representation*, pages 191–236. CSLI Publications, Stanford, California.

[Dreyfus, 1992] Dreyfus, H. (1992). *What Computers Still Can't Do*. MIT Press, Cambridge, Massachusetts.

[Dubois et al., 2001] Dubois, D., Grabisch, M., Prade, H., and Smets, P. (2001). Using the transferable belief model and a qualitative possibility theory approach on an illustrative example: The assessment of the value of a candidate. *International Journal of Intelligent Systems*, 16:1245–1272.

[Dubois and Prade, 1980] Dubois, D. and Prade, H. (1980). *Fuzzy Sets and Systems.* Academic Press, New York.

[Dubois and Prade, 1982] Dubois, D. and Prade, H. (1982). A class of fuzzy measures based on triangular norms. *International Journal of General Systems*, 8:43–61.

[Dubois and Prade, 1988] Dubois, D. and Prade, H. (1988). On fuzzy syllogisms. *Computational Intelligence*, 4:171–179.

[Dubois and Prade, 1991] Dubois, D. and Prade, H. (1991). Updating with belief functions, ordinal conditional functions and possibility measures. In Bonissone, P., Henrion, M., Kanal, L., and Lemmer, J., editors, *Uncertainty in Artificial Intelligence 6*, pages 311–329. North-Holland, Amsterdam.

[Dummett, 1978] Dummett, M. (1978). *Truth and Other Enigmas.* Harvard University Press, Cambridge, Massachusetts.

[Edmonds, 2000] Edmonds, B. (2000). The constructability of artificial intelligence (as defined by the Turing test). *Journal of Logic Language and Information*, 9:419–424.

[Elgot-Drapkin et al., 1991] Elgot-Drapkin, J., Miller, M., and Perlis, D. (1991). Memory, reason, and time: the step-logic approach. In Cummins, R. and Pollock, J., editors, *Philosophy and AI*, chapter 4, pages 79–103. MIT Press, Cambridge, Massachusetts.

[Ellis, 1993] Ellis, J. (1993). *Language, Thought, and Logic.* Northwestern University Press, Evanston, Illinois.

[Ellsberg, 1961] Ellsberg, D. (1961). Risk, ambiguity, and the savage axioms. *Quarterly Journal of Economics*, 75:643–669.

[Englebretsen, 1981] Englebretsen, G. (1981). *Three Logicians.* Van Gorcum, Assen, The Netherlands.

[Englebretsen, 1996] Englebretsen, G. (1996). *Something to Reckon with: the Logic of Terms.* Ottawa University Press, Ottawa.

[Fagin and Halpern, 1988] Fagin, R. and Halpern, J. (1988). Belief, awareness, and limited reasoning. *Artificial Intelligence*, 34:39–76.

[Feigenbaum and McCorduck, 1983] Feigenbaum, E. and McCorduck, P. (1983). *The Fifth Generation : Artificial Intelligence and Japan's Computer Challenge to the world.* Addison-Wesley Publishing Company, Reading, Massachusetts.

[Feyerabend, 1993] Feyerabend, P. (1993). *Against Method.* Verso Books, London, 3rd edition.

[Field, 2001] Field, H. (2001). *Truth and the Absence of Fact.* Oxford University Press, New York.

[Flach and Kakas, 2000] Flach, P. and Kakas, A. (2000). Abductive and inductive reasoning: Background and issues. In Flach, P. and Kakas, A., editors, *Abduction and Induction: Essays on their Relation and Integration*, pages 1–27. Kluwer Academic Publishers, Dordrecht.

[Fodor, 1987] Fodor, J. (1987). *Psychosemantics.* MIT Press, Cambridge, Massachusetts.

[Fodor, 1998] Fodor, J. (1998). *Concepts: where cognitive science went wrong.* Oxford University Press, New York.

[Fox, 2000] Fox, C. (2000). *The Ontology of Language: Properties, Individuals and Discourse.* CSLI Publications, Stanford, California.

[Frege, 1970] Frege, G. (1970). Begriffsschrift, a formula language, modeled upon that of arithmetic, for pure thought. In van Heijenoort, J., editor, *Frege and Gödel: Two Fundamental Texts in Mathematical Logic*, pages 1–82. Harvard University Press, Cambridge, Massachusetts.

[French, 1990] French, R. (1990). Subcognition and the limits of the Turing test. *Mind*, 99:53–65.

[French, 1995] French, R. (1995). *The Subtlety of Sameness: A Theory and Computer Model of Analogy-Making.* MIT Press, Cambridge, Massachusetts.

[Frisch and Haddawy, 1994] Frisch, A. and Haddawy, P. (1994). Anytime deduction for probabilistic logic. *Artificial Intelligence*, 69:93–122.

[Fung and Chong, 1986] Fung, R. and Chong, C. (1986). Metaprobability and Dempster-Shafer in evidential reasoning. In Kanal, L. and Lemmer, J., editors, *Uncertainty in Artificial Intelligence*, pages 295–302. North-Holland, Amsterdam.

[Gaifman, 1986] Gaifman, H. (1986). A theory of higher order probabilities. In Halpern, J., editor, *Theoretical Aspects of Reasoning about Knowledge*, pages 275–292. Morgan Kaufmann, Los Altos, California.

[Gaines, 1978] Gaines, B. (1978). Fuzzy and probability uncertainty logics. *Information and Control*, 38:154–169.

[Gärdenfors and Williams, 2001] Gärdenfors, P. and Williams, M. (2001). Reasoning about categories in conceptual spaces. In *Proceedings of the International Joint Conference on Artificial Intelligence*, pages 385–392.

[Geach, 1968] Geach, P. (1968). *A history of the corruptions of logic: an inaugural lecture.* Leeds University Press, Cambridge.

[Gigerenzer, 1991] Gigerenzer, G. (1991). How to make cognitive illusions disappear: beyond "heuristics and biases". In Stroebe, W. and Hewstone, M., editors, *European Review of Social Psychology, Volume 2*, chapter 4, pages 83–115. John Wiley & Sons Ltd.

[Ginsberg, 1987] Ginsberg, M., editor (1987). *Readings in Nonmonotonic Reasoning.* Morgan Kaufmann, San Mateo.

[Giraud-Carrier and Martinez, 1995] Giraud-Carrier, C. and Martinez, T. (1995). An integrated framework for learning and reasoning. *Journal of Artificial Intelligence Research*, 3:147–185.

[Goertzel and Pennachin, 2006] Goertzel, B. and Pennachin, C., editors (2006). *Artificial General Intelligence.* Springer, New York.

[Goertzel et al., 2003] Goertzel, B., Pennachin, C., Senna, A., Maia, T., and Lamacie, G. (2003). Novamente: An integrative architecture for artificial general intelligence. In *Working Notes of the IJCAI Workshop on Cognitive Modeling of Agents and Multi-Agent Interactions*, Acapulco, Mexico.

[Good, 1965] Good, I. (1965). *The Estimation of Probabilities.* MIT Press, Cambridge, Massachusetts.

[Good, 1983] Good, I. (1983). *Good Thinking: The Foundations of Probability and Its Applications.* University of Minnesota Press, Minneapolis.

[Grosof, 1986] Grosof, B. (1986). An inequality paradigm for probabilistic knowledge. In Kanal, L. and Lemmer, J., editors, *Uncertainty in Artificial Intelligence*, pages 259–275. North-Holland, Amsterdam.

[Grosof, 1990] Grosof, B. (1990). Defeasible reasoning and uncertainty: comments. In Henrion, M., Shachter, R., Kanal, L., and Lemmer, J., editors, *Uncertainty in Artificial Intelligence 5*, pages 61–66. North-Holland, Amsterdam.

[Guha and Lenat, 1990] Guha, R. and Lenat, D. (1990). Cyc: A mid-term report. *AI Magazine*, 11(3):32–59.

[Haack, 1978] Haack, S. (1978). *Philosophy of Logics.* Cambridge University Press, Cambridge.

[Halpern, 1990] Halpern, J. (1990). An analysis of first-order logics of probability. *Artificial Intelligence*, 46:311–350.

[Halpern et al., 2001] Halpern, J. Y., Harper, R., Immerman, N., Kolaitis, P. G., Vardi, M. Y., and Vianu, V. (2001). On the unusual effectiveness of logic in computer science. *The Bulletin of Symbolic Logic*, 7(2):213–236.

[Harman, 1982] Harman, G. (1982). Conceptual role semantics. *Notre Dame Journal of Formal Logic*, 28:252–256.

[Harnad, 1990] Harnad, S. (1990). The symbol grounding problem. *Physica D*, 42:335–346.

[Haussler, 1988] Haussler, D. (1988). Quantifying inductive bias: AI learning algorithms and Valiant's learning framework. *Artificial Intelligence*, 36:177–221.

[Hawkins and Blakeslee, 2004] Hawkins, J. and Blakeslee, S. (2004). *On Intelligence*. Times Books, New York.

[Hayes, 1977] Hayes, P. (1977). In defense of logic. In *Proceedings of the International Joint Conference on Artificial Intelligence*, pages 559–565.

[Hearst and Hirsh, 2000] Hearst, M. and Hirsh, H. (2000). AI's greatest trends and controversies. *IEEE Intelligent Systems*, pages 8–17.

[Hecht-Nielsen, 2005] Hecht-Nielsen, R. (2005). Cogent confabulation. *Neural Networks*, 18(2):111–115.

[Heckerman, 1999] Heckerman, D. (1999). Bayesian learning. In Wilson, R. and Keil, F., editors, *The MIT Encyclopedia of the Cognitive Sciences*, pages 70–72. MIT Press, Cambridge, Massachusetts.

[Hempel, 1943] Hempel, C. (1943). A purely syntactical definition of confirmation. *Journal of Symbolic Logic*, 8:122–143.

[Hofstadter, 1979] Hofstadter, D. (1979). *Gödel, Escher, Bach: an Eternal Golden Braid*. Basic Books, New York.

[Hofstadter, 1985] Hofstadter, D. (1985). *Metamagical Themas: Questing for the Essence of Mind and Pattern*. Basic Books, New York.

[Hofstadter, 1993a] Hofstadter, D. (1993a). From Euler to Ulam: discovery and dissection of a geometric gem. Technical Report 81, Center for Research on Concepts and Cognition, Indiana University, Bloomington, Indiana.

[Hofstadter, 1993b] Hofstadter, D. (1993b). How could a copycat ever be creative? In *Working Notes, 1993 AAAI Spring Symposium Series, Symposium: AI and Creativity*, pages 1–10.

[Hofstadter and FARG, 1995] Hofstadter, D. and FARG (1995). *Fluid Concepts and Creative Analogies: Computer Models of the Fundamental Mechanisms of Thought.* Basic Books, New York.

[Holland, 1986] Holland, J. (1986). Escaping brittleness: the possibilities of general purpose learning algorithms applied to parallel rule-based systems. In Michalski, R., Carbonell, J., and Mitchell, T., editors, *Machine Learning: an artificial intelligence approach*, volume II, chapter 20, pages 593–624. Morgan Kaufmann, Los Altos, California.

[Holland, 1992] Holland, J. (1992). *Adaptation in Natural and Artificial Systems: An Introductory Analysis With Applications to Biology, Control, and Artificial Intelligence.* MIT Press, Cambridge, Massachusetts.

[Hopcroft and Ullman, 1979] Hopcroft, J. and Ullman, J. (1979). *Introduction to Automata Theory, Language, and Computation.* Addison-Wesley, Reading, Massachusetts.

[Horvitz, 1989] Horvitz, E. (1989). Reasoning about beliefs and actions under computational resource constraints. In Kanal, L., Levitt, T., and Lemmer, J., editors, *Uncertainty in Artificial Intelligence 3*, pages 301–324. North-Holland, Amsterdam.

[Hume, 1748] Hume, D. (1748). *An Enquiry Concerning Human Understanding.* London.

[Hutter, 2001] Hutter, M. (2001). Towards a universal theory of artificial intelligence based on algorithmic probability and sequential decisions. In *Proceedings of the 12th European Conference on Machine Learning*, pages 226–238.

[Inhelder and Piaget, 1969] Inhelder, B. and Piaget, J. (1969). *The Early Growth of Logic in the Child.* W. W. Norton & Company, Inc., New York. Translated by E. Lunzer and D. Papert.

[Jeffrey, 1965] Jeffrey, R. (1965). *The Logic of Decision.* McGraw-Hill, New York.

[Johnson, 1997] Johnson, T. (1997). Control in Act-R and Soar. In Shafto, M. and Langley, P., editors, *Proceedings of the Nineteenth Annual Conference of the Cognitive Science Society*, pages 343–348.

[Johnson-Laird, 1983] Johnson-Laird, P. (1983). *Mental Models.* Harvard University Press, Cambridge, Massachusetts.

[Kahneman and Miller, 1986] Kahneman, D. and Miller, D. (1986). Norm theory: comparing reality to its alternatives. *Psychological Review*, 93:136–153.

[Kahneman and Tversky, 1982] Kahneman, D. and Tversky, A. (1982). On the study of statistical intuitions. In Kahneman, D., Slovic, P., and Tversky, A., editors, *Judgment under Uncertainty: Heuristics and Biases*, chapter 34, pages 493–508. Cambridge University Press, Cambridge, England.

[Kamp, 1975] Kamp, J. (1975). Two theories about adjectives. In Keenan, E., editor, *Formal Semantics of Natural Language*, pages 123–155. Cambridge University Press, Cambridge.

[Keynes, 1921] Keynes, J. (1921). *A Treatise on Probability.* Macmillan, London.

[Kitchener, 1994] Kitchener, R. (1994). Semantic naturalism: The problem of meaning and naturalistic psychology. In Overton, W. and Palermo, D., editors, *The Nature and Ontogenesis of Meaning.* Lawrence Erlbaum Associates, Hillsdale, New Jersey.

[Kolmogorov, 1950] Kolmogorov, A. N. (1950). *Foundations of the Theory of Probability.* Chelsea Publishing Company, New York.

[Korb, 1995] Korb, K. (1995). Inductive learning and defeasible infernce. *Journal of Experimental & Theoretical Artificial Intelligence*, 7:291–324.

[Kowalski, 1979] Kowalski, R. (1979). *Logic for Problem Solving.* North Holland, New York.

[Kowalski, 1995] Kowalski, R. (1995). Logic without model theory. In Gabbay, D. M., editor, *What is a Logical System?*, pages 35–71. Oxford University Press.

[Krantz, 1991] Krantz, D. (1991). From indices to mappings: The representational approach to measurement. In Brown, D. and Smith, J., editors, *Frontiers of Mathematical Psychology: Essays in Honor of Clyde Coombs*, Recent Research in Psychology, chapter 1. Springer-Verlag, Berlin, Germany.

[Kugel, 1986] Kugel, P. (1986). Thinking may be more than computing. *Cognition*, 22:137–198.

[Kuhn, 1970] Kuhn, T. (1970). *The Structure of Scientific Revolutions.* Chicago University Press, 2nd edition.

[Kurtonina and de Rijke, 1999] Kurtonina, N. and de Rijke, M. (1999). Expressiveness of concept expressions in first-order description logics. *Artificial Intelligence*, 107:303–333.

[Kyburg, 1983] Kyburg, H. (1983). The reference class. *Philosophy of Science*, 50:374–397.

[Kyburg, 1987] Kyburg, H. (1987). Bayesian and non-Bayesian evidential updating. *Artificial Intelligence*, 31:271–293.

[Kyburg, 1988] Kyburg, H. (1988). Higher order probabilities and intervals. *International Journal of Approximate Reasoning*, 2:195–209.

[Kyburg, 1989] Kyburg, H. (1989). Higher order probabilities. In Kanal, L., Levitt, T., and Lemmer, J., editors, *Uncertainty in Artificial Intelligence 3*, pages 15–22. North-Holland, Amsterdam.

[Kyburg, 1992] Kyburg, H. (1992). Semantics for probabilistic inference. In *Proceedings of the Eighth Conference on Uncertainty in Artificial Intelligence*, pages 142–148.

[Kyburg, 1994] Kyburg, H. (1994). Believing on the basis of the evidence. *Computational Intelligence*, 10:3–20.

[Laffey et al., 1988] Laffey, T., Cox, P., Schmidt, J., Kao, S., and Read, J. (1988). Real-time knowledge based system. *AI Magazine*, 9:27–45.

[Laird et al., 1987] Laird, J., Newell, A., and Rosenbloom, P. (1987). Soar: An architecture for general intelligence. *Artificial Intelligence*, 33:1–64.

[Lakoff, 1987] Lakoff, G. (1987). *Women, Fire, and Dangerous Things: What Categories Reveal about the Mind.* University of Chicago Press, Chicago.

[Lakoff, 1988] Lakoff, G. (1988). Cognitive semantics. In Eco, U., Santambrogio, M., and P., V., editors, *Meaning and Mental Representation*, pages 119–154. Indiana University Press, Bloomington, Indiana.

[Laurence and Margolis, 1999] Laurence, S. and Margolis, E. (1999). Concepts and cognitive science. In Margolis, E. and Laurence, S., editors, *Concepts: Core Readings*, pages 3–81. MIT Press, Cambridge, Massachusetts.

[Lenat, 1995] Lenat, D. (1995). Cyc: A large-scale investment in knowledge infrastructure. *Communications of the ACM*, 38(11):33–38.

[Lenat and Feigenbaum, 1991] Lenat, D. and Feigenbaum, E. (1991). On the thresholds of knowledge. *Artificial Intelligence*, 47:185–250.

[Levesque, 1989] Levesque, H. (1989). Logic and the complexity of reasoning. In Thomason, R., editor, *Philosophical Logic and Artificial Intelligence*, pages 73–107. Kluwer Academic Publishers, Boston.

[Lewis, 1988] Lewis, D. (1988). Desire as belief. *Mind*, 97(387):323–332.

[Littman et al., 1998] Littman, M., Goldsmith, J., and Mundhenk, M. (1998). The computational complexity of probabilistic planning. *Journal of Artificial Intelligence Research*, 9:1–36.

[Lucas, 1961] Lucas, J. R. (1961). Minds, machines and Gödel. *Philosophy*, XXXVI:112–127.

[Lukasiewicz, 1951] Lukasiewicz, J. (1951). *Aristotle's Syllogistic: From the Standpoint of Modern Formal Logic.* Oxford University Press, London.

[Lynch, 1998] Lynch, M. (1998). *Truth in Context.* MIT Press, Cambridge, Massachusetts.

[Marr, 1982] Marr, D. (1982). *Vision: A Computational Investigation into the Human Representation and Processing of Visual Information.* W. H. Freeman & Co., San Francisco.

[McCarthy, 1988] McCarthy, J. (1988). Mathematical logic in artificial intelligence. *Dædalus*, 117(1):297–311.

[McCarthy, 1993] McCarthy, J. (1993). Notes on formalizing contexts. In *Proceedings of the thirteenth international joint conference on artificial intelligence*, pages 555–560.

[McCarthy and Hayes, 1969] McCarthy, J. and Hayes, P. (1969). Some philosophical problems from the standpoint of artificial intelligence. In Meltzer, B. and Michie, D., editors, *Machine Intelligence 4*, pages 463–502. Edinburgh University Press, Edinburgh.

[McDermott, 1987] McDermott, D. (1987). A critique of pure reason. *Computational Intelligence*, 3:151–160.

[McNeill and Freiberger, 1993] McNeill, D. and Freiberger, P. (1993). *Fuzzy Logic.* Simon & Schuster, New York.

[Medin and Ross, 1992] Medin, D. and Ross, B. (1992). *Cognitive Psychology.* Harcourt Brace Jovanovich, Fort Worth.

[Medin and Schaffer, 1978] Medin, D. and Schaffer, M. (1978). A context theory of classification learning. *Psychological Review*, 85: 207–328.

[Michalski, 1993] Michalski, R. (1993). Inference theory of learning as a conceptual basis for multistrategy learning. *Machine Learning*, 11:111–151.

[Minsky, 1985] Minsky, M. (1985). *The Society of Mind.* Simon and Schuster, New York.

[Minsky, 1990] Minsky, M. (1990). Logical vs. analogical or symbolic vs. connectionist or neat vs. scruffy. In Winston, P., editor, *Artificial Intelligence at MIT, Vol. 1: Expanding Frontiers*, pages 218–243. MIT Press, Cambridge, Massachusetts.

[Minsky et al., 2004] Minsky, M., Singh, P., and Sloman, A. (2004). The St. Thomas common sense symposium: Designing architectures for human-level intelligence. *AI Magazine*, 25(2):113–124.

[Mitchell, 1993] Mitchell, M. (1993). *Analogy-Making as Perception: A Computer Model.* MIT Press, Cambridge, Massachusetts.

[Mitchell, 1980] Mitchell, T. (1980). The need for biases in learning generalizations. In Shavlik, J. and Dietterich, T., editors, *Readings in Machine Learning.* Morgan Kaufmann, San Mateo, California. 1990. Originally published as a Rutgers Technical report.

[Murphy and Medin, 1985] Murphy, G. and Medin, D. (1985). The role of theories in conceptual coherence. *Psychological Review*, 92(3): 289–316.

[Neufeld, 1989] Neufeld, E. (1989). Defaults and probabilities: extensions and coherence. In *Proceedings of the First International Conference on Principles of Knowledge Representation and Reasoning.*

[Newborn, 2002] Newborn, M. (2002). *Deep Blue: An Artificial Intelligence Milestone.* Springer Verlag, New York.

[Newell, 1990] Newell, A. (1990). *Unified Theories of Cognition.* Harvard University Press, Cambridge, Massachusetts.

[Newell and Simon, 1976] Newell, A. and Simon, H. (1976). Computer science as empirical inquiry: symbols and search. *Communications of the ACM*, 19(3):113–126.

[Nilsson, 1991] Nilsson, N. (1991). Logic and artificial intelligence. *Artificial Intelligence*, 47:31–56.

[Nilsson, 1998] Nilsson, N. (1998). *Artificial Intelligence: A New Synthesis.* Morgan Kaufmann, San Francisco.

[Nosofsky, 1991] Nosofsky, R. (1991). Typicality in logically defined categories: exemplar-similarity versus rule instantiation. *Memory and Cognition*, 17:444–458.

[Oden, 1977a] Oden, G. (1977a). Fuzziness in semantic memory: choosing exemplars of subjective categories. *Memory and cognition*, 5:198–204.

[Oden, 1977b] Oden, G. (1977b). Integration of fuzzy logical information. *Journal of Experimental Psychology: Human Perception and Performance*, 3:565–575.

[Osherson and Smith, 1981] Osherson, D. and Smith, E. (1981). On the adequacy of prototype theory as a theory of concepts. *Cognition*, 9:35–58.

[Paaß, 1991] Paaß, G. (1991). Second order probabilities for uncertain and conflicting evidence. In Bonissone, P., Henrion, M., Kanal, L., and Lemmer, J., editors, *Uncertainty in Artificial Intelligence 6*, pages 447–456. North-Holland, Amsterdam.

[Palmer, 1981] Palmer, F. (1981). *Semantics.* Cambridge University Press, New York, 2nd edition.

[Pearl, 1988] Pearl, J. (1988). *Probabilistic Reasoning in Intelligent Systems.* Morgan Kaufmann Publishers, San Mateo, California.

[Pearl, 1990] Pearl, J. (1990). Jeffrey's rule, passage of experience, and Neo-Bayesianism. In Kyburg, H., R., L., and G., C., editors, *Knowledge Representation and Defeasible Reasoning*, pages 245–265. Kluwer Academic Publishers, Amsterdam.

[Pearl, 2000] Pearl, J. (2000). *Causality.* Cambridge University Press, Cambridge, UK.

[Peirce, 1931] Peirce, C. (1931). *Collected Papers of Charles Sanders Peirce*, volume 2. Harvard University Press, Cambridge, Massachusetts.

[Penrose, 1994] Penrose, R. (1994). *Shadows of the Mind.* Oxford University Press.

[Piaget, 1960] Piaget, J. (1960). *The Psychology of Intelligence.* Littlefield, Adams & Co., Paterson, New Jersey.

[Piaget, 1963] Piaget, J. (1963). *The Origins of Intelligence in Children.* W.W. Norton & Company, Inc., New York. Translated by M. Cook.

[Poole, 1985] Poole, D. (1985). On the comparison of theories: preferring the most specific explanation. In *Proceedings of the Ninth International Joint Conference on Artificial Intelligence*, pages 144–147.

[Popper, 1959] Popper, K. (1959). *The Logic of Scientific Discovery.* Basic Books, New York.

[Priest et al., 1989] Priest, G., Routley, R., and Norman, J., editors (1989). *Paraconsistent Logic: Essays on the Inconsistent.* Philosophia Verlag, München.

[Putnam, 1981] Putnam, H. (1981). *Reason, Truth and History.* Cambridge University Press, Cambridge.

[Quillian, 1968] Quillian, M. R. (1968). Semantic memory. In Minsky, M., editor, *Semantic Information Processing*, pages 227–270. MIT Press, Cambridge, Massachusetts.

[Rawlins, 1992] Rawlins, G. (1992). *Compared to What?* Computer Science Press, New York.

[Read, 1989] Read, S. (1989). *Relevant Logic: a philosophical examination of inference.* Basil Blackwell, New York.

[Reeke and Edelman, 1988] Reeke, G. and Edelman, G. (1988). Real brains and artificial intelligence. *Dædalus*, 117(1):143–173.

[Rehder, 1999] Rehder, B. (1999). A causal model theory of categorization. In *Proceedings of the 21st Annual Meeting of the Cognitive Science Society*, pages 595–600.

[Reichenbach, 1949] Reichenbach, H. (1949). *The Theory of Probability*. University of California Press, Berkeley, California. Translated by E. Hutten and M. Reichenbach.

[Reiter, 1987] Reiter, R. (1987). Nonmonotonic reasoning. *Annual Review of Computer Science*, 2:147–186.

[Rips, 1995] Rips, L. (1995). The current status of research on concept combination. *Mind and Language*, 10:72–104.

[Roland and Shiman, 2002] Roland, A. and Shiman, P. (2002). *Strategic computing : DARPA and the quest for machine intelligence, 1983-1993*. MIT Press, Cambridge, Massachusetts.

[Rosch, 1973] Rosch, E. (1973). On the internal structure of perceptual and semantic categories. In Moore, T., editor, *Cognitive Development and the Acquisition of Language*, pages 111–144. Academic Press, New York.

[Rosch, 1978] Rosch, E. (1978). Principles of categorization. In Rosch, E. and Lloyd, B., editors, *Cognition and Categorization*, pages 27–48. Lawrence Erlbaum Associates, Hillsdale, New Jersey.

[Rosch and Mervis, 1975] Rosch, E. and Mervis, C. (1975). Family resemblances: studies in the internal structure of categories. *Cognitive Psychology*, 7:573–605.

[Rota, 1989] Rota, G.-C. (1989). The barrier of meaning. In memorium: Stanislaw Ulam. *Notices of the American Mathematical Society*, 36(2):141–143.

[Rumelhart and McClelland, 1986] Rumelhart, D. and McClelland, J. (1986). PDP models and general issues in cognitive science. In Rumelhart, D. and McClelland, J., editors, *Parallel Distributed Processing: Explorations in the Microstructure of Cognition, Vol. 1, Foundations*, pages 110–146. MIT Press, Cambridge, Massachusetts.

[Russell, 1901] Russell, B. (1901). Recent work on the principles of mathematics. *International Monthly*, 4:83–101.

[Russell and Norvig, 2002] Russell, S. and Norvig, P. (2002). *Artificial Intelligence: A Modern Approach*. Prentice Hall, Upper Saddle River, New Jersey, 2nd edition.

[Russell and Wefald, 1991a] Russell, S. and Wefald, E. (1991a). *Do the Right Thing*. MIT Press, Cambridge, Massachusetts.

[Russell and Wefald, 1991b] Russell, S. and Wefald, E. (1991b). Principles of metareasoning. *Artificial Intelligence*, 49:361–395.

[Sapir, 1944] Sapir, E. (1944). Grading: a study in semantics. *Philosophy of Science*, 11:93–116.

[Savage, 1954] Savage, L. (1954). *The Foundations of Statistics*. Wiley, New York.

[Saygin et al., 2000] Saygin, A., Cicekli, I., and Akman, V. (2000). Turing test: 50 years later. *Minds and Machines*, 10(4):463–518.

[Schank, 1991] Schank, R. (1991). Where is the AI? *AI Magazine*, 12(4):38–49.

[Scheutz, 2002] Scheutz, M. (2002). Computationalism — the next generation. In Scheutz, M., editor, *Computationalism: new directions*, pages 1–21. MIT Press, Cambridge, Massachusetts.

[Schweizer and Sklar, 1983] Schweizer, B. and Sklar, A. (1983). *Probabilistic Metric Spaces*. North-Holland, Amsterdam.

[Searle, 1980] Searle, J. (1980). Minds, brains, and programs. *The Behavioral and Brain Sciences*, 3:417–424.

[Segal, 2000] Segal, G. (2000). *A Slim Book about Narrow Content*. MIT Press, Cambridge, Massachusetts.

[Shafer, 1976] Shafer, G. (1976). *A Mathematical Theory of Evidence*. Princeton University Press, Princeton, New Jersey.

[Simon, 1983] Simon, H. (1983). *Reason in Human Affairs*. Stanford University Press, Stanford, California.

[Simon, 1996] Simon, H. A. (1996). *The Sciences of the Artificial.* MIT Press, Cambridge, Massachusetts, third edition.

[Sloman, 2002] Sloman, A. (2002). The irrelevance of Turing machine to artificial intelligence. In Scheutz, M., editor, *Computationalism: new directions*, pages 87–127. MIT Press, Cambridge, Massachusetts.

[Smets, 1991] Smets, P. (1991). Varieties of ignorance and the need for well-founded theories. *Information Sciences*, 57-58:135–144.

[Smith and Osherson, 1984] Smith, E. and Osherson, D. (1984). Conceptual combination with prototype concepts. *Cognitive Science*, 8:337–361.

[Smolensky, 1988] Smolensky, P. (1988). On the proper treatment of connectionism. *Behavioral and Brain Sciences*, 11:1–74.

[Sommers, 1982] Sommers, F. (1982). *The Logic of Natural Language.* Clarendon Press, Oxford.

[Sommers and Englebretsen, 2000] Sommers, F. and Englebretsen, G. (2000). *An invitation to formal reasoning: the logic of terms.* Ashgate, Aldershot.

[Sosa and Tooley, 1993] Sosa, E. and Tooley, M. (1993). Introduction. In Sosa, E. and Tooley, M., editors, *Causation*, pages 1–32. Oxford University Press, Oxford.

[Spiegelhalter, 1989] Spiegelhalter, D. (1989). A unified approach to imprecision and sensitivity of beliefs in expert systems. In Kanal, L., Levitt, T., and Lemmer, J., editors, *Uncertainty in Artificial Intelligence 3*, pages 199–208. North-Holland, Amsterdam.

[Stork, 1997a] Stork, D. (1997a). Computers, science, and extraterrestrials: An interview with Stephen Wolfram. In Stork, D., editor, *HAL's Legacy: 2001's Computer as Dream and Reality*, pages 333–349. MIT Press, Cambridge, Massachusetts.

[Stork, 1997b] Stork, D. (1997b). Scientist on the set: An interview with Marvin Minsky. In Stork, D., editor, *HAL's Legacy: 2001's*

Computer as Dream and Reality, pages 15–30. MIT Press, Cambridge, Massachusetts.

[Strosnider and Paul, 1994] Strosnider, J. and Paul, C. (1994). A structured view of real-time problem solving. *AI Magazine*, 15(2):45–66.

[Sullivan and Cohen, 1985] Sullivan, M. and Cohen, P. (1985). An endorsement-based plan recognition program. In *Proceedings of the National Conference of Artificial Intelligence*, pages 475–479.

[Sun, 1995] Sun, R. (1995). Robust reasoning: integrating rule-based and similarity-based reasoning. *Artificial Intelligence*, 75:241–295.

[Tarski, 1944] Tarski, A. (1944). The semantic conception of truth. *Philosophy and Phenomenological Research*, 4:341–375.

[Thrun and Mitchell, 1995] Thrun, S. and Mitchell, T. (1995). Learning one more thing. In *The Proceedings of the Fourteenth International Joint Conference on Artificial Intelligence*, pages 1217–1223.

[Touretzky, 1984] Touretzky, D. (1984). Implicit ordering of defaults in inheritance systems. In *Proceedings of the National Conference of Artificial Intelligence*, pages 322–325.

[Touretzky, 1986] Touretzky, D. (1986). *The Mathematics of Inheritance Systems*. Pitman Publishing, London.

[Turing, 1950] Turing, A. (1950). Computing machinery and intelligence. *Mind*, LIX:433–460.

[Turksen, 1991] Turksen, I. (1991). Measurement of membership functions and their acquisition. *Fuzzy Sets and System*, 40:5–38.

[Tversky, 1977] Tversky, A. (1977). Features of similarity. *Psychological Review*, 84:327–352.

[Tversky and Kahneman, 1974] Tversky, A. and Kahneman, D. (1974). Judgment under uncertainty: heuristics and biases. *Science*, 185: 1124–1131.

[Tversky and Kahneman, 1983] Tversky, A. and Kahneman, D. (1983). Extensional versus intuitive reasoning: the conjunction fallacy in probability judgment. *Psychological Review*, 90:293–315.

[Valiant, 1984] Valiant, L. (1984). A theory of the learnable. *Communications of the ACM*, 27:1134–1142.

[van Gelder, 1997] van Gelder, T. (1997). Dynamics and cognition. In Haugeland, J., editor, *Mind Design II*, pages 421–450. MIT Press, Cambridge, Massachusetts.

[von Mises, 1981] von Mises, R. (1981). *Probability, Statistics and Truth*. Dover Publications, New York.

[Voorbraak, 1999] Voorbraak, F. (1999). Probabilistic belief change: Expansion, conditioning and constraining. In *Uncertainty in Artificial Intelligence: Proceedings of the Fifteenth Conference (UAI-1999)*, pages 655–662, San Francisco, CA. Morgan Kaufmann Publishers.

[Walley, 1991] Walley, P. (1991). *Statistical Reasoning with Imprecise Probabilities*. Chapman and Hall, London.

[Walley, 1996] Walley, P. (1996). Inferences from multinomial data: learning about a bag of marbles. *Journal of the Royal Statistical Society, Series B*, 58:3–57.

[Wallsten et al., 1993] Wallsten, T., Budescu, D., and Zwick, R. (1993). Comparing the calibration and coherence of numerical and verbal probability judgments. *Management Science*, 39:176–190.

[Wang, 1986] Wang, P. (1986). A reasoning system that can deal with uncertainty. Master's thesis, Peking University, Beijing. In Chinese.

[Wang, 1992] Wang, P. (1992). First ladies and fluid logics. Technical Report 62, Center for Research on Concepts and Cognition, Indiana University, Bloomington, Indiana.

[Wang, 1993a] Wang, P. (1993a). Belief revision in probability theory. In *Proceedings of the Ninth Conference on Uncertainty in*

Artificial Intelligence, pages 519–526. Morgan Kaufmann Publishers, San Mateo, California.

[Wang, 1993b] Wang, P. (1993b). Non-axiomatic logic (version 2.1). Technical Report 71, Center for Research on Concepts and Cognition, Indiana University, Bloomington, Indiana.

[Wang, 1993c] Wang, P. (1993c). Non-axiomatic reasoning system (version 2.2). Technical Report 75, Center for Research on Concepts and Cognition, Indiana University, Bloomington, Indiana.

[Wang, 1994a] Wang, P. (1994a). A defect in Dempster-Shafer theory. In *Proceedings of the Tenth Conference on Uncertainty in Artificial Intelligence*, pages 560–566. Morgan Kaufmann Publishers, San Mateo, California.

[Wang, 1994b] Wang, P. (1994b). From inheritance relation to non-axiomatic logic. *International Journal of Approximate Reasoning*, 11(4):281–319.

[Wang, 1994c] Wang, P. (1994c). On the working definition of intelligence. Technical Report 94, Center for Research on Concepts and Cognition, Indiana University, Bloomington, Indiana.

[Wang, 1995a] Wang, P. (1995a). *Non-Axiomatic Reasoning System: Exploring the Essence of Intelligence.* PhD thesis, Indiana University.

[Wang, 1995b] Wang, P. (1995b). Reference classes and multiple inheritances. *International Journal of Uncertainty, Fuzziness and and Knowledge-based Systems*, 3(1):79–91.

[Wang, 1995c] Wang, P. (1995c). An unified treatment of uncertainties. In *Proceedings of the Fourth International Conference for Young Computer Scientists*, pages 462–467, Beijing.

[Wang, 1996a] Wang, P. (1996a). Heuristics and normative models of judgment under uncertainty. *International Journal of Approximate Reasoning*, 14(4):221–235.

[Wang, 1996b] Wang, P. (1996b). The interpretation of fuzziness. *IEEE Transactions on Systems, Man, and Cybernetics, Part B: Cybernetics*, 26(4):321–326.

[Wang, 1996c] Wang, P. (1996c). Problem-solving under insufficient resources. In *Working Notes of the AAAI Fall Symposium on Flexible Computation*, pages 148–155, Cambridge, Massachusetts.

[Wang, 1998a] Wang, P. (1998a). Grounding the meaning of symbols on the system's experience. In *Working Notes of the AAAI Workshop on the Grounding of Word Meaning: Data and Models*, pages 23–24, Madison, Wisconsin.

[Wang, 1998b] Wang, P. (1998b). Why recommendation is special? In *Working Notes of the AAAI Workshop on Recommender System*, pages 111–113, Madison, Wisconsin.

[Wang, 1999] Wang, P. (1999). A new approach for induction: From a non-axiomatic logical point of view. In Ju, S., Liang, Q., and Liang, B., editors, *Philosophy, Logic, and Artificial Intelligence*, pages 53–85. Zhongshan University Press.

[Wang, 2000a] Wang, P. (2000a). The logic of learning. In *Working Notes of the AAAI workshop on New Research Problems for Machine Learning*, pages 37–40, Austin, Texas.

[Wang, 2000b] Wang, P. (2000b). Non-axiomatic reasoning system (version 4.1). In *Proceedings of the Seventeenth National Conference on Artificial Intelligence*, pages 1135–1136, Austin, Texas.

[Wang, 2000c] Wang, P. (2000c). Unified inference in extended syllogism. In Flach, P. and Kakas, A., editors, *Abduction and Induction: Essays on Their Relation and Integration*, pages 117–129. Kluwer Academic Publishers, Dordrecht.

[Wang, 2001a] Wang, P. (2001a). Abduction in non-axiomatic logic. In *Working Notes of the IJCAI workshop on Abductive Reasoning*, pages 56–63, Seattle, Washington.

[Wang, 2001b] Wang, P. (2001b). Confidence as higher-order uncertainty. In *Proceedings of the Second International Symposium on Imprecise Probabilities and Their Applications*, pages 352–361, Ithaca, New York.

[Wang, 2001c] Wang, P. (2001c). Wason's cards: what is wrong? In *Proceedings of the Third International Conference on Cognitive Science*, pages 371–375, Beijing.

[Wang, 2002] Wang, P. (2002). The logic of categorization. In *Proceedings of the 15th International FLAIRS Conference*, pages 181–185, Pensacola, Florida.

[Wang, 2004a] Wang, P. (2004a). The limitation of Bayesianism. *Artificial Intelligence*, 158(1):97–106.

[Wang, 2004b] Wang, P. (2004b). Problem solving with insufficient resources. *International Journal of Uncertainty, Fuzziness and and Knowledge-based Systems*, 12(5):673–700.

[Wang, 2004c] Wang, P. (2004c). Reasoning in practical situations. In *Proceedings of Knowledge-Based Intelligent Information and Engineering Systems: 8th International Conference*, pages 285–292, Wellington. Published as Lecture Notes in Computer Science, Volume 3215, Springer-Verlag, Heidelberg.

[Wang, 2004d] Wang, P. (2004d). Recommendation based on personal preference. In Zhang, Y., Kandel, A., Lin, T., and Yao, Y., editors, *Computational Web Intelligence: Intelligent Technology for Web Applications*, pages 101–115. World Scientific Publishing Company, Singapore.

[Wang, 2004e] Wang, P. (2004e). Toward a unified artificial intelligence. In *Papers from the 2004 AAAI Fall Symposium on Achieving Human-Level Intelligence through Integrated Research and Systems*, pages 83–90, Washington DC.

[Wang, 2005] Wang, P. (2005). Experience-grounded semantics: a theory for intelligent systems. *Cognitive Systems Research*, 6(4):282–302.

[Wang, 2006a] Wang, P. (2006a). Artificial intelligence: What it is, and what it should be. In *The 2006 AAAI Spring Symposium on Between a Rock and a Hard Place: Cognitive Science Principles Meet AI-Hard Problems*, Stanford, California.

[Wang, 2006b] Wang, P. (2006b). The logic of intelligence. In Goertzel, B. and Pennachin, C., editors, *Artificial General Intelligence*. Springer, New York.

[Wang and Hofstadter, 2006] Wang, P. and Hofstadter, D. (2006). A logic of categorization. *Journal of Experimental & Theoretical Artificial Intelligence*. In press.

[Wason and Johnson-Laird, 1972] Wason, P. and Johnson-Laird, P. (1972). *Psychology of Reasoning: Structure and Content*. Harvard University Press, Cambridge, Massachusetts.

[Whitehead and Russell, 1910] Whitehead, A. and Russell, B. (1910). *Principia mathematica*. Cambridge University Press, Cambridge.

[Whorf, 1956] Whorf, B. (1956). *Language, Thought, and Reality*. MIT Press, Cambridge, Massachusetts.

[Wierzbicka, 1996] Wierzbicka, A. (1996). *Semantics: Primes and Universals*. Oxford University Press, Oxford.

[Wittgenstein, 1999] Wittgenstein, L. (1999). *Philosophical Investigations*. Prentice Hall, Upper Saddle River, New Jersey. Translated by G. Anscombe.

[Wright, 1992] Wright, C. (1992). *Truth and Objectivity*. Harvard University Press, Cambridge, Massachusetts.

[Yang et al., 2001] Yang, Q., Zhang, W., Liu, C., Wu, J., Yu, C., Nakajima, H., and Rishe, N. (2001). Efficient processing of nested fuzzy SQL queries in a fuzzy database. *IEEE Transactions on Knowledge and Data Engineering*, 13:884–901.

[Zadeh, 1965] Zadeh, L. (1965). Fuzzy sets. *Information and Control*, 8:338–353.

[Zadeh, 1972] Zadeh, L. (1972). A fuzzy-set-theoretic interpretation of linguistic hedges. *Journal of Cybernetics*, 2:4–34.

[Zadeh, 1973] Zadeh, L. (1973). Outline of a new approach to the analysis of complex systems and decision processes. *IEEE Transactions on Systems, Man, and Cybernetics*, 3:28–44.

[Zadeh, 1975] Zadeh, L. (1975). The concept of a linguistic variable and its application to approximate reasoning. *Information Sciences*, pages 8:199–249, 8:301–357, 9:43–80.

[Zadeh, 1978] Zadeh, L. (1978). Fuzzy sets as a basis for a theory of possibility. *Fuzzy Sets and System*, 1:3–28.

[Zadeh, 1979] Zadeh, L. (1979). A theory of approximate reasoning. In Hayes, J., Michie, D., and Mikulich, L., editors, *Machine Intelligence*, volume 9, pages 149–194. Halstead Press, New York.

[Zadeh, 1983] Zadeh, L. (1983). The role of fuzzy logic in the management of uncertainty in expert systems. *Fuzzy Sets and System*, 11:199–227.

[Zadeh, 1985] Zadeh, L. (1985). Syllogistic reasoning in fuzzy logic and its application to usuality and reasoning with dispositions. *IEEE Transactions on Systems, Man, and Cybernetics*, 15:754–763.

[Zadeh, 1986a] Zadeh, L. (1986a). Is probability theory sufficient for dealing with uncertainty in AI: a negative view. In Kanal, L. and Lemmer, J., editors, *Uncertainty in Artificial Intelligence*, pages 103–116. North-Holland, Amsterdam.

[Zadeh, 1986b] Zadeh, L. (1986b). Test-score semantics as a basis for a computational approach to the representation of meaning. *Literary and Linguistic Computing*, 1:24–35.

[Zilberstein, 1995] Zilberstein, S. (1995). Operational rationality through compliation of anytime algorithm. *AI Magazine*, 16(2): 79–80.

Index

www.ingramcontent.com/pod-product-compliance
Lightning Source LLC
Chambersburg PA
CBHW060747150926
51501CB00003B/121
* 9 7 8 1 4 0 2 0 5 0 4 4 2 *